COLLOIDAL SYSTEMS
AND INTERFACES

COLLOIDAL SYSTEMS AND INTERFACES

Sydney Ross
Department of Chemistry
Rensselaer Polytechnic Institute
Troy, New York

Ian Douglas Morrison
Webster Research Center
Xerox Corporation
Webster, New York

WILEY

A WILEY-INTERSCIENCE PUBLICATION

JOHN WILEY & SONS
NEW YORK · CHICHESTER · BRISBANE · TORONTO · SINGAPORE

Library of Congress Cataloging in Publication Data:

Ross, Sydney, 1915–
    Colloidal systems and interfaces/Sydney Ross, Ian Morrison.
        p.   cm.

    "A Wiley-Interscience publication."
    Includes bibliographies and index.
    ISBN 0-471-82848-3
    1. Colloids.   2. Surface chemistry.   I. Morrison, Ian.
II. Title.

QD549.R65   1988   87–30529
541.3'45—dc19   CIP
ISBN 0-471-82848-3

Printed in the United States of America

10 9 8 7 6 5 4 3 2

A low man goes on adding one to one,
His hundred's soon hit:
A high man, aiming at a million,
Misses an unit.
BROWNING, *A Grammarian's Funeral*

This book is dedicated to a high man
FREDERICK MAYHEW FOWKES
by two grateful learners

Preface

This book is closely related to a four-day short course on emulsions and dispersions that has been taught by Professor S. Ross and Professor F. M. Fowkes since 1967, joined by Dr. I. D. Morrison in 1985, and by now has gathered some 2000 alumni. The topics discussed in the course are those considered of first importance to those who are confronted by research and development problems in the industrial milieu. This book, like the course, is intended for the industrial chemist or chemical engineer who may not have had a formal university course in colloid and interface chemistry but finds that the nature of the problems that must be solved necessitates the rapid acquisition of some knowledge of that subject. The major step in solving a problem is to define it. If this step is not well considered, the enterprise is sick. We hope to display the armory of concepts and techniques that are available in this discipline, so that investigators may orient their thinking along lines already laid down by the experience of previous workers. Every topic we broach is treated at greater length in monographs and reviews. We do no more than outline its nature, define its terms, explain its elementary concepts, and direct the reader to sources of fuller information. Our book therefore is an index of related topics, by means of which the enquirer, with a specific problem in mind, may hope to find the appropriate context to help formulate it. A great body of organized knowledge is at hand, but many who could use it are only vaguely aware of its existence or are intimidated by its bulk and impenetrability. This book is a guide to those so perplexed.

The behavioral phenomena of foams, emulsions, and suspensions are almost always very complicated. To understand all the details involved in these phenomena requires a more advanced knowledge than what is provided in just the chapters headed "Foams," "Emulsions," and "Suspensions." The explanation of the phenomena is, therefore, materially facilitated by a preliminary account of the general principles of the science.

This field of investigation shares with all other branches of science the potentialities that the electronic chip has bestowed. We are on the threshold of another scientific revolution brought about by technology already available or about to be available. Properties of matter can now be measured with a precision and speed hitherto undreamed of; more recondite properties, once inaccessible, are now within our reach by means of new combinations of hardware and software. A few of these new techniques are described in this book: particle size distributions by quasi-elastic light scattering, dilatational surface elasticity from the damping of ripples, and foam stability by the automatic recording of small pressure differences. Even as we write, new techniques are being developed, soon to be reported in the specialized journals of this field, to remove guesswork from our theories, and to advance our knowledge of phenomena. As the pace of development quickens, so does the rate at which current techniques and even current modes of thought become obsolescent. Today's knowledge may be only of historical interest tomorrow. Even experience may become irrelevant in a mere 20 years, long before a young scientist has reached the end of active life. There is nothing else for it but the prolongation of studenthood throughout one's whole career, by attending short courses and habitual reading. Now more than ever before may it be said that the art is long and life is brief.

Gratefully acknowledged is a grant to one of us (IDM) for books and travel expenses from the James Clerk Maxwell Foundation of Scotland.

SYDNEY ROSS
IAN DOUGLAS MORRISON

Troy, New York
Webster, New York
February 1988.

Acknowledgments

The authors acknowledge permission to use copyrighted material from Paul O. Abbé, Inc. for Figure IE.5; Academic Press, Inc. for Figures IE.3, IIA.10, IIA.13, IIB.1, IIB.2, IIE.9, IIE.10, IIE.11, IIE.13, IIE.14, IIE.15, IIE.19, IIIC.1, IIIC.2, IIIC.3, IVA.1, IVA.2, IVC.8, IVC.16, Table IVC.6; the American Chemical Society for Figures IIA.11, IIA.12, IIG.2, IIIA.1, IVB.4, IVB.5, IVC.1, IVC.2, IVC.3, IVC.4, IVC.12, IVC.17, IVC.18, Tables IB.3, IIA.2, IIA.3, IIA.4, IIA.5, IIA.8, IIA.9, IIA.10, IIA.11; Ann Arbor Science Publishers, Inc. for Figure ID.8; Butterworth for Tables IVA.2, IVA.3; Cahn Instruments, Inc. for Figure IIA.18; Cambridge University Press for Figure IIA.8; Chapman and Hall, Ltd. for Figures ID.1, ID.6; Chemie und Technik Verlagsgesellschaft, mbH for Figures IVA.8, IVA.10; Clay Minerals Society for Figures IIIB.2, IVC.10; Elsevier Science Publishing Co., Inc. for Figures IIA.16, IIE.16, IIE.17, IIG.4, IIG.5, IVC.11, IVC.13, Tables IIG.1, IIIA.3, IVC.5; W. H. Freeman & Co., Inc. for Figure IC.2; Gordon and Breach Science Publishers, S. A. for Figure IIE.18; Hach Co., Inc. for Figure ID.7; D. C. Heath Co., Inc. for Figure IIE.2; IOP Publishers, Ltd. for Table IVC.8; Kinetic Dispersion Corp. for Figure IE.1; R. E. Krieger Co., Inc. for Figure IC.1; Leeds & Northrup Instruments, Inc. for Figure ID.4 (bottom); Longmans, Green & Co., Ltd. for Figure IIG.6; Malvern Instruments, Inc. for Figure ID.4 (top); Marcel Dekker, Inc. for Figure IVA.3, Tables IIIB.1, IIIB.2; the New York Academy of Science for Figure IVC.6; Oxford University Press for Table IVC.8; Premier Mill Corp. for Figure IE.2; Herman E. Ries, Jr. for Figure IID.5; the Royal Society of Chemistry for Figures IIA.17, IIE.6; Scientific American, Inc. for Figures IID.1, IID.4; the Society of Chemical Industry for Figure IVB.2; Sonic Corp. for Figure IE.4; Dr. Dietrich Steinkopft Verlag, GmbH & Co. KG, for Figure IIIB.1; Theorex,

Inc. for Figure IVC.19, Table IIIA.4; Union Process Corp. for Figure IE.6; Van Nostrand-Reinhold for Figure IIE.8; VCH Verlagsgesellschaft mbH for Figure ID.5; John Wiley & Sons, Inc. for Figures IA.1, IB.3, IB.5, IB.7, ID.3, ID.9, IIA.8, IID.6, IVA.5, IVA.9, IVC.5, IVC.9, Tables IB.1, IB.2, ID.1; and Zed Instruments, Ltd. for Figure IVA.6.

Contents

COLLOIDAL SYSTEMS
AND INTERFACES

Introduction

Dispersed-phase systems are two-phase systems in which one phase is subdivided and distributed throughout the other. A suspension is a system of finely divided solid particles in a fluid medium; an emulsion is a system of liquid droplets in a liquid medium; and a foam is a system of gas cells separated by liquid lamellae. These terms are not always so restricted in common usage. A photographic "emulsion" is actually a dispersion of grains of silver halide in a gelatin base. Microbubbles in a liquid may be termed an emulsion. The particles in a suspension, the droplets in an emulsion, and the gas cells in a foam are the dispersed phases. In general, the size of the dispersed unit is least in a suspension, greater in an emulsion, and greatest in a foam. The interfacial area between the dispersed phase and the medium increases as the size of the unit decreases, which means that interfacial properties assume more importance as the size of the unit decreases. The interfacial area per unit mass of uniform cubes in a suspension or of uniform spheres in an emulsion is given by

$$\frac{\text{Area}}{\text{Mass}} = \frac{6}{\rho d}$$

where ρ is the density of the dispersed phase and d is either the edge of the cube or the diameter of the sphere. The increase in specific area with decreasing particle size is illustrated in Table 1 for a dispersed phase with density of $2.00\,\text{g/cm}^3$. The size of the dispersed unit in suspensions, emulsions, and foams ranges from a few centimeters for a foam cell to few nanometers for a colloidal suspension.

The nature of these dispersed systems is determined not only by their specific chemical composition but also by particulate and interfacial properties. These

Table 1   Change of Specific Area with Particle Size

Diameter	Area/g
1 cm	$3\,cm^2/g$
1 mm	$30\,cm^2/g$
$10^{-2}\,mm\,(10\,\mu m)$	$0.3\,m^2/g$
$1\,\mu m$	$3.0\,m^2/g$
$0.1\,\mu m\,(100\,nm)$	$30\,m^2/g$
10 nm	$300\,m^2/g$

properties are the subject matter of colloid and interface science. This book treats particulate properties (Part I) and interfacial properties (Part II) separately.

The word *colloid* was coined by Thomas Graham in 1861 from Greek roots meaning "gluelike" and was based on his observation that glue molecules do not pass through a parchment membrane. Thus *colloid science* is based on the size of the colloidal unit. Many physical properties besides the one observed by Graham are shown by colloidal systems, that is, systems with at least one dimension less than about a micron.

Particulate Properties

The important particulate effects, that is, those that depend on particle size, are optical, rheological (pertaining to flow), and statistical. Common examples of optical effects due to dispersed particles are the colors of the rainbow, the blue of the sky, the crimson of sunsets, the iridescence of peacock feathers, the colors of some gem stones, and the occurrence of natural whites such as in clouds, chalk cliffs, sea foam, milk, teeth, paper, lilies, daisies, powdered glass, and the plumage of swans, sea gulls, and so forth.

Common examples of rheological effects due to dispersed particles are illustrated by lubricating grease, tomato ketchup, paint, ointment, ink, dough, meringue, fire-fighting foam, cornstarch suspension, and modeling clay and in such processes as the churning of cream and the mining of kaolinite. A common example of a statistical property of suspended particles is Brownian motion. This incessant motion arises from the uneven bombardment, by molecules of the medium, of particles small enough to be affected by it.

Interfacial Properties

The important interfacial effects, that is, those that depend on the excess potential energy of surfaces, are capillarity and adsorption. Capillarity refers to any effect resulting from the existence of a tension at a liquid surface. These effects include the rise of a liquid in a capillary, the walking of insects on water, the buoyancy of ducks, the tears of strong wine, the dew drops on spider webs, the breakup of liquid jets, and waterproofing.

Adsorption denotes absorption of a gas or a solute by a surface or an interface. The word is used to imply the action of a surface. Adsorption is a spontaneous process accompanied by reduction of the surface free energy of the adsorbing surface. Common effects of adsorption are the lowering of surface tension of a solvent by adsorbed solutes, the spontaneous emulsification of grease, the scouring of wool, the use of carbon filters to clarify water, beer, and wine, and in gas masks, the use of bone char to decolor sugar solutions, the flocculation of muds at river deltas, and separations by chromatography.

Stability of Dispersed-Phase Systems

Suspensions, emulsions, and foams are known collectively as dispersed-phase systems. Each consists of two distinct phases with a surface or interface between them. But they would not be stable enough to allow any use to be made of them if the two phases were pure components: A condition for their stability is the presence of at least one more component, called the stabilizing agent. The nature of stabilization is, therefore, our next topic of discussion (Part III).

After these general features of dispersed-phase systems, we discuss particulars of the behavior of suspensions, emulsions, and foams (Part IV).

DEFINITIONS AND GLOSSARY OF TERMS

The high and formal discussions of learned men end oftentimes in disputes about words and names; with which (according to the use and wisdom of the mathematicians) it would be more prudent to begin, and so by definitions reduce them to order. Yet even definitions cannot cure this evil in dealing with natural and material things; since the definitions themselves consist of words, and these words beget others: so that it is necessary to recur to individual instances, and these in due series and order. (Francis Bacon, *Novum Organum*, Aphorism LIX)

adsorbate (n)   The species that is adsorbed, which in the case of a solute is termed surface active

adsorbent (n)   The solid that provides a surface for adsorption

adsorption (n)   The spontaneous absorption of molecules from a gas or a solution by a surface or an interface, which may be either liquid or solid; a term coined by Du Bois-Reymond

adsorptive (n)   Synonym for *adsorbate*

agent (n)   (dispersing, emulsifying, profoaming, surface active, etc.) Materials that promote the function described by the qualifying adjective

agglomerate (n)   A collection of primary particles or aggregates joined at their edges or corners, with a specific surface area not markedly different from the sum of the areas of the constituents

aggregate (n)   A group of primary particles joined at their faces, with a specific surface area significantly less than the sum of the area of their constituents. Aggregates are much more difficult to separate into primary parts than are agglomerates

amphipathic (adj)   Combining hydrophilic and lipophilic qualities; a term coined by G. S. Hartley

bubble (n)   A spherical liquid lamella enclosing a gas

Brownian motion (n)   Random thermal motion of suspended particles

coagulum (n)   See *aggregate*

coalescence (n)   The merging of small drops into a single larger drop

colloid particle (n)   A particle, one of whose dimensions is between 10 and 10^3 nm; a term coined by T. Graham

colorimetry (n)   The measurement of the absorption of light by a nonscattering material

comminution (n)   Reduction to minute particles

contact angle (n)   The angle between the tangent to the liquid surface and the tangent to the substrate, measured through the liquid

cosorptive (adj)   Having the same surface activity; a term coined by S. Ross

dilatancy (n)   Increasing resistance to flow with increasing stress; a term coined by O. Reynolds

dispersion (n)   A finely divided solid, liquid, or gas in a continuous medium

dispersion force (n)   A long-range force of attraction between molecules due to mutually induced fluctuations of their electron clouds

electrocratic (adj)   Denotes a dispersed-phase system stabilized by electrostatic repulsion; a term coined by E. Hauser

emulsion (n)   A dispersion of an immiscible liquid in another

floc (n)   A loose structure formed of primary particles, aggregates, and agglomerates

flocculate (v)   To create flocs by adding an agent or by a change of physical conditions

foam (n)   A coarse dispersion of gas in a liquid

gel (n)   A sol with pronounced temperature-dependent viscoelastic properties

homotattic (adj)   An energetically regular solid substrate; a term coined by C. Sanford and S. Ross

hydrophile (n)   A hydrogen-bonding moiety

interfacial phase (n)   A nonisotropic interphase between two extended phases

inverse micelle (n)   A micelle in a nonaqueous medium

inversion (n)   The process of reversing the continuous and discontinuous phases of an emulsion

isaphroic (adj)   Having the same foaming tendency; a term coined by S. Ross

lamella (n)   An extended sheet of liquid

Lewis acid (n)   A substance that can accept an electron pair from a Lewis base to form a covalent bond

lipophile (n)   A non-hydrogen-bonding moiety

liquid crystal (n)   An extended anisotropic micelle

lyophilic (adj)   Having an affinity for the medium

lyophobic (adj)   Lacking affinity for the medium

mesomorphic phase (n)   A micellar structure with one or two extended dimensions: liquid crystals of various types. From the Greek *mesos*, between or intermediate

micelle (n)   The product of spontaneous association of solute molecules

microemulsion (n)   A thermodynamically stable, transparent solution of micelles swollen with solubilizate

miniemulsion (n)   A thermodynamically stable emulsion in the range 100–1000 nm

monodisperse (adj)   All the same size

monolayer (n)   An adsorbate phase one molecule thick; most often refers to an insoluble film on a water substrate

nephelometry (n)   The measurement of the intensity of light scattered at 90°

peptize (v)   To deflocculate, to reverse flocculation, to disperse

Plateau border (n)   The thin vein of liquid at the intersection of three liquid lamellae

polydisperse (adj)   Having a distribution of sizes

protect (v)   Making an electrocratic system less susceptible to electrolyte by adsorbing a hydrophilic polymer

saturated monolayer (n)   A monolayer of maximum surface concentration

sol (n)	A transparent colloidal suspension
solution (n)	A molecular dispersion
sorption (n)	Used technically to indicate a still unknown mechanism of adsorption or absorption; a term coined by J. W. McBain
sorptive (n)	The surface-active molecules in bulk phase
spread (v)	To have no contact angle
substrate (n)	The adsorbing surface, liquid, or solid
surfactant (n)	A surface-active agent; a term coined by F. D. Snell
suspendant (n)	Agent to stabilize a suspension
suspension (n)	Finely divided solid in a liquid medium
tetrahedral angle (n)	The intersection of four Plateau borders
thixotropy (n)	The time dependence of re-forming an internal structure after its destruction by shear
turbidity (n)	The attenuation of transmitted light by light scattering
Tyndall beam (n)	The light scattered by small particles
viscosity (n)	For viscosimetric units see Table IB.2
wet (v)	A liquid wets a solid when it has no contact angle or a contact angle less than $90°$

PART I

PARTICULATES

CHAPTER IA

Optical Properties—Light Scattering

In 1869 Tyndall produced artificial mists (aerosols) and demonstrated that a light beam is made visible only when particles intercept the beam.[1]* He used the absence of the effect as a sensitive test for the cleanliness of air and of solutions, and to this day it remains the simplest test to differentiate between suspensions and solutions. He also found that the scattering power of a given mass of suspended particles increases the larger the particle size. This phenomenon is the most sensitive indicator of the onset of flocculation of sols and of precipitation. He showed that the scattered light at right angles to the incident beam is completely linearly polarized, and this led him to suggest that the light of the sky, which is polarized at right angles to the sun, could be accounted for by scattering in the higher regions of the atmosphere.

No quantitative theory of light scattering was possible before Maxwell's discovery of the electromagnetic nature of light and the formulation of Maxwell's equations. A consequence of the electromagnetic theory of light is that its velocity in a homogeneous medium is related to the dielectric constant and magnetic permeability of the medium. Expressed in absolute units,

$$c = (\varepsilon_0 \mu_0)^{-1/2} \tag{A.1}$$

*Numbered references are given at the end of each part. The references listed in the Bibliography following Part IV are cited in the text by the author's name and the date.

Substituting numerical values for free space gives

$$c = \frac{1}{(4\pi \times 10^{-7}\,\text{W/A-m})(8.9 \times 10^{-12}\,\text{C}^2/\text{N-m}^2)^{1/2}}$$

$$= 3.0 \times 10^8\,\text{m/s}$$

A direct consequence of Maxwell's theory is that all scattering phenomena of light are shown to arise from inhomogeneities in the medium. An electromagnetic wave impinging on matter induces oscillating electric and magnetic moments within atoms, thereby generating secondary electromagnetic waves emitted in all directions, which are observed as scattered light. The scattering center could be an atom or molecule, or a solid or liquid particle. Since the intensity of the scattered light increases with the size of the scattering center, particles have much greater scattering power than atoms or molecules.

Rayleigh applied Maxwell's equations to the propagation of light through a medium containing small spherical particles, treated as discontinuities in refractive index, to obtain the relation (as subsequently corrected)

$$I_u = 8\pi^4 a^6 (r^2 \lambda^4) \left(\frac{n^2 - 1}{n^2 + 2} \right)^2 (1 + \cos^2\theta) \qquad [\text{A.2}]$$

where I_u is the scattered intensity from a single spherical particle of radius a, in the direction θ, at a distance r from the particle, per steradian, when illuminated by unpolarized light of unity intensity of wavelength λ (in that medium), and n is the ratio of the refractive index of the particle to that of the medium. Rayleigh's light-scattering equation has the following important corollaries:

(a) When the refractive index of the particle is the same as that of the medium $n = 1$, no light is scattered.
(b) The intensity of the scattered light increases as the sixth power of the radius of the particle; therefore, in a polydisperse system, the intensity of the scattered light is dominated by the larger particles.
(c) The intensity of the scattered light is inversely proportional to the fourth power of the wavelength of incident light; therefore, blue light is scattered more than red light.

The blue of the sky comes from the scattering of sunlight by air molecules: The blue constituents of the scattered light are more intense than the red. The sky appears blue because scattered blue light comes to us from all quarters of the heavens; in the absence of an atmosphere no scattering takes place and the heavens would appear black except for direct observation of the sun. The yellowish-red of dawning day and of sunsets is caused by transmitted light, from which the blue constituents have been largely eliminated, coming to our eyes directly from the sun or reflected from clouds. If the atmosphere is disturbed, dust particles do not settle during the night; hence the old adage "red sky in the morning is a sailor's warning" betokens windy weather on the way.

Rayleigh's analysis contains the simplifying assumptions that the electric field is uniform throughout the particle and that the particle is transparent. The electric field is uniform or nearly so only when the size of the particle is much less than the wavelength of light. The upper limit of particle radius for the use of the Rayleigh equation is generally taken to be $a/\lambda < 0.05$ or on the order of $a = 20$–30 nm. Most dispersions of practical interest contain larger particles, which are often opaque; hence a more general treatment is needed.

In 1908 Mie developed a more general solution for the scattering of light by spherical particles of any size and any refractive index.[2] His theory was later extended to include cylinders and stratified spheres (Kerker, 1969). The essential result of Mie's treatment is the appearance of many maxima in the intensity of the scattered light as a function of scattering angle; the number and position of these maxima depend on the refractive-index difference and the size of the particle.

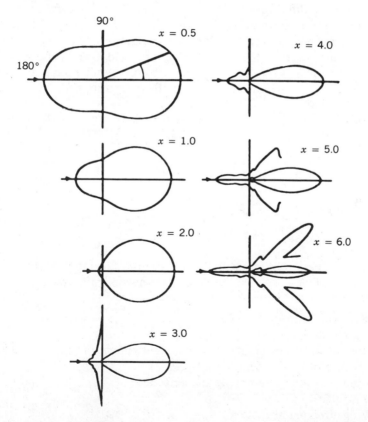

Figure A.1 Angular dependence of scattered intensity for relative refractive index, $n = 1.55$, and size parameter, $x = 2\pi a n/\lambda$, from $x = 0.5$ to 6.0. The arrows indicate the direction of incident light (Barth, 1984, p. 24).

Figure A.1 shows the angular dependence of the scattered intensity for increasing ratios of particle radius to the wavelength of incident light. As the size of the particle increases the scattered light develops lobes of higher intensity at characteristic angles. For the largest particles, bright forward-scattering lobes develop corresponding to the onset of diffraction effects (Fraunhofer diffraction.)

Monodisperse samples of sulfur sols[3] and monodisperse selenium sols[4,5] were used to confirm Mie's theory. If particle properties of a polydisperse system, such as refractive index, distributions of size and shape, and the absorption spectrum are known, the light scattering can be calculated. This possibility offers a practical method to improve the appearance of pigmented films. Efficient computer programs are now available to calculate Mie functions (Bohren and Huffman, 1983). But the inverse problem, that is, to obtain the distribution of size and shape from the data, is impossible, unless the distribution is very narrow, because of the overlapping of maxima and minima from different sizes of particles.

In later publications Rayleigh presented an approximate theory for particles of any size or shape having a refractive-index difference so small that $(n_1 - n_2) a/\lambda \ll 1$. These conditions are well suited to the study of polymers in solution. The observed quantities are the intensity of scattered light as a dual function of concentration and of scattering angle and the refractive index of the solutions as a function of concentration. These data are extrapolated to zero concentration and to zero scattering angle, so that the following limiting relations may be used to obtain the molecular weight M and the radius of gyration R_g.

$$\frac{Kc}{R_\theta} = \frac{1}{M}\left[1 + \frac{16\pi^2 R_g^2}{(3\lambda^2)\sin^2(\theta/2)}\right] \quad \text{as } c \to 0$$

$$\frac{Kc}{R_\theta} = \frac{1}{M} + 2Bc \quad \text{as } \theta \to 0 \qquad\qquad \text{[A.3]}$$

where

$$K = \frac{2\pi n^2}{\lambda_0 N_0}\left(\frac{\partial n}{\partial c}\right)_{T,p}^2 (1 + \cos^2\theta)$$

$$R_\theta = r^2\frac{I_\theta}{I_0} = \text{the Rayleigh ratio for unpolarized}$$
incident light at the scattering angle θ.

n = refractive index of the solution
c = concentration in g/mL
I_θ = intensity of the scattered light from a unit volume of solution at an angle θ
I_0 = intensity of the incident light
r = distance from the sample to the detector,
λ_0 = wavelength in vacuo
N_0 = Avogadro's number
B = virial coefficient

When uniform dispersions of particles of a size approaching the wavelength of light, such as Tyndall's mists and La Mer's sulfur sols, are illuminated with white light, the scattered light appears colored, the color varying with the angle of observation. The variations of the wavelength with the angle of observation are called higher order Tyndall spectra. The colors are more brilliant the closer the dispersion is to being monodisperse. Each size of particle generates a characteristic number of bands of color. Particle size is determined by locating the angles of maximum red (or green) intensity and comparing with computed tables.

CHAPTER 1B
Kinetic Properties—Rheology

1. CLASSIFICATION OF FLOW

a.   Newtonian Flow

The viscosity of a fluid is essentially its internal friction. A body incapable of flow has infinite internal friction; a gas, on the other hand, flows readily as its internal friction is small; liquids are in between. The study of flow, or rheology, provides information about the internal structure of a liquid system; conversely, by controlling the internal structure, flow of a desired quality can be obtained.

As an illustration, consider slicing cold molasses with a knife. The motion of the knife blade is communicated to layers of liquid parallel to the motion of the blade. The internal friction of the liquid diminishes the velocity of this motion even as it is being communicated to layers of liquid lying farther away (Fig. B.1).

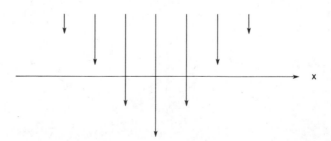

Figure B.1 Velocity profile of liquid flow around a moving knife edge. The central arrow represents the direction and magnitude of the motion of the knife; the parallel arrows represent the direction and magnitude of the motion of the layers of liquid.

14

The shearing stress applied by the knife is the force per unit of surface area of contact; the rate of shear of the liquid is its velocity gradient in a direction normal to that of the knife. Newton described these phenomena in the following words (*Principia*, Prop. LI, Cor. V.):

> If a fluid and an outer cylinder are at rest and an inner cylinder revolves uniformly, a circular motion, communicated to the fluid, will be propagated by decreasing degrees through the fluid to the outer cylinder.

The viscosity coefficient η is defined as the ratio of the shearing stress (F/A) to the rate of shear (dv_y/dx) for one-dimensional flow:

$$\frac{F}{A} = \eta\left(\frac{dv_y}{dx}\right) \qquad\qquad \text{[B.1]}$$

A useful application of this equation is to describe mathematically the flow of a liquid through a cylindrical pipe (Fig. B.2). The rigorous derivation of this problem requires transforming the equations of continuity and motion into cylindrical coordinates (Bird et al., 1960, pp. 42–47). The following simplified derivation leads to the correct result.

Consider the flow of each cylindrical shell (or lamella) of the liquid as equivalent in effect to that of a cylindrical solid slug of the same radius in its place. Let R be the radius of a pipe of length L, and P be the pressure on the end of the pipe. Consider a concentric cylinder of liquid of radius r within the pipe; the applied force

$$F = P\pi r^2 \qquad \text{(dyn)} \qquad\qquad \text{[B.2]}$$

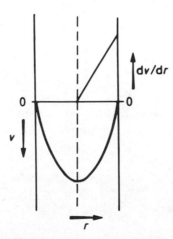

Figure B.2 Parabolic profile of velocity in a cylindrical pipe.

and the tangential area of shear is

$$A = 2\pi r L \quad (\text{cm}^2)$$ [B.3]

hence

$$\frac{F}{A} = \frac{Pr}{2L} \quad (\text{dyn/cm}^2)$$ [B.4]

The rate of shear, or the velocity gradient, is given by

$$\frac{dv}{dx} = -\frac{dv}{dr} \quad (\text{s}^{-1})$$ [B.5]

The negative sign is required since the velocity decreases with increasing r. Equation [B.1] gives

$$\frac{F}{A} = \frac{Pr}{2L} = \eta\left(\frac{dv}{dx}\right) = -\eta\left(\frac{dv}{dr}\right)$$ [B.6]

Integrating this equation gives the variation of v with r, namely,

$$v = \frac{P(R^2 - r^2)}{4L\eta}$$ [B.7]

This equation says that the contour of the velocities of the cylindrical layers of liquid within the cylinder is a paraboloid of revolution.

Our objective is to relate the data of measurements, that is, the rate of flow of liquid out of a pipe corresponding to the pressure forcing it out, to determine the viscosity coefficient. The volume of flow per second, ω, of a cylindrical lamella of liquid is

$$d\omega = [\pi(r + dr)^2 - \pi r^2]v$$ [B.8]

Keeping only the leading terms and substituting Eq. [B.7] gives

$$d\omega = 2\pi r dr (R^2 - r^2)(P/4L\eta)$$
$$= (\pi P/2L\eta)(R^2 r dr - r^3 dr)$$ [B.9]

Therefore

$$\int_0^\omega d\omega = \frac{\pi P R^2}{2L\eta} \int_0^R r dr - \frac{\pi P}{2L\eta} \int_0^R r^3 dr$$ [B.10]

$$\omega = \frac{\pi P R^4}{8L\eta}$$ [B.11]

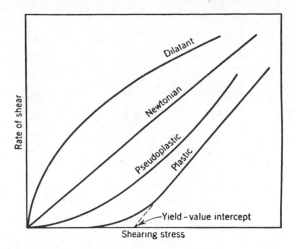

Figure B.3 Schematic rheograms of various types of flow behavior (Fischer, 1950, p. 151).

Equation [B.11] is due to Poiseuille (1844) and relates the rate of flow in cubic centimeters per second to the applied pressure on the liquid flowing through a pipe. Both ω and P are measurable quantities, and from these and the dimensions of the pipe (R and L) the viscosity coefficient can be calculated. The rheogram for the flow of liquid in a pipe may be plotted directly from measured observations as ω versus P, and if the liquid is Newtonian, that is, η constant, a straight line through the origin is the result. The slope of the line is $\pi R^4/8L\eta$. Examples of Newtonian liquids are pure liquids, solutions of solutes smaller than macromolecules, and some polymer solutions.

But many liquids, including some that are common in everyday life, are non-Newtonian. Examples of these are paints, tomato ketchup, cold creams, ointments, and margarine. The flow of these and systems like them can still be described by a plot of shearing stress against rate of shear (the rheogram), but the result is not a straight line through the origin (Fig. B.3).

b. Plastic Flow

The rheogram of a plastic material describes one that does not flow until sufficient stress is applied. The stress at which the flow begins is called the yield point; at stresses larger than the yield point the rate of flow is approximately linear with the shearing stress. Bingham pointed out that the flow behavior of many materials that have a yield point can be approximated on the rheogram as a straight line intercepting the axis at the yield point. Two numerical indices are then available to describe this behavior: the yield point and the plastic viscosity U, derived from the slope of the straight line. Such materials are named "Bingham bodies." Bingham bodies such as spraying lacquer have low yield-point values ($0-10\,\mathrm{dyn/cm^2}$) and low plastic viscosities ($0.1-1.0\,\mathrm{P}$); bodies such as paints or ketchup have medium

Table B.1 Terms Commonly Used to Describe Flow Properties of Consistency of Pigment Dispersion with Approximate Ranges in Absolute Units

Descriptive Terms Frequently Used	Plastic Viscosity		Yield Value		Example
	Magnitude	Range (P)	Magnitude	Range (dyn/cm^2)	
Watery; thin; soupy; highly fluid; nontacky	Low	0.1–1.0	Low	0–10	Spraying lacquer; gravure ink
Creamy	Low	1–5	Medium	50–5,000	Paints; ketchup
Pasty; stiff; buttery; high-consistency; nonfluid; nontacky; short	Low	0.1–5	High	500–5,000	Textile color pastes; (stipple paints); shaving cream; mayonnaise
Fluid; low tack	Medium	5–50	Low	0–1,000	Black news ink
Buttery; stiff; pasty; salve like; short	Medium	5–50	High	1000–10,000	Ointments
Long; molasseslike; tacky; highly viscous	High	100–1000	Medium	100–10,000	Rotary press ink
Long; heavy bodied; high consistency; tacky	High	100–1000	Medium high	1000–30,000	Job press; offset inks
Leathery; tough; rubbery sticky	Very high	10^3–10^7	Very high	—	Resin melts; rubber; asphalt

From Fischer (1950, p. 156).

18

yield-point values (50–500 dyn/cm²) and low plastic viscosities (1–5 P); and bodies such as ointments have high yield-point values (1,000–10,000 dyn/cm²) and medium to high plastic viscosities (5–50 P). Several examples of combinations of yield-point values and plastic viscosities are listed in Table B.1.

The yield point of a Bingham body is a consequence of the destruction of an internal static structure such as is found in a flocculated suspension. On shearing beyond the yield point, the structure may degenerate further, which causes a reduction of the apparent viscosity coefficient, an effect described as "shear thinning."

c. Shear Thinning

Some plastic materials have vanishingly small yield points but still show "shear thinning." The internal structure of such materials is too fragile to withstand even the slightest shearing stress and gradually disintegrates as further shear is applied. These materials, having no appreciable yield point but an apparent viscosity coefficient that continually decreases with applied stress, are sometimes called "pseudoplastic."

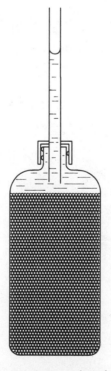

Figure B.4 A mysterious bottle.

d. Dilatancy

A completely different type of flow behavior is displayed by materials whose resistance to flow increases with increased applied stress. Every child who has played with a bucket and spade on the seashore has discovered how difficult it is to stir wet sand. The sand particles are round, approximately all of the same size, and are completely wetted by seawater. The difficulty experienced in stirring this mixture is caused by the obstruction to flow of a closely packed and deflocculated system of particles. Reynolds in 1885 coined the term "dilatancy" to describe this effect. His explanation is that the sand particles, when undisturbed, have settled to a close-packed arrangement and that any disturbance causes rearrangement to a smaller number of nearest neighbors, in which the particles are farther apart. The familiar example that he had in mind was the drying of moist sand when one stands on it. The pressure of the foot expands the underlying sand structure, and the surrounding surface moisture flows into it leaving a visible dry area; on raising the foot, a puddle of water is disclosed as the expanded structure settles back, releasing water as the voids collapse. After a short time the puddle soaks back into the dry area, and all is as before. A simple demonstration of the effect can be made with a plastic bottle with an extended neck containing sand and water (Fig. B.4.) On squeezing the container the net volume increases (this is hard to believe!) and the level of liquid in the neck declines. Try it for yourself!

e. Thixotropy

A feature of some shear-thinning liquid systems is that significant periods of time are required for an internal structure to re-form (heal) after it has been broken by

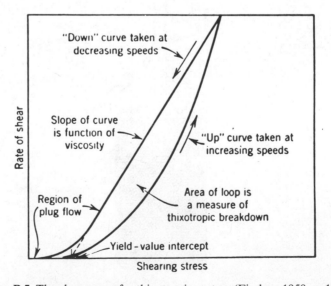

Figure B.5 The rheogram of a thixotropic system (Fischer, 1950, p. 153).

Table B.2   Viscosimetric Units

Property	Symbol	Unit	CGS Units	MKS Units	Dimensions
Viscosity (Newtonian)	η	Poise	dyn-s/cm^2	10^{-1} N-s/m^2	$[ML^{-1}T^{-1}]$
Kinematic viscosity	μ	Stoke	cm^2/s	10^{-4}m^2/s	$[L^2T^{-1}]$
Fluidity	Φ	Rhe	cm^2/dyn-s	10 m^2/N-s	$[M^{-1}LT]$
Plastic viscosity	U	Poise	dyn-s/cm^2	10^{-1} N-s/m^2	$[ML^{-1}T^{-1}]$
Mobility	μ	Rhe	cm^2/dyn-s	10 m^2/N-s	$[M^{-1}LT]$
Yield value	f		dyn/cm^2	10^{-1} N/m^2	$[ML^{-1}T^{-2}]$
Thixotropic breakdown	M		dyn/cm^2-s	10^{-1}N/m^2-s	$[ML^{-1}T^{-3}]$
Rate of shear	$D, dv/dx$		s^{-1}	s^{-1}	$[T^{-1}]$
Shearing stress	F/A		dyn/cm^2	10^{-1} N/m^2	$[ML^{-1}T^{-2}]$
Torque	τ		dyn-cm	10^{-7}N-m	$[ML^{-1}T^{-3}]$
Rate of rotation	Ω		rad/s	rad/s	$[T^{-1}]$

From Fischer (1950, p. 154).

applied stress. The time required may be anything from a few seconds to a few months depending on the nature of the system. Lubricating greases are prone to show this phenomenon. A sample of the grease may be worked energetically, and the rate at which its consistency returns as its structure re-forms is tested periodically. The effect is shown on the rheogram by tracing the variation of the rate of shear as the stress is gradually increased and then reversing the direction of the variation of stress. If the internal structure is re-formed slowly, the reverse rheogram curve lies above the ascending curve, as shown in Fig. B.5. Behavior of this description is called thixotropy. The slow healing of a shear-thinned system is measured conveniently by taking the area of the hysteresis loop.

f. Units and Dimensions

The viscosimetric units and dimensions of the terms described are collected in Table B.2.

2. INSTRUMENTS TO MEASURE VISCOSITY

a. Capillary Viscosimeters

A complete capillary viscosimeter consists of five essential parts: (1) a fluid reservoir, (2) a capillary of known dimensions, (3) a unit to control and measure the applied pressure, (4) a unit to determine flow rate, and (5) a unit to control temperature. Capillary viscosimeters are designed to make use of Poiseuille's equation, hence they are useful only for Newtonian liquids. The applied stress may be the action of gravity on the liquid or it may be created by pressure of a gas to push the liquid through the capillary. The most common instrument of this type is the Ostwald–Ubbelohde viscosimeter (Fig. B.6). A standard liquid of known viscosity and density is used to calibrate the instrument. Since the viscosity of liquids varies with temperature, the instrument is immersed in a constant-temperature bath. The instrument is filled to the level C with liquid, which is then raised by suction above point A and allowed to flow through the capillary. The time t for the liquid to move from A to B is measured. This time is proportional to the ratio of viscosity to density, η/ρ. Comparing the unknown, 1, with the standard, 2, gives

$$\eta_1 = \eta_2 \left(\frac{\rho_1 t_1}{\rho_2 t_2} \right)$$ [B.12]

This instrument is satisfactory for simple liquids and solutions. For different ranges of viscosity, capillaries of different diameters are used; the higher the viscosity, the wider the capillary needed. The capillary viscosimeter is a one-point instrument, so called because it measures a single average applied stress. Other one-point instruments use the time required for a ball, a long needle, or a bubble to pass through the liquid. Instruments that can measure enough points on the rheogram to define its shape are called rheometers rather than viscosimeters. A capillary instrument can be made into a rheometer by measuring various applied stresses.

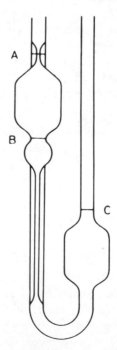

Figure B.6 Capillary viscosimeter.

b. Coaxial Cylindrical Rheometers

Couette or rotational flow occurs in the gap between concentric cylinders. Its advantage over the translational flow in capillary instruments is that measurements may be made continuously. The shear rate is varied by controlling the rate of rotation of the bob or the cup, so that the instrument is capable of providing a rheogram. In bob-and-cup rheometers, the driving force is balanced by retardation by viscous forces. At steady state the rate of rotation is constant and the shear rate nearly uniform; the following relation holds:

$$\tau = 4\pi r^2 \eta h \Omega \qquad\qquad [B.13]$$

where τ is the torque, η is the viscosity coefficient, and Ω is the rate of rotation (radians per second.) The factors r^2 and h are based on the dimensions of the cup and the bob as follows:

$$r^2 = \frac{R_c^2 R_b^2}{R_c^2 - R_b^2} \qquad\qquad [B.14]$$

where R_b and R_c are the radii of the bob and cup, respectively, and h is the height of the bob.

In the original Couette rheometer (1890) and the MacMichael rheometer (1915), the rate of rotation of the cup is the independent variable and the torque is measured. In the latter the torque is measured as "MacMichael degrees" (300° per revolution.) The Haake rheometer measures the torque necessary to drive the bob at constant rotation within a stationary cup.

A widely used instrument is the Brookfield viscosimeter, which consists of a motor-driven bob and a dynamometer to measure the torque. As generally used, the bob has the form of a disk, which is immersed in a large container of liquid. The shear rate is not well defined, and the resulting determination of viscosity is only relative. Its simplicity in use, the wide range of shear rates available, and the ease with which it can be cleaned account for its popularity.

c. Cone-and-Plate Rheometers

The cone-and-plate rheometer (Fig. B.7) consists essentially of a flat plate and a cone, the apex of which barely misses making contact with the plate. The sample of liquid is inserted into the narrow gap between the cone and the plate, where it is retained by capillary forces. The angle of the cone is 179.7°—almost a flat plate itself. The torque resulting from rotating the cone at constant angular velocity is recorded. The design ensures that all portions of the sample are subjected to the same shear rate. Bob-and-cup rheometers do not provide this feature. For Newtonian liquids this difference would not matter as the flow of the liquid is directly proportional to the shearing stress, so for an average shear a proportionately average flow results. Not so for non-Newtonian fluids. If there is a yield

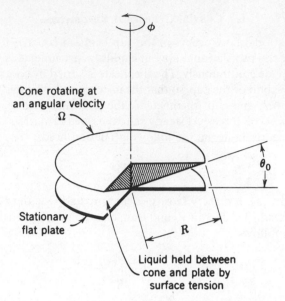

Figure B.7 Schematic diagram of the cone-and-plate rheometer (Bird et al., 1960, p. 99).

point, no flow occurs at shearing stresses less than the yield point. Even at greater shearing stresses the flow is not proportional to the shear. Plug flow and other anomalies begin to appear when non-Newtonian fluids are subjected to shear rates that vary locally. The cone-and-plate rheometer is designed to avoid these anomalies.

The following statement is taken from an article by R. McKennell of Ferranti Ltd., Manchester, England.[6]

Theoretically a cone of half-angle of $\theta_0 = 6°$ gives rise to a departure of only 0.35% from shear-rate uniformity. However, if θ_0 is greater than about 4°, errors may arise due to edge effect and, at higher shear rates, due to temperature rise within the fluid. In practice an angle of only 0.3° is used, with an average gap width of about 0.05 mm, requiring a sample of about 0.1 cc. Cone angles of this order of magnitude simplify the mathematical analysis of non-Newtonian flow data because the whole of the measured sample attains a uniform shear rate and hence a constant apparent viscosity.

Let the rate of rotation $= \Omega$ rad/s; the cone half-angle $= \theta_0$; and the gap width $= w$ at radial distance $= x$ within the sample of radius R. The linear velocity of a point on the cone is proportional to the radial distance. The width of the gap is also proportional to the radial distance. The rate of shear D is given by

$$D = \frac{\text{linear velocity}}{\text{gap width}} = \frac{\Omega x}{w} \quad \text{(cm/s/cm)} \qquad \text{[B.15]}$$

Therefore

$$D = \Omega/\theta_0 \quad s^{-1} \quad \text{(i.e., independent of the distance } x) \qquad \text{[B.16]}$$

Table B.3 Rheological Equations

Property	Couette Rheometer	Cone-and-Plate Rheometer
Viscosity coefficient	$\eta = \tau/(4\pi r^2 h\Omega)$	$\eta = 3\tau/(2\pi R^3 D)$
Rate of shear	$D_{max} = 2\Omega r^2/R_b^2$ $D_{min} = 2\Omega r^2/R_c^2$	$D = \Omega/\Psi$
Plastic viscosity	$U = (\tau - \tau_0)/(4\pi r^2 h\Omega)$	$U = 3(\tau - \tau_0)/(2\pi R^3 D)$
Yield value	$f = U\tau_0/[\ln(R_c/R_b)]$	$f = U\tau_0$
	($\tau_0 =$ extrapolated value of torque for $\Omega = 0$)	

From ref. 6

The torque relations and velocity distribution in the cone-and-plate design are analyzed by Bird et al. (1960, pp. 98–101). Table B.3 shows the fundamental simplicity of the cone-and-plate rheometer compared to the cylindrical rheometer by listing expressions for various rheological properties as a function of the dimensions of the equipment and the experimental values.

d. Viscoelastometers

Fluids that return only partially to their original form when the applied stress is released are called viscoelastic. When stress is suddenly applied and maintained constant, the resulting strain (or change in form) with time is called a *creep curve*. In *stress relaxation* the sample is brought suddenly to a given deformation and the stress required to maintain this deformation is measured as a function of time. Viscoelastic behavior is important in connection with plastics manufacture, performance of lubricant greases, application of paints, processing of foods, and movement of biological fluids. Most of these systems consist of suspended particles or emulsion droplets of colloidal size; the interactions between such particles give rise to their viscoelastic or other non-Newtonian properties. The observed viscoelasticity of a material, therefore, can give information about how its constituents are spatially distributed and the nature of their interactions.

Materials of different types require instruments appropriate to their properties. In materials that flow readily, elastic strains relax quickly. Instruments that use forced sinusoidal deformations (mechanical spectrometers or oscillatory viscoelastometers) are the most useful to determine the relations between stress, deformation, and time. Small deformations are used so that the microstructure is not destroyed during and by the experiment. A suitable rheological test program is based on the identification of critical-flow conditions, both for production and use. A wide choice of instruments is available for materials exhibiting a wide range of properties. About 40 rheometers are described by Walters (1980).

CHAPTER IC

Statistical Properties

1. BROWNIAN MOTION

The most readily observed statistical property of particulates is Brownian motion. Robert Brown, a Scottish botanist, using a simple microscope, discovered (1827) that a dispersion of cytoplasmic granules in water showed lively, random motions. He subsequently established that the same type of motion could be observed with particles of dead plants, glass, and a wide variety of minerals when ground sufficiently small to be temporarily suspended in water. Among his samples was a chip of stone taken by a relic hunter from an Egyptian sphinx. The motion, therefore, has nothing to do with life. That the effect might be due to convection currents or electric charges was also disproved by later investigators. William Prout, an early proponent of Dalton's atomic theory and the author of Prout's hypothesis that atomic weights are multiples of a subatomic *protyle*, was also first in his inquiry: "Are the molecular motions of fluids the cause of those motions which solid particles of matter diffused through them sometimes exhibit?" His remarkably prescient suggestion is now known to be correct.*

In 1905 Einstein noticed the resemblance between Brownian motion and the hypothetical motion of gas molecules according to the kinetic theory of gases.[7] Working on the supposition that size is the essential difference between

---

*Prout's *Bridgewater Treatise*, 4th ed., Bell and Daldy: London, 1870, p. 389. The query does not occur before the 3rd edition of this book. Prout, as an ardent atomist, may even have acquired the idea from Lucretius (c. 95–55 B.C.), who inferred the existence of Brownian motion long before it was discovered (Lucr. *De rer. nat.* **II** 112–141).

suspended particles and gas molecules, Einstein proposed that the mean displacement x of a particle in Brownian motion as a function of time be

$$x = (2Dt)^{1/2} \qquad\qquad \text{[C.1]}$$

where the diffusion coefficient D is given by

$$D = \frac{kT}{3\pi\eta d} \qquad\qquad \text{[C.2]}$$

where η is the viscosity of the medium and d is the diameter of the particle.

Examination of Eq. [C.1] shows that the velocity, x/t, of a particle is a function of time; that is, the observed velocity is larger the shorter the time of observation because a particle changes its direction about 10 million times a second, so that any observed displacement can only be between the beginning and end points of a tortuous path. The rectilinear displacement in a fixed period of time is all that can be determined. From measurements of the diffusion coefficients of molecules in solution, Perrin was able to calculate, by means of Eq. [C.2], the molecular sizes, and from the sizes to calculate Avogadro's number. The agreement of his results with independent determinations was a strong confirmation of the statistical thermodynamics of Boltzmann and showed the connection between Brownian motion and molecular reality, which was Einstein's original hypothesis[8,9] (Nye, 1972). Experimental verifications of Einstein's hypothesis were made independently by Svedberg and Perrin, both of whom obtained Nobel prizes in 1926, one in chemistry and the other in physics.

Nowadays diffusion coefficients are used to determine particle sizes or polymer molecular weights. They are obtained by measuring the degradation of a sharp boundary between a suspension and its serum. The boundary can be obtained by centrifugation and its rate of spread is followed optically. Diffusion constants can also be obtained from quasi-elastic light-scattering data as described in Chapter I.D.

2. COLLIGATIVE PROPERTIES

Colligative properties usually refer to solutions. These properties are osmotic pressure, freezing point depression, boiling point elevation, and lowering of vapor pressure. All these properties depend, to a first approximation, only upon the concentration of dispersed units, whether the units are atoms, molecules, ions, macromolecules, or dispersed particles. The effects with dispersions are significant when particles are less than about 100 nm in diameter and in dispersions free of dissolved species.

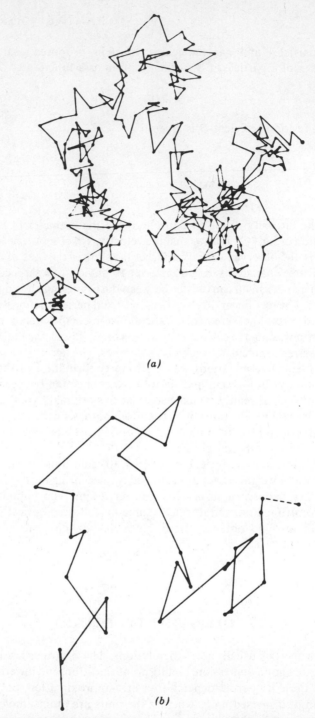

(a)

(b)

Figure C.1 Observed Brownian pathway of a gamboge particle of radius 530 nm. (*a*) Successive positions marked every 30 s, then joined by straight lines having no physical reality whatsoever. (*b*) The same path but only with every tenth position shown and connected by straight lines (Mysels, 1978, p. 104).

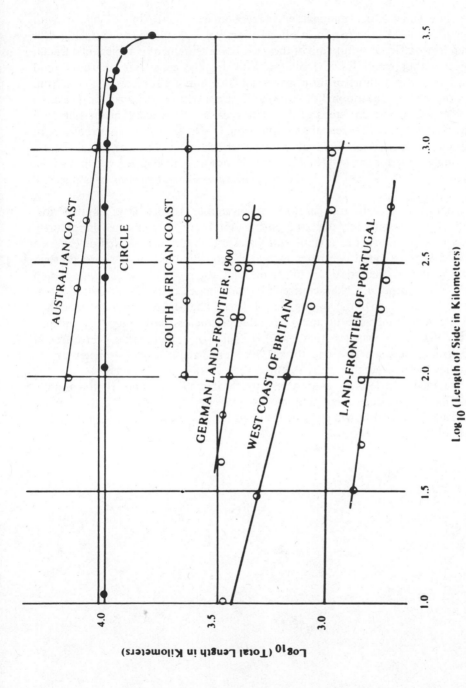

Figure C.2 Richardson's data on the increase of the perimeters of coastlines with the decrease of the size of the step (Mandelbrot, 1983, p. 32).

3. FRACTALS

Measurements of Brownian pathways reveal an ever-increasing irregularity with reduction of the time period of observation. An example is shown in Fig. C.1. Such irregularity as a function of the scale of measurement is typical of a fractal object. Other examples of fractal objects are the coastlines of islands and continents, the outlines of snowflakes and flocs, and the surfaces of solids, from carbon blacks to the moon. For example, if the coastline of Britain were measured as a series of different steps of different size, say, from a kilometer step to a centimeter step, the perimeter so found would be larger the smaller the step. In other words there is no single value for the perimeter of a coastline, just as there is no single value for the length of the pathway of a particle in Brownian motion. The effects of the size of the step on the measurements of various coastlines are given in Fig. C.2.

Figure C.2 reproduces experimental measurements of perimeter length performed on various maps and on a circle, using shorter and shorter steps. Increasingly precise measurements of a circle stabilize very rapidly near a well-determined value; but an irregular perimeter such as the coastline of Britain shows no such limit. Generally, such data are linear on a log-log plot for a wide range of step sizes. These plots are characterized by their slopes. The fractal dimension is defined as the slope of the log-log plot plus one. The perfectly smooth circle then has a fractal dimension of unity; irregular perimeters have fractal dimensions larger than unity (but less than two.) Irregular or rough surfaces have fractal dimensions between two and three. The more irregular the perimeter or the rougher the surface, the larger the fractal dimension.

Fractal dimensions are used to describe the irregularity of floc shapes and the roughness of surfaces. Digitizing microscopes can be programmed to compute the fractal dimension, which is determined by image analysis as a function of magnification.

CHAPTER ID

Size and Surface Area

The particle size distribution of a suspension or emulsion in itself is not a particularly interesting function, rather it is the influence of size on the macroscopic properties such as the optical and rheological properties that is important. A key to choosing the most suitable particle-sizing technique, then, is to pick one that is sensitive to those portions of the size distribution that are most significant to important macroscopic properties. For instance, if the optical properties of a suspension are significant, then using an optical technique to measure size distributions is most likely to yield useful information. In general, a complete description of all the sizes and the shapes of the particles in a suspension or emulsion is unobtainable. At best the usual particle-sizing techniques give a distribution of "equivalent" sized particles, for instance, a distribution of "equivalent" spheres that settle with the same terminal velocity as the particles in the dispersion. Each of the usual particle-sizing techniques is limited to some size range, which might vary with composition. Particles outside that range are either undetected or produce anomalously high signals. For example, a technique based on measuring changes of electrical resistance as particles pass through a sensing zone may not detect the smallest particles. And a few large particles in a suspension can completely overwhelm most light-scattering measurements. Every technique is sensitive to a certain range of particle sizes and hence each technique produces a different distribution of equivalent particles.

The distribution of particle shapes is even more difficult to obtain. The only simple direct determination of shape is through microscopy. A general rule is to look at the sample through a microscope before any particle-sizing technique is tried. The information easily obtainable from microscopic examination of a suspension or emulsion is invaluable. Knowing the approximate shape of the

31

particles being sized can be a useful aid in choosing an appropriate technique.

Two sets of values are necessary to specify a particle size distribution: a measure of size of a particle and a measure of the total quantity of particles with that size. Particle size can be specified by the radius, diameter, or, if a unique one exists, length of side. More likely the measure obtained is one of minimum or maximum chord, fiber length, equivalent diameter from the terminal settling velocity, projected perimeter or diameter of the equivalent circle, or equivalent optical or electrical diameter. The measure of the total quantity of particles of each equivalent size is an additive function such as number, volume, mass, or surface area.

The distribution of particle sizes is sometimes expressed as a histogram, that is, the number (or mass) of particles with equivalent sizes within some range. Some particle-sizing techniques, e.g. sieving, produce size distributions as histograms directly. The distribution of particle sizes can also be expressed as a probability function such that the quantity of particles within a small range of sizes is the area under the probability curve for that range. When particle size distributions are plotted as probability functions, the abscissa is the size and the ordinate is the quantity of particles per unit size for that size. When the ordinate is measured in numbers per unit size, the size distribution is called a number distribution. When the ordinate is measured in mass per unit size, the size distribution is called a mass distribution. For instance, particle sizing by sedimentation produces mass distributions; particle sizing by electrozone detection produces number distributions.

Often the real size distribution of a sample can be adequately described by a best-fit Gaussian (normal) distribution function:

$$\Phi_g(d) = \frac{1}{\sigma\sqrt{2\pi}} \exp\left[-\frac{(d - d_0)^2}{2\sigma^2} \right] \qquad \text{[D.1]}$$

where d is the diameter, d_0 is the mean diameter, and σ is the standard deviation. If the distribution is given as a number density, that is, the number of particles per unit range of particle diameters, it is called a number distribution; and d_0 and σ are the number mean size and the number standard deviation. If the distribution is given as a mass density, that is, the mass of particles per unit range of particle diameters, it is called a mass distribution; and d_0 and σ are the mass mean size and mass standard deviation. If the number distribution is a normal distribution, the mass distribution is not and vice versa.

When a dispersion is obtained by comminution (as is generally the case), then the particle size distribution is often adequately described by the log-normal distribution function:

$$\Phi_{ln}(d) = \frac{1}{\ln\sigma_{ln}\sqrt{2\pi}} \exp\left[-\frac{\ln^2(d/d_0)}{2\ln^2\sigma_{ln}} \right] \qquad \text{[D.2]}$$

where d is the diameter, d_0 is the geometric mean diameter, and σ_{ln} is the gcometric standard deviation. If the distribution is given as a number density, then the distribution is a number distribution, and d_0 and $\sigma_{\ln}$ are the number geometric mean size and the number geometric standard deviation. If the distribution is given as a mass density, it is a mass distribution, and d_0 and $\sigma_{\ln}$ are the mass geometric mean size and the mass geometric standard deviation. If the number distribution is log-normal, then the mass distribution is so too, with the same geometric standard deviation; and vice versa. The relation between the geometric mean for a number distribution, d_n, and the geometric mean for a mass distribution, d_w, is

$$\ln d_n = \ln d_w - 3 \ln^2 \sigma_{\ln} \qquad [\text{D}.3]$$

When reporting an average size and standard deviation for a particle size distribution, it is clearly important to state whether the distribution is normal or log-normal, number or mass. For a thorough discussion of the statistics of small particles, including graphical techniques, see Herdan (1953).

Errors in sampling procedure are most significant when sizing dry powders. In typical dispersions the particle sizes are so small and the samples so easily mixed that sampling is usually not a problem. For a detailed discussion of sampling techniques and considerations see Kaye (1981).

In the following section we discuss various commercially available techniques to measure particle size, dividing the techniques into groups corresponding to the approximate range of sizes appropriate to each: coarse particles ($> 2 \,\mu$m), fine particles (0.5–2 μm), and ultrafine particles ($< 0.5 \,\mu$m). The size limitations are, of course, not exact but serve as a convenient guide. Detailed monographs devoted to particle-sizing techniques are by Allen (1981), Barth (1984), Jelínek (1970), Kaye (1981), and Stockham and Fochtman (1977). Appendix J includes some names and addresses of manufacturers of particle-sizing equipment.

1. PARTICLE-SIZING TECHNIQUES FOR COARSE PARTICLES (2 μm AND ABOVE)

a. Optical Microscope

The microscope is the basic instrument to study particulates, as it is the only method by which individual particles are observed in situ. It is the first instrument of choice if the material is within its range, as it is accepted as the reference technique. The evidence of the microscope has a high degree of credibility. As well as giving the particle size, it provides information about shape, crystal structure, whether the particles are primary or aggregated, and whether different compounds are in the sample. Care has to be taken to obtain a representative sample, both in what is put under the microscope and in what is taken from the image, and to count a statistically significant number of particles. The short depth of field adds to the sampling problem. Modern image analyzers are capable of doing the counting and subsequent calculations automatically.

The optical microscope will often detect but cannot measure particles below the ~ 2-μm limit because of diffraction. Specimen contrast often limits the utility of the optical microscope, although special methods such as phase contrast, dark-field illumination, and Nomarski interference are often used to advantage. Specimen contrast for nonabsorbing particles decreases as the refractive index of the particles approaches that of the medium. Optical methods of all types fail for nonabsorbing particles when the refractive index of the particle and the dispersing medium match. To get optimum results proper techniques must be used: See McCrone and Delly (1973) for practical information. This reference also gives detailed techniques for particle identification as well as size and shape analysis.

b.   Sieves

A powder can be split into fractions of known size by using a stack of sieves of different mesh size. Standard woven-wire sieves are made with openings, called the mesh, in the range of 37–5660 μm. Micromesh sieves, made by photoetching or electroforming, extend the range downwards to 5 μm. Sieving can be "wet" or "dry" according to the medium. "Sifting" is the term used to denote horizontal agitation of the powder over the sieve surface. Rotation and tapping are the usual forms of agitation to reduce "blinding" of sieve apertures.

Figure D.1 represents schematically the stages of sieving. A known mass of powder is placed on the upper (coarsest mesh) sieve. The mass of powder is lifted either by air or water pressure or by agitation. Gradually the smaller particles work their way down the stack until some final distribution of masses on the various sieves is obtained. The mass of powder left on each sieve as a function of the mesh is the size distribution histogram.

For irregularly shaped particles the apparent size of the particle varies with its aspect ratio and the length of time spent sieving. Ideally, irregularly shaped particles are caught on sieves with a mesh size just smaller than the second largest dimension. (Consider a needle: It will ultimately pass through a sieve whose mesh is just greater than its diameter.)

Sieve surfaces "blind" when particles are jammed into the apertures of the sieve. To minimize blinding, the sieves are vibrated vigorously. Too vigorous a vibration can cause particle attrition and hence a misleading size distribution. The American Society for the Testing of Materials (ASTM) specifies standard sieve sizes and procedures (Table D.1).

To overcome particle–particle agglomeration, pancaking, or electrostatics, powders are often sieved with water. The water can be used as a spray to prevent blinding of the sieves. The disadvantages of using water are that the smaller mesh sizes offer considerable resistance to water and suspended fines are difficult to recover from large volumes of wash water.

Numerous commercial equipments are available with combinations of shaking, tapping, sonic vibration, rocking, and air and water sprays. Some manufacturers are listed in Appendix J.

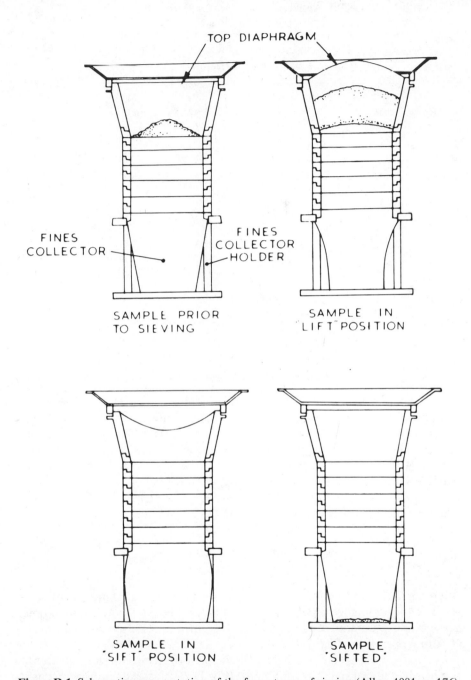

TOP DIAPHRAGM

FINES
COLLECTOR

FINES
COLLECTOR
HOLDER

SAMPLE PRIOR
TO SIEVING

SAMPLE IN
'LIFT' POSITION

SAMPLE IN
'SIFT' POSITION

SAMPLE
'SIFTED'

Figure D.1 Schematic representation of the four stages of sieving (Allen, 1981, p. 176).

35

**Table D.1 Aperture Dimensions for Wire-Woven Sieves
Conforming to the U.S. Sieve Series**

Screen Number	Width of Opening (μm)	Screen Number	Width of Opening (μm)
3.5	5660	40	420
4	4760	45	350
5	4000	50	297
6	3360	60	250
7	2830	70	210
8	2380	80	177
10	2000	100	149
12	1680	120	125
14	1410	140	105
16	1190	170	88
18	1000	200	74
20	840	230	62
25	710	270	53
30	590	325	44
35	500	400	37

From Kaye (1981, p. 59).

c. Elutriators, Impactors, and Cyclones

Particles of different size can be separated by allowing them to settle down a column of liquid that itself is being pumped upward. When the Stokes sedimentation velocity of a particle is greater than the velocity of the upward-moving stream (i.e., for large particles), then the particles settle. This process, called elutriation, separates the particles into two populations. Those with Stokes diameters larger than some value will settle, and those with smaller diameters will be carried out of the column. This "cut-size" diameter d_c is expressed by a rearrangement of the Stokes equation [D.8]:

$$d_c = \left[\frac{28 \eta v}{[(\rho_2 - \rho_1)g]} \right]^{1/2} \qquad \text{[D.4]}$$

where η is the fluid viscosity, v is the average velocity of the fluid motion, $(\rho_2 - \rho_1)$ is the density difference between the solid and the fluid, and g is the gravitational acceleration. Numerous air- or water-based elutriators are available for manufacturing operations (Kaye, 1981, p. 227).

An impactor consists of a nozzle and a flat plate (the collector) located near the nozzle exit and normal to the fluid flow (Fig. D.2). A cascade impactor consists of several impactors in series. The stream containing the particles is deflected by the collector: The heavier particles cannot follow the stream lines because of their

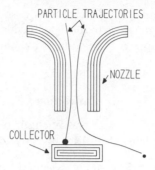

Figure D.2 A single-stage impactor showing the particle trajectories for a large particle (terminating on the collector) and for a small particle (continuing out of the impactor).

inertia and are deposited on the collector; hence the impactor classifies particles according to the aerodynamic or hydrodynamic size. In the cascade impactor the jet velocity is increased at each succeeding stage to trap particles of progressively smaller size. The instrument is calibrated with standards to provide the measurement of particle size. It is useful for on-line determinations but is less accurate than sieving. In general, sampling is a problem because large flowing streams may not be homogeneous. The collectors are sometimes greased and rotated to hold collected powder and to prevent uneven buildup on the collector. Numerous models are available commercially.

d.   Hegman Fineness-of-Grind Gauge

The Hegman gauge is a precision-machined steel plate with a center channel tapered from a slight to a deep depression. A sample of the wet dispersion is

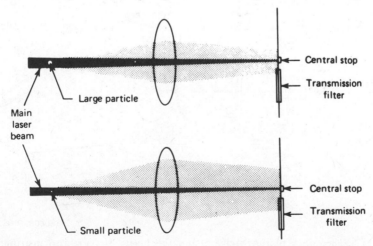

Figure D.3 A schematic representation of Fraunhofer diffraction for two different particle sizes (Barth, 1984, p. 175).

placed in the tapered channel and leveled with a doctor blade. Particle size is determined by holding the gauge up to the light and identifying the point at which coarse particles begin to appear. The standard channel is tapered from 0 to 5 thousandths of an inch.

e.   Diffraction

Particles significantly larger than the wavelength of incident light (usually a He–Ne laser of wavelength 632.8 nm) act as tiny lenses and diffract light, a

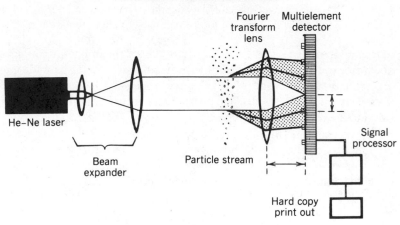

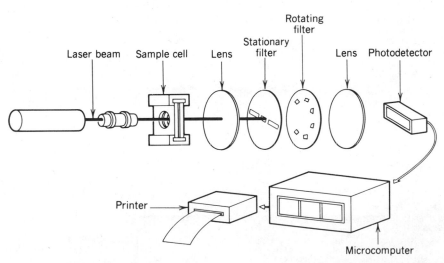

Figure D.4 Schematics of the Malvern and the Leeds & Northrup instruments for measuring Fraunhofer diffraction (Courtesy of Malvern Instruments, Ltd., Malvern, Worcester, UK, and Leeds & Northrup, North Wales, PA.)

phenomenon called Fraunhofer diffraction. The intensity of light scattered is proportional to the particle volume, and the scattering angle is inversely proportional to the diameter (Fig. D.3). Neither the intensity of the scattered light nor the angle of scattering is sensitive to particle shape. The technique is to measure the intensity of light as a function of angle near the forward direction (low angles.) The light scattered from all the particles in the forward direction is blocked and not analyzed. Two commercial instruments are available for the measurement. In the Microtrac Analyzer from Leeds & Northrup, the scattered light at low angles is sampled by passing it through a series of rotating optical filters with holes at different radial positions (Fig. D.4). The intensity of light scattered at 13 different scattering angles is measured. The Malvern Fraunhofer diffraction instrument uses a series of 15 concentric photosensitive rings to measure intensity as a function of scattering angle (Fig. D.4). The intensity of light as a function of scattering angle and size of particle is represented by Airy functions. The data collected are deconvoluted by least-squares analysis to give volume distributions. Both instruments provide particle size histograms of flowing dispersions either in liquids or in the air (Barth, 1984, Chapters 5 and 6), and are appropriate for continuous monitoring of flocculation or comminution.

2.   PARTICLE-SIZING TECHNIQUES FOR FINE PARTICLES (0.5–2 μm)

a.   Sedimentation

Sedimentation methods depend on an application of the Stokes law, which may be derived as follows. At constant velocity of fall, the frictional force F_f is proportional to the radius a, the velocity v, and the viscosity η of the medium. The proportionality constant for conditions of laminar flow was derived by Stokes as equal to 6π. Hence

$$F_f = 6\pi a v \eta \qquad [\text{D.5}]$$

The gravitational force F_g acting on a sphere of radius a is

$$F_g = \tfrac{4}{3}\pi a^3 (\rho_2 - \rho_1) g \qquad [\text{D.6}]$$

Under conditions of steady state

$$F_f = F_g \qquad [\text{D.7}]$$

therefore $6\pi a v \eta = \tfrac{4}{3}\pi a^3 (\rho_2 - \rho_1) g$ or

$$v = \frac{2a^2 (\rho_2 - \rho_1) g}{9\eta} \qquad [\text{D.8}]$$

Equation [D.8] is called Stokes law. If the particle is not a sphere, then a is the radius of the equivalent sphere, and the size distribution is given in terms of equivalent spherical radii.[10]

The commonly used techniques start with a homogeneous sample of particles dispersed in a liquid. The general method is to sample the mass of particles at different heights at different times, so as to determine the time for a particle to fall from the liquid surface to a known depth in the liquid. Initially the composition of a volume of suspension at any depth remains constant, with as many particles falling into the volume as falling out. Eventually, however, the largest particles are lost from the suspension above the test depth. The velocity of those particles is

$$v = h/t \qquad\qquad [D.9]$$

where the velocity v can be converted to an equivalent Stokes radius by

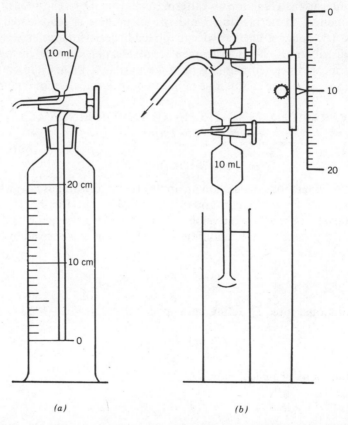

(a)   (b)

Figure D.5 Two modifications of the Andreason sedimentation pipette. In (a) the cylinder is filled with homogeneous suspension, the sampling pipette inserted, and the level h of the suspension recorded. Samples of the suspension are removed from the 0-cm level by means of an aspirator at a time t. Any decrease in concentration of suspension is due to the loss of particles with the Stokes sedimentation velocities greater than h/t. In (b) the sampling pipette is mounted on a movable arm so that the suspension can be sampled at various depths as well as at various times (Jelínek, 1970, p. 77).

Eq. [D.8]. The particles can be detected in various ways. In the Andreason pipette, samples of the suspension at a determined height are drawn off at various times and analyzed for percent solids (Fig. D.5). Any decrease in the percent solids at a time t is due to a loss of particles with Stokes velocity greater than h/t. In the Cahn sedimentation balance the increase in mass of particles deposited on a balance pan is monitored continuously. The mass on the pan increases linearly with time until the largest particles have all settled to the depth of the pan. The mass on the pan then increases more and more slowly corresponding to the deposition of only smaller and smaller particles. In the Micromeritics SediGraph or the Quantachrome Microscan, the solids concentration is measured continuously by the absorption of X rays. The advantage of using X rays as the light source is that the wavelength of X rays is so short that scattering and diffraction effects are insignificant for particles in this size range. The mass of particles in the beam is determined by Beer's law of absorption. The disadvantage is that the particles must absorb X rays. Many biological and polymeric materials cannot be sized by this method. The data are most directly presented as a cumulative particle size distribution. To decrease the time required for the analysis, the detector can be moved steadily up the sedimentation column. All these sedimentation techniques produce particle size distributions as mass distributions.

Sedimentation techniques are limited at the lower sizes by thermal gradients and Brownian motion, both of which disturb steady sedimentation. For narrow tubes, wall effects are significant, especially if the walls are not vertical (Kaye, 1981, p. 102). The analysis depends on each particle falling independently, which is only true for dilute suspensions. The upper limit of concentration is of the order of 1–2% by volume but must be established in each case by making measurements at lower concentrations. Clusters of particles, the effect of crowding, tend to settle as a unit and hence appear to be larger than they really are.

b.   Individual Particle Sensors

The following techniques depend on isolating and measuring individual particles, by running a dilute dispersion through a small sensing zone. Differences between methods depend on how the size of each particle is measured. The Coulter counter uses the change in electrical resistance of the sensing zone due to the presence of a particle, thus simultaneously detecting and measuring the size of the particle. It is a popular and well-established automatic instrument for this size range; its major limitation is that its use is confined to aqueous systems. It has the advantage that the change in resistance is directly related to the volume of the particle. The general name for such an instrument is resistazone counter. A careful examination of the shape of the electrical transient caused by a particle moving through the sensing volume enables the equipment to reject automatically signals from the presence of multiple particles in the sensing zone as well as to give some information about the shape of the particles by characteristics of the lead and the trail edge of the signal.

Pacific Scientific markets particle size analyzers that detect and measure the size of a particle in the sensing volume by its absorption of visible light. The general name for such an instrument is photozone counter (Fig. D.6). This sensing technique is not restricted to aqueous systems; it may be used for nonaqueous dispersions and for aerosols. These instruments are not used for particles less than one micron in diameter, that is, near the wavelength of visible light, as the absorption is then not a simple function of volume because of scattering.

The lower size limit to the sensitivity of the resistazone and photozone counters is essentially set by the dimensions of the sensing volume. To detect smaller particles, smaller sensing volumes are used. The limitation is that too small a sensing volume can be plugged by the larger particles. The commercially available equipment has built-in algorithms and mechanical devices to minimize the effects of plugging the sensing zone.

Other particle counters are designed to size individual particles as they flow through a small section of an illuminating beam. A particle crossing the illuminated measuring volume generates a pulse of scattered light. The pulse height determines the particle size and the pulse width determines its velocity. The Spectrex particle counter uses a scanning laser beam to sample a dilute suspension rather than passing it through a small sensing zone. The intensity of the forward-scattered light from individual particles is analyzed to give the

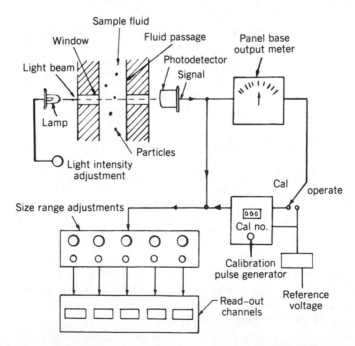

Figure D.6 The optical sensing and the flow pattern of a dispersion in a typical photozone counter. In the resistazone counter the lamp and the photodetector are replaced by electrodes (Allen, 1981, p. 92).

particle size. The output of the instrument after a few minutes of scanning is a histogram of the number of particles as a function of diameter.

All the techniques to sense individual particles have a fixed number of sensing channels or thresholds. The spacing of the channels is set by the operator, either by changing the sensing volume or resetting electronic limits, hence the data are easily represented by size histograms. All these methods respond to individual particles, hence the size distributions are number distributions.

c. Mercury Porosimetry

Mercury porosimetry is routinely used to evaluate pore volume and pore size distribution of porous and particulate materials in powder form. The essential idea is to use a nonwetting liquid, such as mercury, and to measure the volume of mercury that is forced into pores as a function of pressure. The radius of the pore is calculated by Laplace's equation [IIA.73] from the applied pressure, the surface tension of mercury, and an estimate of the mercury–solid contact angle. The pressure required to push mercury into void spaces depends on particle size and packing; therefore, void-volume distribution curves for granular or powdered materials, which are related to their size distributions, are readily obtained while the porosity of the powder is being measured. The pressure at which mercury enters interstitial or void space is much less than that required to penetrate porous particles, and so the effect of particle size is readily distinguished from that of pores. Lowell and Shields (1984) give a detailed description of mercury porosimetry.

3. PARTICLE-SIZING TECHNIQUES FOR ULTRAFINE PARTICLES ($< 0.5\ \mu m$)

a. Transmission and Scanning Electron Microscopes

These methods share with the optical microscope the advantages of direct observation and pictorial representation (imaging.) The major disadvantages are the effects of high vacuum and the high temperature created by impingement of the electron beam on the sample, which may alter its state. The preparation of the sample and its imaging are tedious and many pictures have to be taken. The limited depth of field of electron microscopes and artifacts due to taking microtome cross sections introduce sampling errors. In order to be observed materials have to be opaque to the electron beam, especially compared to other constituents in the dispersion, such as stabilizers or polymer matrices. The technique is both equipment intensive and labor intensive and so is expensive to run. The degree of magnification in the electron microscope is unknown, so that reference materials are required for calibration. In spite of these defects, the electron microscope gives the most reliable information about submicron particles.

b. Turbidity

Turbidity, or cloudiness, is the most apparent optical effect to detect the presence of particles in a medium (the Tyndall effect). Turbidity is important in water treatment and purification, in monitoring bacterial growth in microbiological research, quality control for foods, milk, cheese, and beverage manufacture, determination of sulfates and silica concentrations, and sensing the onset of precipitation or nucleation. The turbidity of a dispersion is measured by the attenuation of light passing through a thickness d:

$$I = I_0 \exp(-\tau d) \qquad\qquad [D.10]$$

where I_0 is the intensity of the incident beam, I is the intensity of the attenuated beam, and τ is the turbidity coefficient. Turbidity measurements are useful to size those particles that do not absorb light and are less than 30–40 nm in diameter, that is, Rayleigh scatterers. All the light lost from the incident beam is by scattering. The total light scattered out of the incident beam by a single particle can be calculated by integrating the Rayleigh equation, Eq. [A.2], over all angles. If the dispersion is monodisperse, the following relation, which gives particle volume V as a function of measured turbidity τ, is obtained (Kerker, 1969, p. 326):

$$V = \frac{[(n^2 + 2)/(n^2 - 1)]^2 \lambda^4 \tau \rho}{(24\pi^3 c)} \qquad\qquad [D.11]$$

where n is the ratio of the refractive indices of the dispersed phase to that of the medium, λ is the wavelength of the light in the medium, ρ is the density of the particles, and c is the concentration of the suspension in grams per cubic centimeters. Equation [D.11] shows that flocculation of a fixed mass of dispersed material, as measured by V, can be followed by turbidity measurements, at least within the limits of Rayleigh scattering. Generally, white light rather than monochromatic light is used and the sample is polydisperse; nevertheless, turbidity gives a light-scattering mean volume. Rayleigh scattering increases as the sixth power of the particle size, so that large particles greatly influence the average size.

Commercially available turbidimeters are calibrated in turbidity units (TU). Originally, one TU was equal to the turbidity caused by 1 ppm of a standard suspended silica.

Turbidity is often measured today by applied nephelometry, which measures light scattered by particles at right angles to the incident beam (Fig. D.7). The standard is one nephelometric turbidity unit (NTU). Standards are available from equipment manufacturers. These are aqueous suspensions of formazin, a polymer formed by the condensation reaction between hydrazine sulfate and hexamethylenetetramine.

The turbidimetric method may also be applied in conjunction with Mie's theory of light scattering for uniform spherical particles less than 2000 nm in

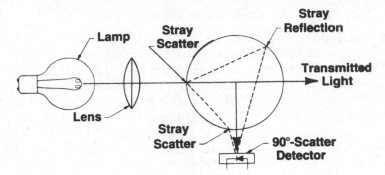

Figure D.7 The optics of an apparatus used to measure particle size by nephelometry. The light scattered by the particles in the sample cell is detected at 90°. (Courtesy of Hach Co., Loveland, CO.)

diameter. The particle size is obtained by matching the measured angular dependence of the intensity of scattered light with tabular or graphical computations. Polydisperse systems, other than very narrow distributions, cannot be treated in this way, as the mathematical difficulties of deconvoluting the data are insuperable.

c.   Laser Light Scattering

Lasers are preferred for light-scattering studies as they are monochromatic, stable and intense. Particle size distributions of suspended particles are measured by the scattering of laser light in either of two modes: The first is to measure the average intensity of the scattered light as a function of scattering angle; the second is to measure the time dependence of the scattered light at a fixed scattering angle (quasi-elastic light scattering.) For monodisperse particles either mode of measurement is sufficient to determine particle size; for polydisperse suspensions, however, with sizes greater than a few hundredths of a micron, the convolution of overlapping angle-dependent scattering functions makes the analysis of the angle-dependent intensity too difficult. In this range of sizes the use of quasi-elastic scattering is preferred.

For particles smaller than a few hundredths of a micron the method of analysis is due to Zimm and is generally described in terms of molecular weight measurements rather than particle size measurements. The essence of Zimm's method is to measure the intensity of light as a function of concentration and scattering angle and to extrapolate the data to zero concentration and zero angle, from which the average molecular weight and z-average radius of gyration can be obtained. The average molecular weight and the radius of gyration can be combined to give information about the width of the size or weight distribution (Kerker, 1969, 432f). Commercial equipment capable of simultaneous multiangle detection and commercial equipment for very low angle detection, both optimized for Zimm analysis, are available. (See Chapter IA).

d. Quasi-Elastic Light Scattering

Light scattering as developed by Rayleigh and Mie measures the time-average intensity. In the 1960s it was noticed that the time fluctuations of the intensity could be used to obtain information about the motion of dispersed particles or macromolecules in solution. The motion is Brownian movement. The light scattered by a moving particle is Doppler shifted by a tiny amount with respect to the incident frequency. Each moving particle scatters light at a slightly different frequency depending on its velocity with respect to the incident beam. A photodetector senses the light scattered from a number of particles and generates a time-varying signal with a scale of the order of milliseconds or even microseconds, depending on the velocity of the particles. Brownian motion is inversely proportional to particle size. The time dependence of the fluctuating intensity therefore is a function of particle size. The autocorrelation function of these fluctuations in intensity has a simple dependence on particle size. Commercial instruments take the input from the photomultiplier tube and calculate the autocorrelation function versus delay time automatically.

For a monodisperse system of spherical particles in Brownian motion, the normalized autocorrelation function $g^{(1)}(\tau)$, is an exponential decay:

$$g^{(1)}(\tau) = \exp(-\Gamma\tau) \qquad [\text{D.12}]$$

where τ is the delay time of the autocorrelation function and Γ is the decay constant given by

$$\Gamma = D\kappa^2$$

$$\kappa = (4n\pi/\lambda)\sin\frac{\theta}{2} \qquad [\text{D.13}]$$

where n, λ, η, and θ are the refractive index of the medium, wavelength of the light, viscosity of the medium, and scattering angle, respectively. The diffusion coefficient D is inversely proportional to particle size, as shown by the Einstein equation [C.2].

For a polydisperse system of spherical particles in Brownian motion, the normalized autocorrelation function, $g^{(1)}(\tau)$, is an integral of exponentials:

$$g^{(1)}(\tau) = \int_0^\infty F(\Gamma)\exp(-\Gamma\tau)\,d\Gamma \qquad [\text{D.14}]$$

where $F(\Gamma)$ is the normalized distribution of decay constants. The mathematical problem is to invert the integral to obtain the distribution of decay constants. Each decay constant corresponds to a particle size through its dependence on the diffusion coefficient. The problem has been solved by the method of cumulants,[11] by the method of regularization,[12] and by a nonnegatively constrained optimization.[13,14] The technique of quasi-elastic light scattering is a more practicable procedure to determine particle size distribution than the classical

Rayleigh–Mie procedure, as the simplicity of Eq. [D.14] results in a more tractable mathematical problem. Commercial instruments are available to measure the autocorrelation function and to do the complete mathematical analysis of the data.

The great advantage of all light-scattering techniques is that particle size measurements can be made in any solvent, as the sample need only be held in a transparent container. The disadvantages of all light-scattering techniques are that dispersions must be dilute and that the effects of particle shape (other than spherical) are not easily estimated.

e.   Centrifugation

Gravitational sedimentation as a technique to measure particle size is limited to coarse particles, but centrifugation allows the range to be extended to smaller sizes. It is also useful for larger particles whose density does not differ much from that of the dispersion medium. The efficiency E of a centrifuge with rotor arm of length r centimeters, moving at ω radians per second, in terms of multiples of terrestrial gravitation is

$$E = \frac{r\omega^2}{g} \qquad [D.15]$$

Efficiency thus depends on the angular acceleration, $r\omega^2$. Centrifuge efficiency should be in the range of 100–20,000. An ultracentrifuge, driven by compressed air, of which part drives an air turbine and another part forms an air bearing, is in the range of 200,000–400,000, depending on the radius r.

The commercial instrument in most common use to measure particle sedimentation by centrifugation is the Joyce–Loebl disk centrifuge (Fig. D.8).[15] The rotor is a transparent, hollow, plastic disk that is filled, while spinning, with the suspension medium, or serum. The suspension is then injected as a second concentric layer situated nearer the center. Particles settle toward the circumference and are detected by the absorption of light at a known distance from the starting line. The output of the photodiode appears on a chart recorder, and the particle size distribution is obtained by data reduction.

In a centrifuge particles do not sediment at constant velocity as in gravitational sedimentation, but they accelerate as they move farther and farther from the axis of rotation. The time t required for a particle of diameter d to move from an initial axial position r_0 to a final axial position r is

$$t = \frac{18\eta \ln(r/r_0)}{(\rho_2 - \rho_1)\omega^2 d^2} \qquad [D.16]$$

where η is the viscosity of the fluid, $(\rho_2 - \rho_1)$ is the density difference between the particle and the fluid, and ω is the angular velocity of the centrifuge. The time

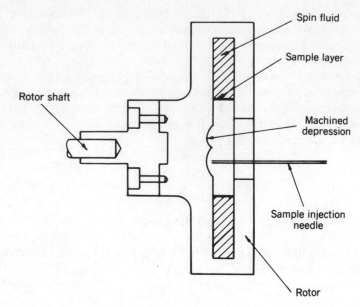

Figure D.8 The Joyce–Loebl disk centrifuge. The sample is injected on top of the liquid inside a spinning disk. A stationary optical sensor near the outside edge of the spinning disk detects changes in light absorption (Stockham and Fochtman, 1977, p. 81).

required for the analysis can be adjusted by varying the angular velocity. The method is absolute and needs no calibration. The disadvantage of a transparent plastic disk is that all solvents cannot be used. A variation is to use matched glass tubes mounted in an opaque disk with apertures to allow light to pass.

f. Hydrodynamic Chromatography

When a suspension is injected into a stream of liquid flowing through a column packed with small solid spheres, the particles separate according to size, the largest particles eluting first. This phenomenon can be understood by considering the velocity profile of liquid flowing through a small channel. The liquid flow is slowest at the walls of the channel. The smaller a particle, the closer it can approach the walls of the channel, and the slower it will move on the average. The largest particles, which are confined by their size to the center of the channel, actually move through the packed column faster than the *average* fluid velocity (Fig. D. 9).

   The equipment for hydrodynamic chromatography (HDC) is essentially the same as used in liquid chromatography, the major difference being the use of inert packing. If the particles are attracted to the packing, the possibility of separation based on size is lost. The complexity of phenomena in HDC requires that calibration standards be used. Commercially available equipment uses ultraviolet absorption to detect eluting particles. Because particles eluting from a

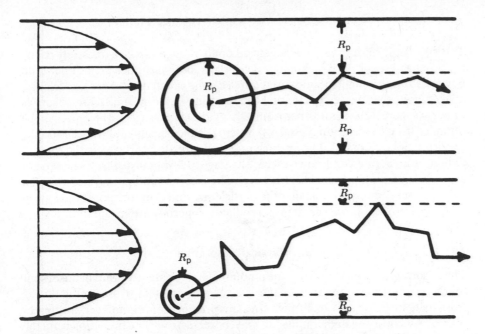

Figure D.9 The rheological model of separation by hydrodynamic chromatography. The average velocity of the larger particles confined to the center of the tube is greater than the average velocity of the smaller particles, which can diffuse closer to the wall where the fluid flow is less. R_p is the particle radius (Barth, 1984, p. 279).

column are detected optically, the measured absorption must be corrected for scattering effects. The commercial unit is designed for aqueous dispersions.

g. Sedimentation Field-Flow Fractionation

Sedimentation field-flow fractionation is a particle-sizing technique that combines the flow separation used in hydrodynamic chromatography and the size separation in centrifugation. The sample (generally $5 \mu L$ of a 1% suspension) is injected into a spinning annular channel through which liquid is flowing. The larger particles are thrown to the outer edge of the spinning channel by centrifugal force. The liquid at the outer edges of the channel flows more slowly than that in the center, so that the larger particles at the outside move more slowly than the smaller particles in the center. By adjusting the flow rate and the centrifugal force, the separation of the particles by size can be optimized. To optimize size separation the flow rates and the centrifugal forces are varied independently. The mass of particles as a function of size is measured by an ultraviolet detector on the exit port of the channel. The method needs no calibration as the laminar flow of liquid in the channel and the centrifugal forces can all be calculated. The engineering challenge is to form tight seals between the spinning channel and the stationary inlet and outlet ports. The commercial equipment is designed for aqueous dispersions.

4. SURFACE AREA BY ADSORPTION

The specific surface area of a powdered solid is an indirect expression of the particle size. Adsorption techniques deliver the specific surface area rather than particle size. Although specific surface area can be calculated from a particle size distribution, the converse cannot be done. But if the particles are taken as uniform spheres, cubes, or other isometric solids, an average particle size can be obtained from the relation $d = 6/(\rho A)$, where ρ is the density and A is the specific surface area (area/mass). The average radius obtained in this way is the ratio of volume to area, or d^3/d^2, known as a 3:2 average to distinguish it from the ratio of area to diameter, known as a 2:1 average. The determination of surface area of a solid is more accurate the smaller the particle size and is more appropriate when interfacial properties rather than particulate properties are in question.

a. Physical Adsorption

The adsorption isotherm relates the amount of adsorbate per gram of adsorbent to each equilibrium pressure of the gas to constant temperature (Fig. D.10). Adsorbate molecules at small pressures are present as a monolayer; at greater pressures second and higher molecular layers are formed. The transition from the first to the second layer is usually accompanied by a sharp decrease in the binding energy of the adsorbate, as the first layer is directly held by the substrate whereas succeeding layers are farther away and in direct contact only with adsorbed molecules. This transition, therefore, is often marked by a change of slope of the adsorption isotherm. The inflection point of the isotherm occurs approximately at the pressure where the substrate is covered with a fully compressed monolayer, designated n_m. A fully compressed monolayer is an artificial concept, which is never actually attained, because succeeding layers form before the first is saturated. The adsorption isotherm may be determined either by gradual addition of gas (see Ross and Olivier, 1964) or by continuous flow of an adsorbable gas in an inert carrier gas (see Lowell and Shields, 1984).

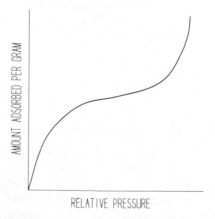

Figure D.10 A simple multilayer adsorption isotherm typical of nitrogen at 77 K.

A theory designed to evaluate n_m from adsorption data was proposed by Brunauer, Emmett and Teller.[16] It is usually referred to as the BET theory. It supposes that the first layer of adsorbate is laid down according to the Langmuir equation:

$$p = \frac{K\theta}{1-\theta} \qquad [D.17]$$

where p is the equilibrium pressure, θ is the degree of coverage, given by n/n_m, where n is the number of moles of adsorbate per gram of adsorbent at pressure p, and K is a constant. The following assumptions underlying this equation:

(a) The adsorption is only monomolecular
(b) The adsorption is localized on specific sites
(c) The energy of adsorption on the solid substrate is uniform and characteristic
· of the solid and the adsorbate
(d) No adsorbate–adsorbate attraction exists

The BET theory goes on to postulate that simultaneous Langmuir-type adsorptions take place in successive layers. Further assumptions are required to simplify the mathematical treatment, namely, that the energies of adsorption in the second and succeeding layers of adsorbate are also uniform, and that all of these are equal to the heat of liquefaction of the adsorbate. The final expression of the BET theory is

$$n = \frac{n_m C p}{(p_0 - p)[1 + (C-1)(p/p_0)]} \qquad [D.18]$$

where p_0 is the saturation vapor pressure of the adsorbate, n_m and C are constants, which are determined from the data either graphically or by a linear least-squares fit: n_m has the same significance in the BET theory as it has in the Langmuir theory.

The BET theory is the basis of a method to measure the specific surface area of a solid, using the value of n_m and Eq. [D.25]. This method has been experiment-ally shown to be sufficiently accurate for most practical purposes and has successfully met a requirement for a rapid estimate of particle size. By so doing, it has made possible great advances in both theoretical and practical studies of surface chemistry and physics. We must not, however, confuse the triumphs of the BET theory in this practical application with its claims as a model of the adsorption process—in this latter respect, it is obviously far from satisfactory. The Langmuir model with its unjustifiable assumptions is the basis of the BET model, and the additional assumptions introduced in the BET derivation merely add further improbabilities to the total picture. Nevertheless the following admonition (A. W. Rücker, Presidential Address to the British Association meeting at Glasgow, September, 1901) is pertinent:

From the practical point of view, it is a matter of secondary importance whether our theories and assumptions are correct, if only they guide us
to results in accord with facts.... By their aid we can foresee the results of
combinations of causes which would otherwise elude us.

At low temperatures a more realistic model than that of Langmuir is to assume
that the adsorbate behaves as a nonideal two-dimensional gas. The virial
equation of state of such a gas is

$$\frac{\pi A}{kT} = 1 + \frac{nB}{A} + \frac{n^2 C}{A^2} + \cdots \tag{D.19}$$

where π is the two-dimensional (spreading) pressure, A/n is the area per molecule,
k is the Boltzmann constant, and B and C are the second and third virial
coefficients. The adsorption isotherm equation can be obtained from any
equation of state by means of the Gibbs theorem in the form

$$\int d\ln p = \frac{1}{kT} \int \frac{A}{n} d\pi \tag{D.20}$$

Solving Eq. [D.19] for $d\pi$ and then substituting in Eq. [D.20] gives

$$\ln p = \ln\frac{n}{A} + \frac{2nB}{A} + \frac{3n^2 C}{2A^2} + \cdots + \ln K \tag{D.21}$$

where $\ln K$ is the integration constant. Equation [D.21] has been shown to
describe the adsorption isotherms of argon on graphitized carbon black in the
monolayer region.[17]

Few substrates, of which graphitized carbon black is one, are sufficiently
uniform in terms of their adsorptive energies for adsorption of a gas to proceed
according to the description given by Eq. [D.21]. Heterogeneity of adsorptive
energies is the rule rather than the exception. Unless substrate heterogeneity is
taken into account in the analytical treatment of the adsorption isotherm, the
model lacks an essential feature and the evaluation of n_m and consequently of the
specific surface area, is open to doubt.

The variation in adsorptive potential causes the density of the adsorbed
molecules (molecules per unit area) to vary across the surface; hence the
adsorbate cannot be considered as a single surface phase. An analysis can
proceed, however, if the adsorbate is treated as a set of independent surface
phases, each with its own homogeneous density. This entails that the substrate be
regarded as composed of different unisorptic patches. (A liquid surface is
structureless and isoenergetic; a perfect crystalline surface is homotattic; both are
unisorptic.) A realistic model of an heterogeneous substrate can be based on the
following premises:

(a) The whole adsorptive substrate is made up of homotattic patches, each with a different adsorptive energy for the adsorbate, and on each of which monolayer adsorption can be described by an adsorption isotherm equation, such as Eq. [D.21], which has been established for an homotattic substrate.

(b) Elemental patches are filled simultaneously, though not to the same density, subject to the condition that the adsorbed phase on each patch has the same chemical potential. These elemental patches are assumed to be sufficiently large that boundary effects are unimportant.

(c) The number of moles adsorbed on the whole heterogeneous substrate is obtained by summing the individual values of the number of moles adsorbed on each patch, n_i, over all the patches.

$$n(p) = \sum_i n_i(p) = \sum_i A_i f\left(\frac{p}{K_i}\right) \qquad \text{[D.22]}$$

where A_i is the area of the ith patch and $f(p/K_i)$ is found by solving Eq. [D.21] for n_i/A_i. If the heterogeneity of the substrate can be approximated by a continuous distribution function, then

$$n(p) = A \int dK \, P(K) f\left(\frac{p}{K}\right) \qquad \text{[D.23]}$$

where A is the specific surface area (m^2/g); $P(K)$ is the probability of a patch having the Henrys law constant K. The distribution of the K's is a description of the heterogeneity of adsorptive energies of the substrate. Simple techniques for fitting experimental data to Eq. [D.22] or [D.23] are available (CAEDMON.)[18]

b.   Adsorption from Solution

Adsorption from solution as a procedure to measure specific surface area is much simpler than any technique that requires vacuum apparatus and is sufficiently rapid to make it attractive for routine determinations, quick estimates, or quality control. Adsorption from solution is a common phenomenon, taking place in dyeing, tanning, decolorization, lubrication, catalysis, flotation, and detergency. Frequently used as adsorbates for determination of specific surface area are fatty acids, dyes, and molecules containing a radioactive isotope. In using this method, two quantities are to be ascertained: the monolayer capacity of the substrate and the area of the molecule in a close-packed monolayer. Both these determinations are subject to error, even more than in gas adsorption (Kipling, 1965, Chapter 17). Giles and Nakhwa[19] list the attributes that would be required for a solute to give reliable results for specific surface area:

(a) It should be highly polar to ensure adsorption by polar solids.
(b) It should have hydrophobic properties to permit adsorption by nonpolar solids.

(c) It should be a small molecule, preferably planar.
(d) It should not be highly surface active, as micelle formation at the surface is undesirable.
(e) It should be colored for ease of analysis.
(f) It should be readily soluble in water for convenience in use but also soluble in nonpolar solvents so that it can be used with water-soluble solids.

Giles et. al.[20] give details of a number of dyes recommended for surface area determinations, Table D.2. When used to examine a wide variety of oxide, carbon, and halide surfaces, ranging in area from 2 to $100 \, m^2/g$, these dyes give results lying within 10% of those obtained by nitrogen gas adsorption.

The adsorption isotherms of these dyes on many solids appear to be Langmuir type, Eq. [D.17],

$$\frac{n}{n_m} = \theta = \frac{c_2/K}{1 + c_2/K}$$ [D.24]

The evaluation of n_m is obtained from a plot of $1/n$ versus $1/c_2$, which has an intercept of $1/n_m$ and a slope of K/n_m. The surface area of the adsorbent can then be calculated from the value of n_m, by assuming that the adsorption is confined to a single molecular layer and that each adsorbed molecule covers an area determined by further assumptions about the molecular orientation of the adsorbate in the saturated monolayer. One may assume, for instance, that the adsorbate molecules pack as they do in their crystalline form, taking some

Table D.2 Details of Recommended Dyes[a]

Name	C I. No.	Molecular Area, Flat $\sigma_0, (\text{Å}^2)$	Aggregation Number N	Molar Extinction Coefficient ($\times 10^{-4}$)	λ_{max} (nm)
Anionic Dyes					
Orange II	15510	120	3.0	2.4	480
Naphthalene Red J.	15620	150	3.6	2.1	500
Cationic Dyes					
Methylene Blue BP	52015	120	2.0	4.0	610
Crystal Violet BP	42558	225	3.6	3.6	590
Victoria Pure Lake Blue BO	42598	270	9.0	6.8	620

[a]The molecular areas of these dyes are based on molecular cross-sections obtained by calibration against BET measurements with nitrogen gas.

principal plane as their mode of two-dimensional arrangement on the substrate; or, alternatively, one can derive the molecular volume from the density of the adsorbate in its liquid form, and take the two-thirds power of that volume as the effective cross-sectional area of each molecule. A factor for square or for hexagonal packing is also included to give a reasonable estimate of σ_0, the covering area per molecule. More and more frequently the method of dye adsorption is calibrated against the application of the BET theory to nitrogen gas adsorption, using an adsorbent selected to give an answer for the molecular area that lies within 10% of the value obtained by one of these calculations based on crystalline form or bulk density of the adsorbate. With this much agreement, the answers forthcoming are considered satisfactory. The specific surface of the adsorbent is then derived as

$$A = n_m \sigma_0 N_0 \qquad \qquad [D.25]$$

where N_0 is Avogadro's number.

In spite of the uncertainties in the technique, its extreme simplicity makes it the method of choice for many practical problems. The major experimental consideration is to be able to adjust the concentration of dye and the concentration of particles in suspension so that detection is sensitive enough at the initial and final concentrations. Once the proper ranges have been established (usually by trial and error) the method is quite satisfactory and may be used for process control.

CHAPTER IE

Processing Methods for Making Emulsions and Suspensions

Processing methods are of two types: those that emulsify or that pull agglomerates apart by shear forces and those that comminute aggregates by fracture. Different equipment is used for each process. Equipment to generate shearing forces need only provide sufficient energy to attenuate an immiscible liquid within another or to separate agglomerates. Comminution requires higher energy input to break tightly bound aggregates or to shatter coherent solids.

The generation of high-shear forces requires narrow gaps, or high rates of flow, or both. The following are examples of equipment in which one or the other of these modes to generate high-shear forces is used. A Banbury mixer, used to blend carbon and other fillers into rubber or plastics, functions at low speed with a loading of high millbase viscosity. The operation of a colloid mill, used to emulsify liquids, depends on flow through a narrow gap. A Kady mill, used to disperse powders in liquids, functions at high speed with a loading of low viscosity (Fig. E.1).

The breaking of aggregates requires impact, which is favored when unhindered by viscous resistance. The fineness of the grind depends on the size of the grinding media: the smaller the media, the finer the grind. Ball or pebble mills are rotated on a shaft or by rollers and the impacts are driven by gravity. Finer media require stirring.

Table E.1 lists high-shear mills in order of increasing viscosity of the millbase and high-impact mills in decreasing order of the size of the grinding media. Appendix K lists the names and addresses of some manufacturers of processing equipment.

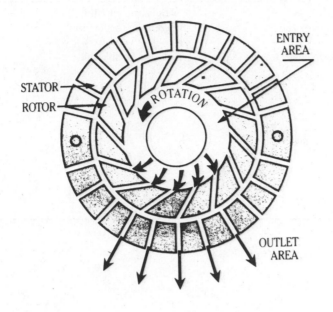

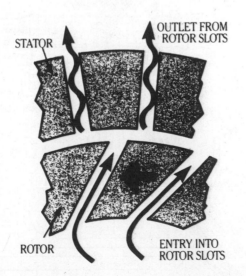

Figure E.1 A Kady mill rotor and stator. (Courtesy of the Kinetic Dispersion Corp.)

57

Table E.1 Classification of Milling Equipment

High-Shear Mills (increasing viscosity)	High-Impact Mills (decreasing media size)
VariKinetic dispersers Kady Mills	Ball and pebble mills
Colloid mills Microfluidizer Homogenizer Sonolator Ultrasonic mills	Attritor
Three-roll mills	Sand mills Dyno-Mill
Banbury mixer	

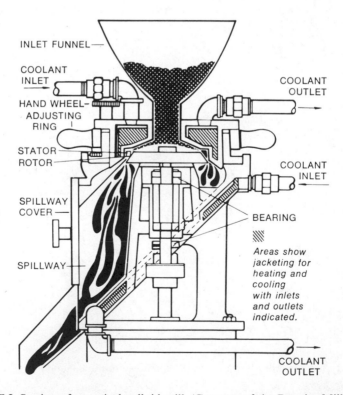

Figure E.2 Section of a vertical colloid mill. (Courtesy of the Premier Mill Corp.)

1. HIGH-SHEAR MILLS

Gaulin's VariKinetic disperser features variable-pitch impeller vanes, the angle of which can be adjusted while the unit is operating. The tip of the vanes moves at 4000–5000 ft/min. Kady mills are continuous mixers using a high-speed rotor turning within a labyrinth stator in which particles are accelerated to velocities in excess of 10,000 ft/min.

Colloid mills (Fig. E.2) operate by sucking liquids through a narrow gap between a high-speed rotor and a stator. The rotor is dynamically balanced and can rotate at speeds of 1,000–20,000 rpm. The gap between the rotor and stator surfaces is adjustable down to a thousandth of an inch. The rotor is sometimes grooved. The liquids are recirculated and the stator is water cooled.

In the Microfluidizer two streams impact at high velocity and high pressure in precisely defined microchannels within the interaction chamber. The process pressure can be varied from 500 to 20,000 psi; the process stream accelerated to velocities of 1500 ft/s. Fine droplets with a narrow size distribution are produced by a combination of shear, turbulence, impact, and cavitation forces. The equipment may be used to produce emulsions, suspensions, and foams.

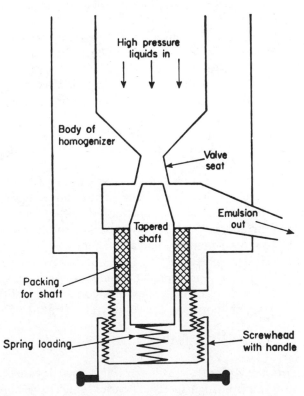

Figure E.3 Section of a single-stage homogenizer (Sherman, 1968, p. 11).

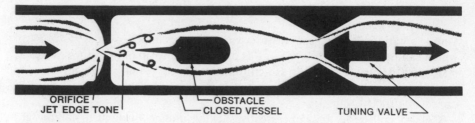

ORIFICE
JET EDGE TONE   OBSTACLE
CLOSED VESSEL   TUNING VALVE

Figure E.4 Schematic diagram of the Sonolator. (Courtsey of the Sonic Corp.)

Homogenizers produce emulsions by forcing the mixture at high pressure (up to 12,000 psi) through a small orifice against a spring-loaded plunger (Figure E.3). The velocity of the liquid through a small gap creates high-shear forces. The lack of moving parts makes it preferable to the colloid mill. A familiar application is to reduce the size of fat globules in milk.

A Sonolator is a variation of the homogenizer (Figure E.4). A jet of liquid at pressures between 200 and 2000 psi is pumped through an orifice against a bladelike obstacle in the jet stream. The turbulent flow of the liquid causes the blade to resonate at ultrasonic frequencies. A high level of cavitation, turbulence, and shear is the result. The Sonolator can be tuned while in operation to a peak of acoustic intensity.

Ultrasonic activators convert conventional 60-Hz line frequency to 20,000 Hz. The high frequency is fed to an electrostrictive element, which converts the signal to mechanical vibrations in tips of various shapes called horns. The tip of the horn is immersed in the liquid in which the ultrasonic vibrations cause cavitation. The units are primarily laboratory instruments. For continuous use the vibrating horns must be cooled.

A three-roll mill is a set of rolls rotating in opposite directions with a small clearance between the rolls. High-shearing action is exerted on the agglomerate, causing it to break up. High viscosity of the loading is important for this type of mill; therefore, percent solids is kept as high as possible. Its advantages are that it handles viscous materials such as printing inks. The disadvantage is that it is open to the air and so cannot be used with volatile solvents.

In the Banbury mixer or mill two kneading arms or rotors encased in a mixing chamber rotate in opposite directions and at different speeds. They are so shaped that the plastic mixture is pressed against the walls of the chamber, forming a wedge during the kneading operation. The wedge is continuously formed and sheared, while the motion of the rotors ensures good mixing of the batch. The chamber is heated to mill polymers above their glass transition temperatures.

2. IMPACT MILLS

The Micronizer is a dry-process machine in which particles are fluidized in two opposed streams of air from high-speed jet nozzles, which project particle against particle at high kinetic energies. The nozzles are precisely aimed—within fractions of a second in degrees of arc. The carrier fluid pressure and feed materials are equivalent on both sides. The mixture is classified by being blown into a vortex. The smaller particles follow the streamlines of the air and exit with it; the larger particles are recirculated until ground small enough to escape. The equipment is well suited to break up soft solids such as carbon blacks, molybdenum disulfide, and polymers.

A ball mill is any rolling mill in which steel or iron balls are used as the grinding medium (Fig. E.5). The cylinder is usually made of steel. A pebble mill uses flint

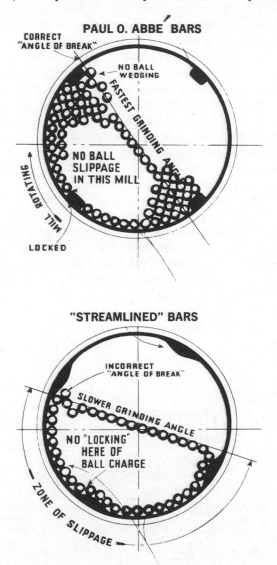

Figure E.5 Milling action in a ball mill. (Courtesy of Paul O. Abbé, Inc.)

pebbles or porcelain balls as the grinding medium and the inside of the cylinder is lined with some nonmetallic substance. The cylinder is rotated by a shaft or on rollers. For efficient grinding, the speed, the amount and size of grinding medium, the amount of loading, and its viscosity (if a wet process) are all adjusted so that the top layers of balls form a cascading, sliding stream moving faster than the lower layers, thus causing a grinding action between them. If the viscosity is too high or the rotation too rapid, the balls are carried all the way around the cylinder without any grinding action. Since the grinding depends on the cascading motion, the balls have to be dense and large, $\frac{1}{4}$–2 in. diameter.

In all stirred ball mills smaller grinding media can be used than in gravity-driven ball mills, resulting in a finer grind. In the Attritor the grinding media are moved by a series of staggered horizontal rods attached to a central shaft (Fig. E.6). Typically the grinding media are $\frac{1}{8}$–$\frac{1}{2}$ in. steel shot. By the motion of the rotating rods, both shear fields and impacts are generated, leading to more efficient power consumption and shorter grinding times. The regularity of the motion allows a process to be scaled up from pilot to manufacturing size. Recirculation of the dispersion can be provided, as can cooling for prolonged grinding. Temperature control and constant rate of stirring make stirred mills scalable.

Sand mills consist of several impeller disks attached to a shaft rotating at speeds of 5,000–12,000 rpm. The standard grinding medium is Ottawa sand, 20–30 mesh (0.6–1 mm). This is the smallest grinding medium in use and so makes the finest dispersions. The dispersion is pumped into the bottom of the sand mill and is drained from the top in a continuous flow often resulting in an average residence time of only a few minutes. Recirculating the dispersion from a large holding tank through the sand mill promotes efficient size reduction by providing intermittent milling action. Frequently several sand mills are connected in series, each succeeding mill filled with finer grinding media.

A variation of the small-media stirred mills is the Dyno-Mill. The working

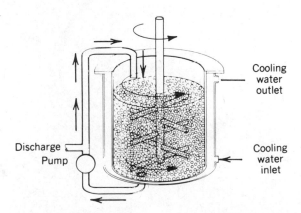

Cooling water outlet

Discharge Pump

Cooling water inlet

Figure E.6 The Attritor—a stirred ball mill. (Courtesy of Union Process, Corp.)

system is completely enclosed, jacketed for temperature control, and the grinding chamber is mounted horizontally. Agitator disks, connected to a fast-rotating shaft, make the grinding beads move in a circle at high speed. The beads are made of ceramic, steel, or glass with diameters of 0.1–3 mm. The loading is moved through the grinding chamber by means of a variable feed pump.

3. CHEMICAL PROCESSING AIDS

The production of emulsions in a high-shear mill requires the presence of an emulsifying agent, since, if the interface is not stabilized, coalescence soon occurs. The emulsifying agent is usually added to the system before emulsification. Surface-active solutes contribute to the mechanical breakdown of agglomerated particles by promoting internal wetting. Also, by reducing the viscosity of the system and keeping the millbase fluid, surface-active solutes aid the action of ball and sand mills. Another way in which a surface-active solute enhances milling action is an effect discovered by Rehbinder and his colleagues,[21-23]. The Rehbinder effect is the reduction of strength of solids by the action of surface-active solutes. Originally this effect was interpreted simply as a lowering of surface energy by adsorption and hence reduction of the work required to produce new interface; but it must also include the creeping of surface-active substances along grain boundaries or dislocations while the body is under stress. Examples of the effect occur with materials as diverse as pure metals and various types of rock under the action of rock drills or during crushing and grinding.

The various effects introduced by the presence of a surface-active solute in wet milling are: (a) stabilizing new interfaces by adsorption, (b) wetting of the constituent particles (primary or aggregate) of an agglomerate with elimination of air, (c) reduction of viscosity by preventing flocculation, and (d) the Rehbinder effect. All these effects operate simultaneously and hitherto have proved impossible to differentiate in practice.

These effects of surface-active solutes all depend on adsorption at the liquid–solid interface. The rate of adsorption is frequently slow, particularly with polymers (Section IV.C.1d), so that intermittent milling is often found to be more efficient and sometimes more effective than continuous action. After some fresh interface is created, a quiescent period of time is allowed for adsorption to take place before more interface is created.

References for Part I

1. Tyndall, J. On the blue colour of the sky, the polarization of skylight, and on the polarization of light by cloudy matter generally, *Phil. Mag.* **1869**, [5], *37*, 384–394.
2. Mie, G. Beitrage zur optik trüber medien, speziell kolloidaler metallösungen (Contributions to the optics of diffusing media, particularly of colloidal metals in solution), *Ann. Phys. (Leipzig)* **1908**, [4], *25*, 377–445.
3. La Mer, V. K. Nucleation in phase transitions, *Ind. Eng. Chem.* **1952**, *44*, 1270–1277.
4. Dauchot, J.; Watillon, A. Optical properties of selenium sols I. Computation of extinction curves from Mie equations, *J. Colloid Interface Sci.* **1967**, *23*, 62–72.
5. Watillon, A.; Dauchot, J. Optical properties of selenium sols II. Preparation and particle size distribution, *J. Colloid Interface Sci.* **1968**, *27*, 507–515.
6. McKennell, R. Cone-plate viscometer, *Anal. Chem.* **1956**, *28*, 1710–1714.
7. Einstein, A. "Über die von der molekularkinetischen theorie der wärme geforderte bewegung von in ruhenden flüssigkeiten suspendierten teilchen" (On the movement of small particles suspended in a stationary liquid demanded by the molecular-kinetic theory of heat), *Ann. Phys. (Leipzig)* **1905**, [4], *17*, 549–560; also (Engl. transl.) *Investigations on the Theory of Brownian Motion*, Einstein, A.; Methuen: London, 1926, 1–18.
8. Kerker, M. The Svedberg and molecular reality, *Isis* **1976**, *67*, 190–216.
9. Kerker, M. Brownian movement and molecular reality prior to 1900, *J. Chem. Ed.* **1974**, *51*, 764–768.
10. Siebert, P. C. "Simple sedimentation methods, including the Andreason pipette and the Cahn sedimentation balance," in *Particle Size Analysis*; Stockman, J. D.; Fochtman, E. G., Eds.; Ann Arbor Science: Ann Arbor, MI, 1977; pp. 45–55.
11. Koppel, D. E. Analysis of macromolecular polydispersity in intensity correlation spectroscopy: The method of cumulants, *J. Chem. Phys.* **1972**, *57*, 4814–4820.

12. Provencher, S. W.; Hendrix, J.; De Maeyer, L.; Paulussen, N. Direct determination of molecular weight distributions of polystyrene in cyclohexane with photon correlation spectroscopy, *J. Chem. Phys.* **1978**, *69*, 4273–4276.

13. Grabowski, E. F.; Morrison, I. D. "Particle size distributions from analyses of quasi-elastic light-scattering data," in *Measurement of Suspended Particles by Quasi-Elastic Light Scattering*, Dahneke, B., Ed.; Wiley: New York, 1983; pp. 199–236.

14. Morrison, I. D.; Grabowski, E. F.; Herb, C. A. Improved techniques for particle size determination by quasi-elastic light scattering, *Langmuir* **1985**, *1*, 496–501; Herb, C. A.; Berger, E. J.; Chang, K.; Morrison, I. D.; Grabowski, E. F. Using quasi-elastic light scattering to study particle size distributions in submicrometer emulsion systems, *A.C.S. Symp. Ser.* 1987, **332**, 89–104.

15. Puretz, J. "Centrifugal particle size analysis and the Joyce-Loebl disc centrifuge," in *Particle Size Analysis*; Stockman, J. D.; Fochtman, E. G., Eds.; Ann Arbor Science: Ann Arbor, MI, 1977; pp. 77–87.

16. Brunauer, S.; Emmett P. H.; Teller, E. Adsorption of gases in multimolecular layers, *J. Am. Chem. Soc.* **1938**, *60*, 309–319.

17. Morrison, I. D.; Ross, S. The second and third virial coefficients of a two-dimensional gas, *Surf. Sci.* **1973**, *39*, 21–36.

18. Sacher, R. S.; Morrison, I. D. An improved CAEDMON program for the adsorption isotherms of heterogeneous substrates, *J. Colloid Interface Sci.* **1979**, *70*, 153–166.

19. Giles, C. H.; Nakhwa, S. N. Studies in adsorption. XVI. The measurement of specific surface areas of finely divided solids by solution adsorption, *J. Appl. Chem.* **1962**, *12*, 266–273.

20. Giles, C. H.; D'Silva, A. P.; Trivedi, A. S. Use of dyes for specific surface measurement, *Surf. Area Determination, Proc. Int. Symp., 1969* **1970**, 317–329.

21. Rehbinder, P. A.; Lichtman, V. Effect of surface active media on strains and rupture in solids, *Proc. Int. Congr. Surf. Act., 2nd, 1957,* **1957**, *3*, 563–582.

22. Rehbinder. P. A. Formation and aggregative stability of disperse systems, *Colloid J. USSR* (Engl. Transl.) **1958**, *20*, 493–502.

23. Shchukin, E. D.; Rehbinder, P. A. Formation of new surfaces in the deformation and destruction of a solid in a surface-active medium, *Colloid J. USSR* (Engl. Transl.) **1958**, *20*, 601–609.

PART II
INTERFACES

CHAPTER IIA

Capillarity of Pure Liquids

The word *capillarity* refers to phenomena in which surface tension plays a part and not merely to those associated with capillary tubes. It is an old term: What earlier writers such as Freundlich meant by "capillary chemistry" would now be expressed as "the physics and chemistry of interfaces." But the old term denotes a branch of physics for which no other term equally succinct is available.

1. MOLECULAR THEORY OF SURFACE TENSION

The surface of a liquid is in a condition of tension, the most prominent evidence of which is its tendency to contract in area. The sphere has the least surface for a given volume; hence liquid drops are spherical. The spherical shape of a raindrop or the circular arc of a rainbow testify to the existence of a tension at the surface of water, which makes each drop minimize its surface area. The symmetry of the rainbow is a consequence of raindrops being spheres, and so, indirectly, is evidence of tension in the surface of water.

The surface tension of a liquid can be traced to forces of attraction between its molecules. Evidence for such intermolecular attractive forces is afforded by the very existence of the liquid state itself. Given the presence of these forces, molecules at or near the surface are subject to a net force toward the denser phase; and being mobile, a characteristic of the liquid phase, molecules respond to this force by moving from the region of the surface into the interior. Within a very short time, but probably not less than a few milliseconds after a new surface is created,[1] the surface region is depleted of molecules to such an extent that a lower density prevails there than in the bulk liquid. This process of depletion of the

69

surface region by migration of molecules into the interior is soon brought to an end by a countermovement of diffusion in the opposite direction, from the higher density bulk to the lower density surface region. Thus a dynamic equilibrium is rapidly established in which the rate of migration out of the surface is balanced by an equal and opposite rate of diffusion into the surface, with the lower density at the surface maintained as a time average.[2]

The potential energy between two molecules as a function of the distance between them is represented in Fig. A.1. A useful mathematical model for this pair potential is the 6–12 potential, in which the attraction potential varies with the inverse sixth power of the distance of separation, and the repulsion potential varies with the inverse twelfth power of the distance of separation, that is,

$$U(r) = 4\varepsilon[(s/r)^{12} - (s/r)^6] \qquad\qquad [\text{A.1}]$$

where $U(r)$ is the net potential dependent on the separation r and s is a characteristic distance. Equation [A.1] has a minimum potential energy, $U(r_0)$, that can be evaluated from $dU/dr = 0$, giving

$$r_0 = 2^{1/6}s$$
$$U(r_0) = -\varepsilon \qquad\qquad [\text{A.2}]$$

In the surface region of lower density any two molecules are on the average farther apart than a corresponding pair in the bulk; consequently the pair-potential energy in the surface region is greater than in the bulk. The excess potential energy of the surface region is the source of the tension, for the mobility

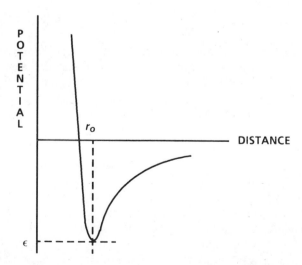

Figure A.1 The potential energy between two molecules as a function of the distance between them.

of molecules in the liquid state allows them to lose their excess potential energy by moving from the surface region into the bulk phase; this motion results in a spontaneous contraction of the surface to its minimum possible area, which, incidentally, is not always a spherical surface, as the minimizing tendency must accommodate to the action of other forces, particularly gravity. It follows that any expansion of the surface is resisted, that is, requires work, the work required being what is necessary to create the excess potential energy per unit area of additional surface. The work to create new surface is proportional to the extent of new surface created:

$$W = \sigma \, \Delta A \qquad\qquad [A.3]$$

where ΔA is the area of additional surface and σ is a proportionality constant, which has the units newtons/meter (or dynes/centimeter in the CGS system). The work W is said to be done against "the surface tension of the liquid," and the phrase provides a useful designation for the proportionality constant σ.

The quantity σ varies from one liquid to another, depending on the strength of the attractive forces between their molecules. Liquids with relatively weak forces create less of a density difference at their surfaces, and so less of an excess potential energy, requiring less work to extend their surfaces. Those with stronger forces would require more work and have a correspondingly larger value of σ. The relative strength of the intermolecular forces in liquids is reflected in their boiling points and vapor pressures. We should expect, therefore, that volatile liquids of low boiling point would have low values of σ and liquids of low volatility and high boiling point would have larger values of σ. This expectation is confirmed generally, with many exceptions, by reference to the data collected in Table A.1.

Table A.1   Surface Tensions and Boiling Points of Some Pure Liquids

Liquid	T (°C)	σ (mN/m)	Boiling point (°C)
Perfluoropentane	20	9.9	
Perfluoroheptane	20	11.0	
Perfluorohexane	20	11.9	
Pentane	20	16.0	36.1
Diethyl ether	20	17.0	34.5
Polydimethylsiloxanes			
Tetramer	20	17.6	
Dodecamer	20	19.6	
Hexane	20	18.4	68.7
Heptane	20	20.1	98.4
Octane	20	21.1	125.7
Ethanol	20	22.3	78.3

(*Table Contd.*)

Table A.1 *(Contd.)*

Liquid	T (°C)	σ (mN/m)	Boiling point (°C)
Methanol	20	22.5	64.7
Methyl isobutyl ketone	20	23.6	116.5
Ethyl acetate	20	23.7	77.1
1-Propanol	20	23.7	97.2
Decane	20	23.7	174.9
Acetone	25	24.0	56.3
Methyl ethyl ketone	25	24.0	79.6
1-Butanol	20	24.6	117.7
Cyclohexane	20	25.2	80.7
Dodecane	20	25.3	216.3
1, 1, 1-Trichloroethane	20	25.6	74.0
1-Pentanol	20	25.6	137.8
1-Octanol	20	26.1	195.2
Tetrahydrofuran	25	26.4	66.0
Carbon tetrachloride	25	26.4	76.7
Chloroform	25	26.7	61.2
Acetic acid	20	27.4	117.9
Methylene chloride	20	27.8	39.7
Cellosolve	25	28.2	135.6
Toluene	20	28.5	110.6
Benzene	20	28.9	80.1
Octanoic acid	20	29.2	239.9
Methyl cellosolve	15	31.8	124.6
Oleic acid	20	32.8	360 (decomposes)
p-Dioxane	15	34.4	101.3
Cyclohexane	20	34.5	155.6
Bromobenzene	20	36.5	155.9
Benzaldehyde	20	38.8	178.9
Aniline	20	42.9	184.4
Nitrobenzene	20	43.9	210.8
Ethylene glycol	20	46.5	197.3
Methylene iodide	20	50.8	182 (decomposes)
Glycerol	20	63.3	290
Water	20	72.9	100.0
Water	30	71.3	100.0
Water	60	67.0	100.0
Mercury	20	484	356.6

From Jasper, 1972; Riddick and Bunger, 1970.

2.   MOLECULAR THEORY OF INTERFACIAL TENSION

When two liquids, such as ether and water, that are only partially soluble in one another, are superposed, a distinct surface of separation, known as the liquid–liquid interface, persists. The liquid–liquid interface, like the liquid–vapor surface, is also in a state of tension. The convention has arisen that the tension of the former is called "interfacial tension" to distinguish it from the "surface tension" of the latter. An experiment by the Belgian physicist, J. A. F. Plateau illustrates the reality of the interfacial tension. Water and olive oil were the two immiscible liquids that he used. He added enough alcohol to the water to bring its density to the same value as that of the oil. When he then added a quantity of oil, it neither floated nor sank to the bottom, but remained suspended as a single spherical mass in the water. If no interfacial tension had been operative, the oil would have retained the irregular shape first formed on being poured into the water; but it did not do so; its surface contracted to that of a sphere, which is proof positive that a tension acts at the oil–water interface to contract the interfacial area to its minimum.

Girifalco and Good[3] have indicated how interfacial tension is related to the surface tensions of the contacting liquids. Just as surface tension depends on the strength of the intermolecular forces, the interfacial tension also depends on intermolecular forces, but three different types of force come into play: homomolecular forces in each of the two liquids and heteromolecular forces between the two liquids. Instead of evaluating these forces in fundamental terms, a simpler method is to express them in terms of the surface tensions to which they give rise. To estimate the interfacial tension from the two surface tensions, they adopted a suggestion of Berthelot, who estimated the attraction between unlike gas molecules, α_{12}, as the root mean square of the attraction between like molecules (α_{11} and α_{22})[4,5].

$$\alpha_{12} = (\alpha_{11}\alpha_{22})^{1/2} \qquad\qquad \text{[A.4]}$$

By analogy, Girifalco and Good[3] set up the corresponding ratio for the work of adhesion between two phases in terms of the work of cohesion of each phase, leading to

$$W_{12}^{adh} = \Phi(W_1^{coh} W_2^{coh})^{1/2} \qquad\qquad \text{[A.5]}$$

where (Section II.A.5d)

$$W_1^{coh} = 2\sigma_1 \qquad W_2^{coh} = 2\sigma_2 \qquad\qquad \text{[A.6]}$$

$$W_{12}^{adh} = \sigma_1 + \sigma_2 - \sigma_{12} \qquad\qquad \text{[A.7]}$$

and where Φ is a characteristic constant, which was introduced to confer generality and which would be evaluated for various types of systems.

Rearranging these equations gives

$$\sigma_{12} = \sigma_1 + \sigma_2 - 2\Phi(\sigma_1 \sigma_2)^{1/2} \qquad [A.8]$$

Equation [A.8] shows how interfacial tension is related to the surface tensions of the contacting liquids for a simple model of homo- and heteromolecular interactions. It is referred to as the Girifalco–Good equation.

3. DISPERSION FORCE CONTRIBUTIONS TO INTERFACIAL TENSION

Fowkes modified this model by introducing the concept that the effects of different types of intermolecular attraction could be expressed separately in terms of their contribution to the observed surface tension.[6] Thus, for instance, in water the intermolecular attractive forces are the London dispersion forces (formerly called van der Waals forces) and hydrogen bonding (Chapter III.A). The observed surface tension of water may be expressed in terms of the contributions made to it by each of these forces:

$$\sigma_w = \sigma_w^d + \sigma_w^H \qquad [A.9]$$

where σ_w is the observed surface tension, and σ_w^d and σ_w^H are the two contributions to σ_w made by the dispersion forces and by hydrogen bonding, respectively. Hydrogen bonding is a special case of a more general type of interaction, namely, that between a Lewis base and a Lewis acid, giving σ^{ab} as an alternative designation for σ^H and also making it applicable to liquid pairs where the acid–base interaction is not one of hydrogen bonding. Liquid metals introduce another type of bonding, the metallic bond, whose contribution to the surface tension is designated σ^m. The surface tension of mercury would be written as

$$\sigma_{Hg} = \sigma_{Hg}^d + \sigma_{Hg}^m \qquad [A.10]$$

Between molecules of saturated aliphatic hydrocarbons the only interaction is that of the London dispersion forces, therefore

$$\sigma_{hc} = \sigma_{hc}^d \qquad [A.11]$$

London dispersion forces exist between any two molecules; consequently the term σ^d enters into every expression for the observed tension. For some liquids, as we have seen, it is the only term.

When a saturated aliphatic hydrocarbon, such as octane, is superposed on water, the net density lowering of the interfacial region is the sum of that arising from the homomolecular water attraction and from the homomolecular hydrocarbon attraction, but is partially relieved by the heteromolecular attractions between water and hydrocarbon. The net difference of density is directly proportional to the tension that develops there, so that either the strength of the intermolecular forces or the density difference at the interface that is the direct outcome of these forces may be expressed in terms of surface tension. By using values of surface tension for this purpose, the argument does not need to be taken all the way back to calculations of intermolecular forces, with a great gain in simplicity. The expression for the interfacial tension, σ_{12} between water 1 and hydrocarbon 2, is therefore,

$$\sigma_{12} = \sigma_1 + \sigma_2 - W_{12} \qquad\qquad [A.12]$$

where W_{12} is a work term dependent on the degree of heterointeraction at the interface between molecules of 1 and those of 2.

Heteromolecular forces are often evaluated as a root mean square of the homomolecular forces (Chapter III.A). Continuing to use values of surface tension in the place of homomolecular forces, and restricting the type of interaction between 1 and 2 to that of their dispersion forces only, gives

$$W_{12} = 2(\sigma_1^d \sigma_2^d)^{1/2} \qquad\qquad [A.13]$$

The factor 2 is introduced to take into account both 1–2 and 2–1 dispersion force interactions. Combining Eqs. [A.12] and [A.13] gives

$$\sigma_{12} = \sigma_1 + \sigma_2 - 2(\sigma_1^d \sigma_2^d)^{1/2} \qquad\qquad [A.14]$$

Equation [A.14], the Fowkes equation, can be tested with known values of

Table A.2   Determination of σ_{Hg}^d for Mercury (mN/m at 20°C)

Hydrocarbon	σ_2	σ_{12}	σ_{Hg}^d
n-Hexane	18.4	378	210
n-Octane	21.8	375	199
n-Nonane	22.8	372	199
Benzene	28.85	363	194
Toluene	28.5	359	208
o-Xylene	30.1	359	200
m-Xylene	28.9	357	211
p-Xylene	28.4	361	203
n-Propylbenzene	29.0	363	194
n-Butylbenzene	29.2	363	193
		Average	200 ± 7

From ref. 6.

surface and interfacial tension for a number of hydrocarbons with mercury. For hydrocarbons, Eq. [A.11] applies, so there is only one unknown in Eq. [A.14], namely, σ_{Hg}^d the dispersion force component of the surface tension of mercury.

Table A.2 shows the determination of σ_{Hg}^d by Eq. [A.14] based on data for surface and interfacial tensions of 10 different hydrocarbons with mercury at 20°C. The average value is 200 ± 7 mN/m. A similar use of Eq. [A.14], using the surface and interfacial tensions of hydrocarbons against water, is shown in Table A.3 for the determination of $\sigma_{H_2O}^d$. The average value is 21.8 ± 0.7 mN/m.* A more extensive list of interfacial tensions of liquids against water is given in Table A.4.

These values of $\sigma_{H_2O}^d$ and σ_{Hg}^d can be used to calculate the interfacial tension between water and mercury, assuming that the interaction between these two liquids is solely one of dispersion forces.

$$\sigma_{H_2O/Hg} = \sigma_{H_2O} + \sigma_{Hg} - 2(\sigma_{H_2O}^d \cdot \sigma_{Hg}^d)^{1/2} \qquad \text{[A.15]}$$

Therefore,

$$\sigma_{H_2O/Hg} = 72.8 + 484 - 2(21.8 \times 200)^{1/2} = 424.8 \text{ mN/m}$$

The calculated value of 424.8 mN/m with a standard deviation of 4.4 mN/m agrees with the best measured values of 426–427 mN/m. This result supports the assumptions underlying the derivation. For water, it shows that the dispersion forces between the molecules are 30% of the total attractive forces, the remainder being hydrogen bonding. It also shows that the interaction between water and mercury is almost entirely dispersion force interaction and therefore that the dipole image forces are comparatively weak.

Equation [A.12] points out that the magnitude of an interfacial tension depends on the degree of heteromolecular interaction: If the term W_{12} is small, the interfacial tension is large; if the term W_{12} is large, the interfacial tension is small; if the term W_{12} is greater than the sum of the two surface tensions, the calculated interfacial tension is negative, that is, no interface can exist and the two liquids are miscible. While surface tension is increased by the degree of homomolecular attraction, interfacial tension is diminished by the degree of heteromolecular attraction. Compare, for example, cyclohexane and benzene: Their surface tensions are about the same, but their interfacial tensions against water are very different. The former value is 50.2 mN/m and the latter value is 35 mN/m. Clearly, on the basis of Eq. [A.12], cyclohexane has less heteromolecular interaction with water than has benzene. This conclusion may be explained by the presence of π orbitals in benzene, which have electron-donor interactions with water. This type of interaction is lacking between cyclohexane and water.

---

*In later work[7] Fowkes suggests that more reliable values of σ^d for liquids and solids are to be obtained from branched or cyclic alkanes rather than from n-alkanes used as test liquids, as the latter have small anisotropic polarizabilities whereas the former have none.

Table A.3 Determination of $\sigma^d_{H_2O}$ for Water (mN/m at 20°C)

Hydrocarbon	σ_2	σ_{12}	$\sigma^d_{H_2O}$
n-Hexane	18.4	51.1	21.8
n-Heptane	20.4	50.2	22.6
n-Octane	21.8	50.8	22.0
n-Decane	23.9	51.2	21.6
n-Tetradecane	25.6	52.2	20.8
Cyclohexane	25.5	50.2	22.7
Decalin	29.9	51.4	22.0
White oil (25°)	28.9	51.3	21.3
		Average	21.8 ± 0.7

From ref. 6.

Table A.4 Selected Liquid–Liquid Interfacial Tensions against Water

Liquid	Temperature (°C)	σ_{12} (mN/m)	W^{adh} (mN/m)
Hexane	20	51.1	40.1
Heptane	20	50.2	42.7
Octane	20	50.8	43.4
Decane	20	51.2	45.3
Carbon tetrachloride	20	45.0	54.2
Toluene	20	36.1	65.2
Benzene	20	35.0	66.7
Chloroform	20	31.6	67.9
Bromobenzene	20	38.1	71.2
Methylene chloride	20	28.3	72.6
Methylene iodide	20	48.5	75.0
Diethyl ether	20	10.7	79.1
Ethyl acetate	20	6.8	89.7
Oleic acid	20	15.7	89.9
1-Octanol	20	8.5	90.4
Nitrobenzene	20	25.7	91.0
Octanoic acid	20	8.5	93.5
1-Pentanol	20	4.4	94.0
1-Butanol	20	1.8	95.6
Benzaldehyde	20	15.5	96.1
Aniline	20	5.8	109.9
Mercury	25	427	130

From ref. 3.

By the same token, Eq. [A.14] is inadequate to evaluate the interfacial tension between bezene and water because it lacks a term to take into account the additional heteromolecular attraction beside that of the dispersion forces. On the other hand, the interfacial tension between mercury and water, which might seem more complex, is well described by Eq. [A.14], as the capability of water to hydrogen bond is not reciprocated by the mercury and the capability of mercury to amalgamate with other metals finds no response with water: Only dispersion forces are common to both and constitute their sole mode of heteromolecular attraction.

The underlying assumption of Fowkes' approach is that the components of the surface tension or of the work of adhesion are linearly related; that is, that the whole property can be subdivided into fractions, that each fraction can be treated separately, and that they can then be recombined to reconstitute the whole property. This would assume that each type of molecular interaction is without effect on any other type: that, for example, the simultaneous actions of dispersion forces and hydrogen bonding on any given surface site are mutually independent. Such a question cannot be decided on theoretical grounds alone: it is essentially a question of how matter behaves, and can only be decided by an appeal to experimental data.

4. ACID–BASE CONTRIBUTIONS TO INTERFACIAL TENSION

Dispersion forces alone do not account for all the attraction between bodies across their interface. Different approaches to nondispersion forces have been proposed. The success of the treatments by Girifalco and Good[3] and by Fowkes[6] of the dispersion force interactions, using a geometric mean of the homomolecular interaction to obtain the heteromolecular, was at first merely imitated in form without reference to the nature of the molecular forces. For example, the work of adhesion between two liquids was given by

$$W_{12} = 2(\sigma_1^d \sigma_2^d)^{1/2} + 2(\sigma_1^p \sigma_2^p)^{1/2} \qquad [A.16]$$

where the superscripts d and p refer to dispersion and polar forces, respectively. Wu[8] found better agreement by substituting the harmonic mean for the geometric mean, but still without theoretical justification.

Fowkes holds that all such treatments are in error: Keesom and Debye forces, which can be approximated as the geometric mean of polar forces, are revealed by calculation to be small (Table IIIA.2). The remaining polar contribution, namely, the acid–base interaction (including hydrogen bonding as a special case), is significant where it exists but cannot be represented by a geometric mean. For instance, if the two immiscible liquids are both acidic or both basic, very small polar interactions take place, and the work of adhesion is given by the dispersion force interaction alone. If one material is a Lewis acid and the other a Lewis base (i.e., electron acceptor or proton donor and electron donor or proton acceptor, respectively), then the reaction between them across an interface has a profound effect on the interfacial tension. For example, the interfacial tension between a Lewis base and water is lowered appreciably by a decrease in the pH of the aqueous phase; and the interfacial tension between a Lewis acid and water is lowered appreciably by an increase in pH of the aqueous phase. Indeed, how the interfacial tension varies with the pH of the aqueous phase indicates whether the water-immiscible organic liquid is a Lewis acid or a Lewis base or neither.[9]

A number of empirical treatments of Lewis acid–base reactivity have been developed to predict equilibrium constants and reaction rates. These are based on empirical linear correlations between the logarithms of the equilibrium constants or of the rate constants of one series of reactions and those of another related series. An early and well-known example of a linear free-energy relation is the Hammett equation, originally developed to predict equilibrium constants for the reactions of substituted aromatic acids, but which turns out to be reasonably successful whether the substrates are attacked by electrophilic, nucleophilic, or free-radical reagents, at least within a given reaction series. Three empirical reactivity treatments, designed particularly for acid–base reactions, are the hard–soft acid–base (HSAB) principle of Pearson, the donor–acceptor number (DN–AN) scales of Gutmann, and the $E\&C$ equation of Drago (Jensen, 1980). Of these the $E\&C$ treatment is readily adapted to quantify the contribution of acid–base reactions at interfaces to the total work of adhesion. The DN–AN scales bear a general resemblance to the $E\&C$ equation, but are not yet in as suitable a form to calculate works of adhesion. The HSAB principles are only qualitative but provide guidelines for conditions where other approaches are unavailable.

Drago[10–12] proposes the following relation to calculate the enthalpy of an acid–base reaction:

$$-\Delta H_{ab} = E_a E_b + C_a C_b \qquad\qquad [A.17]$$

where ΔH_{ab} is the enthalpy of adduct formation per mole, E_a and C_a are empirically determined parameters for the acid, and E_b and C_b are empirically determined parameters for the base.

The magnitudes of the E parameters can be interpreted as measures of the susceptibility of the molecule for electrostatic interaction; the magnitudes of the C parameters can be interpreted as measures of the susceptibility for covalent interactions (similar to the concept of "hard" and "soft" acids and bases). Equation [A.17] accurately correlates over 280 enthalpies of formation. Representative values of the acid–base constants from Drago[10] are listed in Table A.5, which, when substituted in Eq. [A.17], give enthalpies in kilojoules per mole.

A similar list of E and C parameters of acidic and basic polymers could be obtained from measured heats of solution, determined calorimetrically, in solvents whose E and C values are already known. The long time required to dissolve a polymer is a disadvantage of this method. A more convenient method is to measure the infrared spectral shifts of the stretching frequencies of the active

Table A.5 Drago E and C Parameters for a Variety of Molecular Acids and Bases

Acids	C_a	E_a
Iodine	2.05	2.05
Iodine monochloride	1.697	10.43
Thiophenol	0.405	2.02
p-tert-Butylphenol	0.791	8.30
p-Methylphenol	0.826	8.55
Phenol	0.904	8.85
p-Chlorophenol	0.978	8.88
tert-Butyl alcohol	0.614	4.17
Trifluoroethanol	0.922	7.93
Pyrrole	0.603	5.19
Isocyanic acid	0.528	6.58
Sulfur dioxide	1.652	1.88
Antimony pentachloride	10.49	15.09
Chloroform	0.325	6.18
Water	0.675	5.01
Methylene chloride	0.02	3.40
Carbon tetrachloride	0.00	0.00

Bases	C_b	E_b
Pyridine	13.09	2.39
Ammonia	7.08	2.78
Methylamine	11.41	2.66
Dimethylamine	17.85	2.23
Trimethylamine	23.6	1.652
Ethylamine	12.31	2.80
Diethylamine	18.06	1.771
Triethylamine	22.7	2.03
Acetonitrile	2.74	1.812
p-Dioxane	4.87	2.23
Tetrahydrofuran	8.73	2.00
Dimethyl sulfoxide	5.83	2.74
Ethyl acetate	3.56	1.994
Methyl acetate	3.29	1.847
Acetone	4.76	2.018
Diethyl ether	6.65	1.969
Isopropyl ether	6.52	2.27
Benzene	1.452	1.002
p-Xylene	3.64	0.851

From ref. 10–12.

groups of either polymer or solvent. The greater the enthalpy of interaction between the solvent and the polymer, the greater is the spectral shift. Both the total enthalpy of interaction and the total spectral shift are separated into contributions from the dispersion force interaction and the acid–base interaction. The dispersion force component of the spectral shift is calculated from the dispersion force component of the surface tension of the solvent; and the enthalpy of acid–base interaction is then calculated from the remaining component of the spectral shift. The E and C parameters for the polymer are obtained by repeating this procedure with several solvents. Ultimately, E and C parameters of various acidic and basic polymers will be available analogously to those of the solvents listed in Table A.5.[13]

Using the enthalpy of adduct formation, Fowkes[14] proposed that the work of adhesion, including acid–base interactions and other polar interactions (Keesom and Debye forces), be approximated by

$$W_{12} = W_{12}^d - FN_{ab}\,\Delta H_{ab} + W_{12}^p$$

or

$$W_{12} = \sigma_1 + \sigma_2 - \sigma_{12} = 72.8 + 28.9 - 35.0 = 66.7\,\text{mJ/m}^2 \qquad [\text{A.18}]$$

where N_{ab} is the moles of acid–base pairs per unit area and where the constant F therefore converts the enthalpy per unit area into surface free energy (F is near unity since enthalpy is the major part of the free energy for these strong interactions) and the last term is usually small. For example at the benzene–water interface, $N_{ab} = 2 \times 10^{14}$ acid–base pairs per square centimeter (based on a molecular area for benzene of $0.5\,\text{nm}^2$); and, using Drago values from Table A.5, Eq. [A.18] gives $W_{12}^{ab} = 19.9\,\text{mJ/m}^2$. The total work of adhesion of benzene on water is given by

$$W_{12} = \sigma_1 + \sigma_2 - \sigma_{12} = 72.8 + 28.9 - 35.0 = 66.7\,\text{mJ/m}^2$$

That part of the work of adhesion due to dispersion force interaction is

$$W_{12}^d = 2(\sigma_1^d \sigma_2^d)^{1/2} = 2(22.0 \times 28.9)^{1/2} = 50.4\,\text{mJ/m}^2$$

The difference between W_{12} and W_{12}^d is close to the acid– base part of the total work of adhesion; that is, $66.7 - 50.4 = 16.3\,\text{mJ/m}^2$, to compare with the value calculated above of $19.9\,\text{mJ/m}^2$. This result attests to the success of Fowkes' treatment of works of adhesion as a simple combination of only the two major types of molecular interaction: dispersion forces and acid–base forces.

5.   THERMODYNAMICS OF SURFACES AND INTERFACES

Systems in which interphase zones are significant require the introduction of terms for interfacial area and interfacial tension into their thermodynamic descriptions, analogous to the usual volume and pressure terms used for gaseous processes. If the surface tension is expressed by σ and the surface area by A, the work done on extending a surface against the force of its surface tension is σdA. The analogy to the work term for the compression of a gas, namely, pdV, is complete except for the sign of the work term: Since the spontaneous action of a gas is expansion and the spontaneous action of a surface is contraction, the signs of the work terms are opposite. The first and second laws of thermodynamics are therefore summarized for open systems by the expression

$$dU = T\,dS - p\,dV + \sigma\,dA + \sum \mu_i\,dn_i \qquad [A.19]$$

Similarly, the definition of enthalpy applied to interfacial systems requires the addition of a term σA, that is,

$$H = U + pV - \sigma A \qquad [A.20]$$

A complete listing of the thermodynamic functions, differential equations, and Duhem equations are given in Table A.6. Thermodynamic quantities are expressed in work units: in the CGS system, ergs; and in the SI system, joules. The expression for the differential Helmholtz function, dF, leads to

$$\left(\frac{\partial F}{\partial A}\right)_{V,T,n_i} = \sigma \qquad [A.33]$$

This result establishes that the surface tension, which is a physical property of a liquid surface as real as the tension in a stretched elastic band, also has the

Table A.6   Thermodynamic Functions for Interfacial Systems[a]

$U = TS - pV + \sigma A + \sum \mu_i n_i$	[A.21]
$dU = T\,dS - p\,dV + \sigma\,dA + \sum \mu_i\,dn_i$	[A.22]
$H = U + pV - \sigma A$	[A.23]
$dH = T\,dS + V\,dp - A\,d\sigma + \sum \mu_i\,dn_i$	[A.24]
$F = U - TS$	[A.25]
$dF = -S\,dT - p\,dV + \sigma\,dA + \sum \mu_i\,dn_i$	[A.26]
$G = H - TS = \sum \mu_i n_i$	[A.27]
$dG = -S\,dT + V\,dp - A\,d\sigma + \sum \mu_i\,dn_i$	[A.28]
$S\,dT - V\,dp + A\,d\sigma + \sum n_i\,d\mu_i = 0$	[A.29]
$S\,dT + p\,dV - \sigma\,dA + \sum n_i\,d\mu_i = 0$	[A.30]
$-T\,dS - V\,dp + A\,d\sigma + \sum n_i\,d\mu_i = 0$	[A.31]
$-T\,dS + p\,dV - \sigma\,dA + \sum n_i\,d\mu_i = 0$	[A.32]

[a] U is the internal energy, H is the enthalpy, F is the Helmholtz free energy, and G is the Gibbs free energy.

meaning of the Helmholtz function per unit area at constant volume, temperature, and composition. When expressed as a tension, the appropriate units of σ are force/unit length; when expressed as a free energy, the appropriate units are energy/unit area. The measured surface tension of water at 20°C is 73 mN/m; therefore, the Helmholtz free energy is 73 mJ/m² at the same temperature, volume, and composition.

For a one-component system containing an interface, the entropy per unit area at constant pressure and composition is (from Eq. [A.29])

$$\frac{S}{A} = -\left(\frac{\partial \sigma}{\partial T}\right)_p \qquad [A.34]$$

Combining Eq. [A.34] with [A.21] for a closed system gives, for the surface.

$$\frac{U}{A} = \sigma - T\left(\frac{\partial \sigma}{\partial T}\right)_p \qquad [A.35]$$

The internal energy per unit area, U/A, is a useful thermodynamic function to compare surface properties of members of an homologous series as it is less temperature dependent than the surface tension.

We shall now use the free-energy functions listed in the Table A.6 to describe a few simple processes with closed systems (i.e. constant composition) of significant interphase areas.

a.   Coalescence of Droplets

The free-energy change (Helmholtz) for the process of coalescence of two droplets (Fig. A.2) at constant volume, temperature, and composition is

$$\Delta F = F_{\text{final}} - F_{\text{initial}} = \sigma(A_{\text{final}} - A_{\text{initial}}) \qquad [A.36]$$

The area decreases as drops coalesce; hence the expression for ΔF is negative and the coalescence is therefore spontaneous. The stability of aerosols and emulsions depends on the inhibition of coalescence, which requires that the thermodynamic expressions for the free energy of coalescence include additional terms of opposite sign. These terms may refer to the free energy of solvation or to the free energy of desorption of a third component (Section IV.A.3).

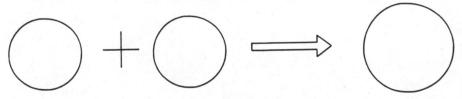

Figure A.2 The coalescence of two droplets.

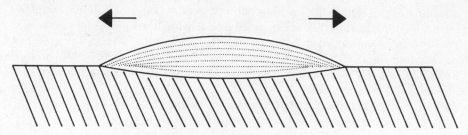

Figure A.3 The spreading of one liquid on another.

b.   Spreading of One Liquid on Another

The Helmholtz free-energy change *per unit area* for the process of liquid 2 spreading on the surface of liquid 1 (Fig. A.3) at constant volume, temperature, and composition is

$$\Delta F = F_{\text{final}} - F_{\text{initial}}$$
$$= (\sigma_2 + \sigma_{12} - \sigma_1) \qquad \text{[A.37]}$$

If this expression is negative, liquid 2 spreads spontaneously; if it is positive, liquid 2 sits as a lens on the surface of liquid 1. If the drop of liquid 2 is small enough, it will still remain on the surface even though its density is greater than that of liquid 1, held there by the predominance of surface over gravitational forces.

The spreading coefficient is defined as equal to the negative of the Helmholtz free energy of spreading per unit area: $S = -\Delta F$. Therefore

$$S_{(\text{of } 2 \text{ on } 1)} = \sigma_1 - \sigma_2 - \sigma_{12} \qquad \text{[A.38]}$$

If the spreading coefficient is positive, liquid 2 spreads spontaneously on liquid 1; if the spreading coefficient is negative, liquid 2 sits as a lens on the surface of liquid 1.

c.   Encapsulation of One Liquid by Another

Spontaneous encapsulation of one liquid by another requires spreading of the medium on the dispersed phase. Two spreading processes are possible: that of liquid 2 on 1 and that of liquid 1 on 2. The corresponding coefficients are

$$S_{(\text{of } 2 \text{ on } 1)} = \sigma_1 - \sigma_2 - \sigma_{12}$$

$$S_{(\text{of } 1 \text{ on } 2)} = \sigma_2 - \sigma_1 - \sigma_{12}$$

Therefore

$$S_{(\text{of } 2 \text{ on } 1)} + S_{(\text{of } 1 \text{ on } 2)} = -2\sigma_{12} \qquad \text{[A.39]}$$

This equation tells us that *both* spreading coefficients cannot be positive; one may be positive and the other negative; or both may be negative. The diagrams published by Kelvin (Fig. A.4) show two liquids for which both spreading coefficients are negative. No matter what their relative proportions may be, encapsulation of one of these liquids by the other is never achieved. If the spreading coefficient of one of these liquids on the other had been positive, it would have encapsulated the other; the reverse type of encapsulation would then never occur.

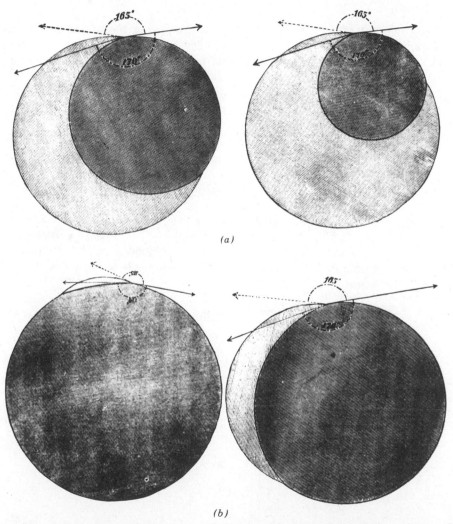

(a)

(b)

Figure A.4 The constancy of the contact angles is independent of the relative volumes of two immiscible liquids with negative spreading coefficients on each other (W. Thomson, Lord Kelvin, 1889).

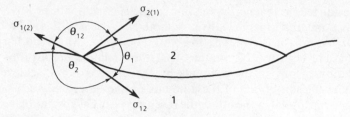

Figure A.5 Neumann's triangle of surface forces at each point on the line at which three fluid phases meet.

Kelvin's diagrams also illustrate the constancy of the angles of contact between the two liquids, independent of their relative amounts. The relation between the angles and the surface and interfacial tensions is given by Neumann's triangle of forces (Fig. A.5):

$$\sigma_1 = \sigma_2 \cos(\pi - \theta_{12}) + \sigma_{12} \cos(\pi - \theta_2) \qquad \text{[A.40]}$$

or

$$\frac{\sigma_1}{\sin\theta_1} = \frac{\sigma_2}{\sin\theta_2} = \frac{\sigma_{12}}{\sin\theta_{12}} \qquad \text{[A.41]}$$

d.   Works of Adhesion and Cohesion

The processes of adhesion and cohesion (Fig. A.6) may be described in terms of their opposites, namely, the processes of separation. The work of adhesion per unit area is the work done on the system when two condensed phases (1 and 2), forming an interface of unit area, are separated reversibly to form unit areas of each of the 1 and 2 surfaces.

$$W^{\text{adh}} = \sigma_1 + \sigma_2 - \sigma_{12} \qquad \text{[A.42]}$$

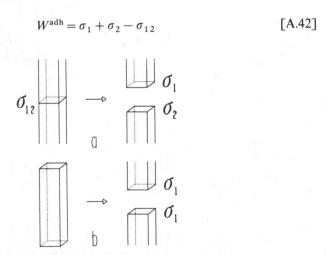

Figure A.6 The processes of (*a*) adhesion and (*b*) cohesion.

The work of adhesion is the Helmholtz free energy of separation. The work of cohesion is the work of adhesion when liquids 1 and 2 are the same. In that case σ_{12} is nonexistent and

$$W^{coh} = 2\sigma_1 \qquad [A.43]$$

The work of adhesion given by Eq. [A.42] refers to the adhesion between two liquids or a liquid and a solid but does not measure the strength of the adhesive interface between two solid surfaces. At the dry contact, between a polymer and a metal for example, electric charge is exchanged, typically 10^3–$10^4\,esu/cm^2$ (Section IV.C.6). This small charge does not contribute significantly to the interfacial free energy σ_{12}, nevertheless Derjaguin et al.[15,16] have shown that even such a small surface charge may contribute over a hundred times as much adhesive energy as the dispersion components when two such surfaces are separated. The reason is that the charges in the polymer film are immobile and work must be done to separate them from their countercharges in the metal. Although the electric force across the interface is weak, it extends to a greater distance than molecular forces; therefore, the work against the electric field, due to the trapped charges, is greater than the work done against the molecular forces. This and other aspects of adhesion are reviewed elsewhere (Lee, 1975, 1980, 1984; Mittal, 1983).

e.   Free Energy of Emersion

The process of emersion is shown in Fig. A.7. The process as measured is one of immersion, but for consistent thermodynamic equations, it is calculated as the negative process, emersion. The Helmholtz free energy *per unit area* of emersion at constant temperature, volume, and composition is

$$\Delta F_{emersion} = \sigma_{sv} - \sigma_{sl} \qquad [A.44]$$

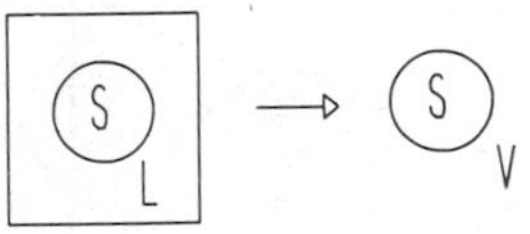

Figure A.7 The process of emersion.

6.   THERMODYNAMICS OF LIQUIDS IN CONTACT WITH SOLIDS

The first equation of capillarity describes the thermodynamics of a liquid in contact with a solid (Fig. A.8). The equilibrium condition for a sessile (i.e., sitting) drop on a solid substrate was developed in 1805 by Thomas Young[17] by equating the horizontal components of the forces acting at each unit length of the

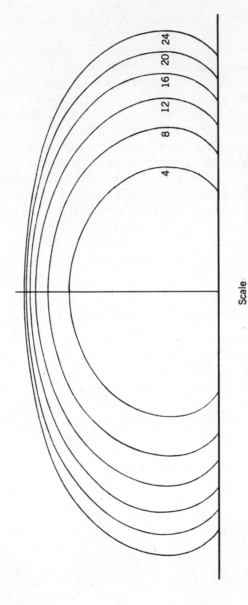

Scale

0 .01 .02 .03 .04 .05 .06 .07 .08 .09 .10 .11 .12 .13 .14 .15 .16 .17 .18 .19 .20 inch

Figure A.8 The forms of six sessile drops of mercury, weighing 4, 8, 12, 16, 20, and 24 grains (1 grain equals 64.8 mg), on a clean glass substrate, as measured by Bashforth in 1882. The contact angle of 140° is the same for each drop. From Bashforth and Adams (1883).

88

periphery of the three-phase contact line (Fig. A.9):

$$\sigma_{sv} = \sigma_l \cos \theta + \sigma_{sl} \qquad\qquad [A.45]$$

Although recent literature frequently refers to this relation as the Young–Dupré equation, no support for this designation is obtained from any adequate review of the classical work in this field, such as that of Bakker (1928). If any modification is justified, it would seem more suitable to refer to the relation as the Young–Gauss equation.[18]

The contact angle is defined as the angle between the tangent to the liquid surface and the tangent to the solid surface at any point of contact on the three-phase line measured through the liquid. Young's equation supposes that the contact angle is solely defined by the energies of the three surfaces in contact along a line and does not vary with the size of the drop or the profile of the substrate. If the liquid spreads without limit over the surface, there is no finite angle of contact; if the liquid does not spread, the angle can vary from 0° to 180°. The contact angle is readily measured by means of a goniometer if the substrate of the solid is sufficiently plane and sufficiently extensive, which is not always the case. The derivation in terms of forces is not rigorous, as it assumes that the surface energy of a solid substrate or of a solid–liquid interface is associated with a tension. Only at the liquid–gas surface or the liquid–liquid interface is the presence of a tension beyond doubt. We may avoid introducing these assumptions by discussing the equilibrium of the contact angle in terms of surface and interfacial energies.

Each unit area of a liquid has a potential energy numerically equal and dimensionally equivalent to the surface tension. A solid surface, because of its lack of mobility, does not develop a tension in the same way as a liquid surface; nevertheless, we may conceive that a part of the energy of a solid is proportional to its surface area and that this surface potential energy is thermodynamically equivalent in a solid surface or in a solid–liquid interface to the surface energy

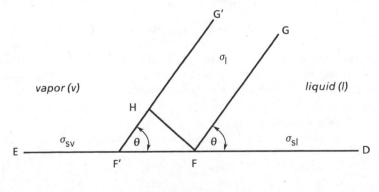

Figure A.9 The liquid–solid–vapor contact line. The three components are acting on a line of unit length, located at F and normal to the page.

measured by the surface tension of a liquid surface. With this concept we consider the equilibrium of a liquid, in contact with its vapor, resting on a solid substrate and not subject to any other forces than those of the surface and interfacial tensions.

In Fig. A.9 let s represent the solid, l the liquid, and v the vapor; ED is the surface of the solid, FG the tangent to the liquid surface at a point of contact on the periphery of the line of contact of the three phases s, l, and v. The contact angle θ is the angle GFD in the diagram. The angle is characteristic of the nature of the three phases in contact, and if the system is in equilibrium, then its potential energy is a minimum in this position. Now let the surface GF come into the position $G'F'$ parallel to GF: The angle of equilibrium is not affected by the displacement of GF to $G'F'$. This displacement of the edge of the liquid covers an differential area dA of the surface between v and s, replacing it with the same area of interface between l and s; and also extends the surface area of the liquid between l and v by an amount proportional to $F'H$. Hence the total change of Helmholtz free energy is

$$dF = dA(-F_{sv} + F_{sl} + \sigma_1 \cos \theta) \qquad [A.46]$$

At equilibrium, the free energy per unit area is a minimum; hence

$$F_{sv} - F_{sl} - \sigma_1 \cos \theta = 0$$

or

$$F_{sv} = F_{sl} + \sigma_1 \cos \theta \qquad [A.47]$$

This derivation is correct for some geometries such as flat surfaces; it is not a general proof. Consider the displacement of a liquid in a capillary tube: The liquid–vapor surface does not change in area. In 1960 F. P. Buff[19] finally derived the Young–Dupré equation by statistical thermodynamics in terms of surface free energies where

$$F_{sv} = F_{s0} - \pi_e \qquad [A.48]$$

and π_e is the reduction of the free energy of the solid substrate by the adsorbed vapor of the liquid. This quantity is sometimes referred to as the equilibrium spreading pressure. The equilibrium spreading pressure of an insoluble monolayer on a liquid surface is established when the spread monolayer is in equilibrium with its bulk phase.

For a liquid drop on a solid, the *final*, or equilibrium, spreading coefficient is

$$S = F_{sv} - F_{sl} - \sigma_1 \qquad [A.49]$$

For a finite angle of contact, Eq. [A.47] may be introduced to give

$$S = \sigma_1(\cos \theta - 1) \qquad [A.50]$$

The work of adhesion of a liquid to a solid substrate is

$$W^{\text{adh}} = F_{\text{sv}} - F_{\text{sl}} + \sigma_1 \qquad [\text{A.51}]$$

The corresponding equation for a finite contact angle is

$$W^{\text{adh}} = \sigma_1(\cos\theta + 1) \qquad [\text{A.52}]$$

and under the same conditions the free energy of emersion is

$$\Delta F^{\text{emersion}} = \sigma_1 \cos\theta \qquad [\text{A.53}]$$

7. DEGREES OF LIQUID–SOLID INTERACTION

The degree of interaction of a liquid with a solid substrate may be expressed by any one of a number of different but not independent quantities, each one representing the change in surface free energy per unit area for some process significant in the phenomenology of surfaces. These quantities are the spreading coefficient, the free energy of separation, and the free energy of emersion, all of which have already been defined.

The complete range of behavior is shown in Fig. A.10. The field is divided into three parts: (1) nonfinite contact angle, (2) contact angles less than 90°, and (3) contact angles larger than 90°. The behavior of the first region is associated with high-energy substrates and that of the third region with low-energy substrates. In the first region the liquid spreads spontaneously on the substrate, which can hold a uniform film of liquid that is stable at any thickness. In the second region the liquid "wets" but does not spread; it disproportionates to make an acute angle of

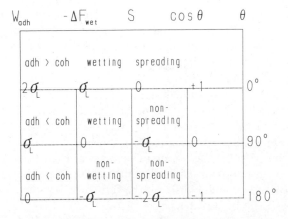

Figure A.10 The degrees of liquid–solid interaction.

Table A.7   Contact angles π_e and W^{adh} for Selected Systems

System	θ(deg)	π_e(mJ/m^2)	W^{adh}(mJ/m^2)
Water on hydrated silica	0	316	462
Water on gold	0		145.6
Water on silica	47	0	122.4
Water on graphite	85.7	15	93.3
Water on polyethylene	105	0	54.0
Water on paraffin wax	110	0	47.9
Water on Teflon®	115	0	42.0

contact between an adsorbed film in equilibrium with a thick lens or sessile drop. In the third region the liquid neither spreads nor wets the substrate: it forms a sessile drop, with an obtuse angle of contact, in equilibrium with the unwetted substrate. The chart of Fig. A.10 is a schematic report of the variation of the contact angle with the extent of liquid–solid interaction. High-energy solid surfaces interact strongly with liquids, either spreading spontaneously or giving low contact angles; the lower the contact angle, the greater the surface energy of the substrate. Table A.7 lists the contact angles of water on various solids. Water is seen to adhere more strongly (larger values of the work of adhesion) to inorganic surfaces and to have very strong adhesion to hydrated silica (where the surface is not SiO_2 but SiOH.) A bibliography of contact angle literature is available.[20]

8.   DISPERSION ENERGIES OF SOLID SUBSTRATES

Extensive series of measurements of contact angles of various liquids on low-energy polymer substrates were reported by W. A. Zisman and his co-workers at the Naval Research Laboratory, who found an empirical linear relation between the cosine of the contact angle and the surface tension of the liquid of the sessile drop (Fig. A.11). The extrapolation of the line to $\cos \theta = 1$ gives the "critical surface tension" σ_c of the substrate, that is, the surface tension of a liquid that just spreads on that substrate. The term *critical* is used because any liquid on the Zisman plot whose surface tension is greater than σ_c makes a finite contact angle with the substrate. The Zisman plot has no theoretical basis, although values of σ_c are useful empirical values to characterize relative degrees of surface energy of polymer substrates. Table A.8 lists Zisman's values of the critical surface tension of various polymeric substrates. Zisman's empirical prediction fails for liquids that form hydrogen bonds or acid–base interactions with the substrate. These liquids would spread spontaneously on the substrate, but their surface tensions would not necessarily be less than σ_c.

Many years later Fowkes worked out a theoretical treatment of the

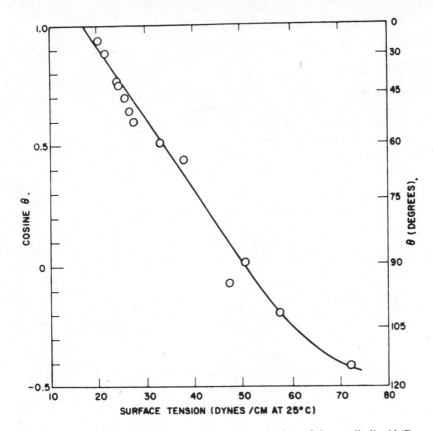

Figure A.11 Zisman plot of $\cos\theta$ versus the surface tension of the sessile liquid (Bernett, M. K., Zisman, W. A., Wetting properties of tetrafluoroethylene and hexafluoropropylene copolymers, *J. Phys. Chem.* **1960**, *64*, 1292–1294).

interactions across interfaces, which sufficiently accounts for the contact angles of all liquids, including hydrogen-bonding liquids and liquid metals. The basic concept is to separate the net interaction into dispersion force and electron donor–acceptor contributions (Chapter II.A.3–4). His treatment follows.

The Young–Dupré equation for the contact angle θ of the liquid l on a plane solid substrate s is

$$F_{sv} = \sigma_l \cos\theta + F_{sl} \qquad \text{[A.47]}$$

where

$$F_{sv} = F_{s0} - \pi_e \qquad \text{[A.48]}$$

For solid–liquid systems interacting by dispersion forces only, Eq. [A.14], originally developed to calculate the interfacial tension between two liquids, can be used to obtain F_{sl}. Rewritten in terms of the surface free energy per unit area of

Table A.8   Critical Surface Tensions of Various Polymeric Substrates

Polymeric substrate	σ_c (mN/m 20°C)
Polymethacrylic ester of perfluorooctanol	10.6
Polyhexafluoropropylene	16.2
Polytetrafluoroethylene	18.5
Polytrifluoroethylene	22
Polyvinylidene fluoride	25
Polyvinyl fluoride	28
Polyethylene	31
Polytrifluorochloroethylene	31
Polystyrene	33
Polyvinyl alcohol	37
Polymethyl methacrylate	39
Polyvinyl chloride	39
Polyvinylidene chloride	40
Polyethylene terephthalate	43
Polyhexamethylene diamide	46

From ref. 21.

the substrate and the surface tension of the liquid, Eq. [A.14] becomes

$$F_{sl} = F_{s0} + \sigma_1 - 2(F_{s0}^d \sigma_1^d)^{1/2} \qquad [A.54]$$

Rearranging and introducing Eq. [A.48] gives

$$F_{sv} - F_{sl} = -\sigma_1 + 2(F_{s0}^d \sigma_1^d)^{1/2} - \pi_e \qquad [A.55]$$

Using Eq. [A.47] gives

$$\sigma_1 \cos \theta = -\sigma_1 + 2(F_{s0}^d \sigma_1^d)^{1/2} - \pi_e$$

or

$$\cos \theta = -1 + \frac{2(F_{s0}^d \sigma_1^d)^{1/2}}{\sigma_1} - \frac{\pi_e}{\sigma_1} \qquad [A.56]$$

A plot of $\cos \theta$ versus $(\sigma_1^d)^{1/2}/\sigma_1$ should give a straight line originating at $\cos \theta = -1$ and with a slope of $2(F_{s0}^d)^{1/2}$. The data of Zisman and his co-workers treated in this way are shown in Fig. A.12. In this plot the values of π_e are neglected as the adsorption of vapors by solids of low surface energy is slight.

Figure A.12 testifies to the merits of Fowkes' theory. These data refer to a series of liquids of known surface tension on a variety of low-energy solid substrates. Most of the liquids are hydrocarbons for which σ_1^d is equal to σ_1; for other liquids such as water or mercury, σ_1^d can be obtained from measurements of

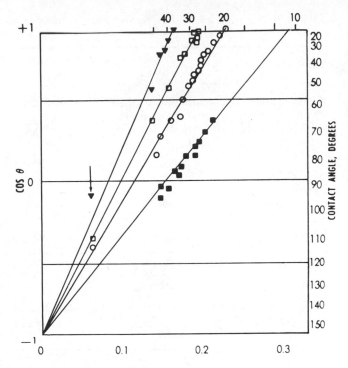

Figure A.12 Fowkes' plot of Zisman's data to obtain surface free energy of solids.[6]

interfacial tension as already described in Section II.A.3. When θ, σ_1^d and σ_1 are known, Eq. [A.56] with $\pi_e = 0$ gives F_{s0}^d; when θ, σ_1 and F_{s0}^d are known, Eq. [A.56] with $\pi_e = 0$ gives σ_1^d. A useful test liquid to obtain F_{s0}^d is methylene iodide, CH_2I_2, for which $\sigma_1 = \sigma_1^d = 50.8 \, \text{mN/m}$ at $20°C$. Its large surface tension leads to high contact angles, which can be measured more accurately. To obtain F_{s0}^d with this liquid, a single measurement of contact angle and the application of Eq. [A.56] is sufficient. Table A.9 reports some values of F_{s0}^d obtained from contact angle measurements such as are shown in Fig. A.12.

Table A.9 reports values of F_{s0}^d obtained from measurement of the contact angle when the properties of the liquid are known. Conversely, the dispersion force contribution to the surface tension of polar organic liquids can be determined from contact angle measurements on low-energy reference solids.[6] Some values are reported in Table A.10.

Equation [A.56] is applicable to obtain F_{s0}^d when the liquid makes a discrete contact angle with the solid substrate, which occurs when the substrate has relatively small surface energy. On substrates of relatively large surface energy, especially when they are clean, most liquids, liquid metals excepted, spread readily with no discrete angle of contact. The values of F_{s0}^d may then be found by a different method.[6]

If, as is held by many (Section II.A.10), the spreading coefficient in the absence

Table A.9   Values of F_{s0}^d of Low Surface Energy Solids from Contact Angles at 20°C

Name	F_{s0}^d
Perfluorododecanoic acid monolayer on Pt	10.4
Perfluorodecanoic acid monolayer on Pt	13.1
Polyhexafluoropropylene (HFP)	18.9
Polytetrafluoroethylene (TFE)	19.5
n-$C_{36}H_{74}$ wax crystal	21.0
n-Octadecylamine monolayer on Pt	22.1
Paraffin wax	25.5
Polytrifluoromonochloroethylene (Kel-F®)	30.8
Polyethylene	35
Polystyrene	44

From ref. 6.

Table A.10   Values of σ_1^d (mN/m at 20°C) for Polar Organic Liquids from Contact Angle Measurements on Reference Solids

Liquid	σ_1	σ_1^d ($\pm$ std. dev.)
Tricresyl phosphate	40.9	39.2 ± 4
α-Bromonaphthalene	44.6	47 ± 7
Trichlorobiphenyl	45.3	44 ± 6
Methylene iodide	50.8	48.5 ± 9
Glycerol	63.4	37.0 ± 4
Formamide	58.2	39.5 ± 7
Polydimethylsiloxane	19.0	16.9 ± 0.5

From ref. 6.

of a specific interaction, is never positive at equilibrium, its largest attainable value is zero. Observed examples of positive spreading are ascribed to nonequilibrium conditions. The final spreading coefficient is given by

$$S = F_{sv} - F_{sl} - \sigma_1 = F_{s0} - \pi_e - F_{sl} - \sigma_1 \qquad \text{[A.49]}$$

Putting $S = 0$ gives

$$\pi_e = F_{s0} - F_{sl} - \sigma_1$$

Substituting Fowkes' Eq. [A.54] for solid–liquid systems interacting by dispersion forces only,

$$F_{sl} = F_{s0} + \sigma_1 - 2(F_{s0}^d \sigma_1^d)^{1/2} \qquad \text{[A.54]}$$

gives

$$\pi_e = 2(F_{s0}^d \sigma_1^d)^{1/2} - 2\sigma_1 \qquad [A.57]$$

By means of Eq. [A.57] the value of F_{s0}^d can be determined if the surface tension of the spreading liquid and its equilibrium spreading pressure π_e are both known. The value of π_e is determined from the vapor adsorption isotherm, as described in Section II.A.11. Values of F_{s0}^d of metals and other high-energy solid substrates have been determined by this method and are included in Table A.11.

The numerical value of F_{s0}^d is probably the best index of the wettability of a solid by any liquid. Substrates with high values are readily wetted and those with low values are not. High-energy substrates readily adsorb ambient impurities that lower their surface potential. A familiar example is a water-wettable glass surface that soon becomes nonwettable merely on exposure to laboratory or household air. Stainless steel when clean is water wettable but rapidly becomes water repellent by chemisorption of fatty-acid vapors or aerosols.

Higher energy substrates contain additional potential for interactions, such as donor–acceptor interactions. These depend on both the substrate and the liquid; and can be calculated from the enthalpies of adduct formation as described in Section II.A.4. The Drago parameters (or parameters for any other treatment of

Table A.11   Values of F_{s0}^d Obtained from π_e Measurements for Adsorbed Vapors

Adsorbent	Adsorbate	Temperature	π_e (mJ/m^2)	F_{s0}^d (mJ/m^2)
Polypropylene	Nitrogen	78 K	12	26
	Argon	90 K	13	28.5
Graphite	Nitrogen	78 K	51	123
	n-Heptane	25°C	63,56,58	132,115,120
Copper	n-Heptane	25°C	29	60
Silver	n-Heptane	25°C	37	74
Lead	n-Heptane	25°C	49	99
Tin	n-Heptane	25°C	50	101
Iron	n-Heptane	25°C	53	108
	Argon	90 K	47	106
	Nitrogen	78 K	40	89
Ferric oxide	n-Heptane	25°C	54	107
Anatase (TiO$_2$)	n-Heptane	25°C	46	92
	Butane	0°C	43	89
	Nitrogen	78 K	56	141
Silica	n-Heptane	25°C	39	78
Stannic oxide	n-Heptane	25°C	54	111
Barium sulfate	n-Heptane	25°C	38	76

From ref. 6.

Table A.12   Acidic and Basic Solid Surfaces

Acidic	Basic
Post-chlorinated polyvinyl chloride	Polycarbonates
Asphalt components	Polymethylmethacrylate
Silica	Calcium carbonate
	Glass

donor–acceptor interactions) are obtained by microcalorimetry or FTIR spectroscopy (Section IV.C.1). These may also be characterized by comparing the adsorption of acid and basic polymers from solution.[14] Basic polymers are adsorbed more strongly on acid surfaces than on basic surfaces, and vice versa. Table A.12 records some typical results.

Some polymeric solids are formed by crystallization, but they are generally formed by freezing into various metastable glassy states. The surface properties of glassy materials are often studied by means of the temperature dependences of the surface free energy or of the specific heat. The temperature dependence of the free energy has a discontinuity at the melting point but not at the temperature of the glass transition; the heat capacity has a discontinuity at the temperature of the glass transition.[22] Monographs on the characterization of polymer surfaces are available.[22] (Wu, 1982).

9.   FLOATING PARTICLES

Many finely divided minerals float on water, even though they may be denser than water. Gold dust, finely divided tungsten, molybdenum, silicon, sulfur, talc, arsenic trioxide, coal, graphite, galena, molybdenite, zinc blende, pyrites, and many other substances have this property. Even massive bodies, such as an oil-coated needle, will float on water. An uncut diamond weighing 0.05 g (specific gravity = 3.5) floated readily on water and remained floating for hours. The requirement for the effect is a contact angle of more than 90° for water against the solid particle: A vertical component of the surface tension equal to $\sigma \sin \theta$ acts upward on every unit length of the perimeter of contact S of the water on the solid; then if $S\sigma \sin \theta > mg$, where m is the apparent mass of the particle, it will float. By reducing the surface tension of the water, the contact angle is reduced, the particle is wetted, and sinks.

The foregoing principles have been used in various ways, sometimes without being understood. Dr. Mathew Hay's test for jaundice is to dust flowers of sulfur over the urine of a patient; on normal urine sulfur floats, while the presence of bile salts causes the sulfur to sink. Again, in the East Indies the natives engaged in gold washing, who are in the habit of chewing the betel nut, spit into the washing water to prevent finely divided gold from floating away.[23] The natural protective oils on the plumage of birds, by ensuring a high contact angle, prevent air from being displaced by water, thus insulating them against the cold and permitting, for example, ducks and sea gulls to float on water. A method used to destroy a plague of starlings was to spray the flock with a solution of a powerful wetting agent; the next rain shower exposed them fatally to the cold. Water flies can walk on water, bouyed up by the high contact angle; mosquito larvae suspend themselves under the surface of stagnant water by means of a hydrophobic pseudopod penetrating the surface.

Industrial processes that make use of these principles are the separation of metal-bearing ores from the gangue by flotation and the production of fluorinated polymers, such as PTFE (polytetrafluoroethylene), which float on the surface of the solution when the polymerization is completed.

10. INITIAL AND FINAL SPREADING COEFFICIENTS

High-energy substrates, unlike low-energy substrates, must be carefully protected from inadvertent contamination on contact with the atmosphere, which, especially in a laboratory, is likely to be poluted with organic vapors. The surface energy of a substrate is reduced by adsorption; indeed, adsorption would not occur unless this were so. Some striking examples of the spontaneous reduction of surface free energy on exposure to the atmosphere were demonstrated by Quincke and by Devaux. A drop of water flashes across the surface of freshly cleaved mica, but on a mica surface exposed to the atmosphere for 20 min, a drop of water remains sessile. Devaux mounted a rose petal above the surface of freshly distilled mercury on which talc had been dusted to act as an indicator. Volatile odor from the petal was adsorbed by the mercury surface, as indicated by the withdrawal of the talc, leaving a bare spot. The surface energy of carefully purified water is so great that a freshly exposed surface is instantly contaminated. The surface of glass is another high-energy substrate readily contaminated on exposure to ordinary air.

For dry benzene on a clean water (or glass) surface, the initial spreading coefficient is

$$S_{(b \text{ on } a)} = S_{b/a} = \sigma_a - \sigma_b - \sigma_{a'b'} \qquad \text{[A.58]}$$

$$= 72.8 - 28.9 - 35.0 = 8.9 \, \text{mJ/m}^2$$

That is, if the area of the spread benzene increases by one square meter and that of the clean surface of the water decreases by the same amount, there is a decrease in free energy of 8.9 mJ, which is considerable. This indicates that benzene should spread readily. However, since the benzene soon becomes saturated with water and the water ultimately becomes saturated with benzene, the surface tensions change to give the final spreading coefficient as follows:

$$S'_{(b' \text{ on } a')} = S'_{b'/a'} = \sigma_{a'} - \sigma_{b'} - \sigma_{a'b'} \qquad [A.59]$$

$$= 62.2 - 28.8 - 35.0 = -1.6 \,\text{mJ/m}^2$$

where the prime terms denote mutually saturated solutions. The interfacial phase is held to be mutually saturated immediately on contact.

The negative value of the final spreading coefficient indicates that benzene will not spread over the surface of water if the liquids are mutually saturated. In every case mutual saturation reduces the surface tensions, so that the final spreading coefficients are always less than initial spreading coefficients. When a drop of benzene is placed on a water surface, it spreads rapidly over the whole available area, but in a few seconds the film retracts to a lens, with a finite contact angle. This behavior conforms to the calculations of initial and final spreading coefficients given above.

Some theorists maintain that the final spreading coefficient at true equilibrium is never positive (Rowlinson and Widom, 1982, p. 216), based on an argument by Gibbs (1906, p. 258) to the effect that the surface tension of a solution of b in a, $\sigma_{a(b)}$, can be no greater than the sum of the tensions of the surface phase, namely, that against air, $\sigma_{b(a)}$, and that against the solution, $\sigma_{b(a)/a(b)}$. The surface phase is considered by Gibbs to be a solution of a in b. It may be very thin, but cannot be as thin as a monolayer, as surface tension is a macroscopic property. A layer thick enough to have different tensions against its two adjoining phases is called a duplex film. Expressed as an equation, Gibbs' argument is

$$\sigma_{b(a)} + \sigma_{b(a)/a(b)} \geqslant \sigma_{a(b)} \qquad [A.60]$$

An equality in Eq. [A.60], which Gibbs anticipated would describe the usual interface between a and b, is now known as Antonoff's rule. Neither Gibbs's anticipation nor Antonoff's rule is well established in practice. The inequality itself states that the spreading coefficient is never positive. Experiments, however, consistently indicate small positive values of the spreading coefficient for some systems[24] for long periods of time. The source of the contradiction may be the requirement for true equilibrium at the surface, for this may take an indefinite time. A supreme example of the persistence of a positive spreading coefficient is the spreading pressure measured on the film balance, which if enough time (eons) were allowed would always equal zero, by dissolution of the monolayer and its subsequent adsorption on the other side of the float.

The following analysis of spreading of a liquid on a solid substrate leads to the same conclusion as that of Gibbs for the fluid–fluid interface. Let the free energy of the solid substrate in a vacuum be F_{s0} and in equilibrium with a vapor be F_{sv}. When the substrate is in equilibrium with the saturated vapor, the intercalation of a liquid film of macroscopic thickness (energy $F_{sl} + \sigma_l$) between the solid surface and the vapor would not decrease the total surface free energy of the system since the liquid is in equilibrium with its own vapor. Therefore[25]

$$F_{sv} = F_{sl} + \sigma_l \qquad [A.54]$$

since the interfaces represented on both sides of the equation are in equilibrium with the same vapor phase, and therefore are in equilibrium with each other. This equality substituted in Young's equation

$$F_{sv} = F_{sl} + \sigma_l \cos \theta \qquad [A.47]$$

leads to $\cos \theta = 1$ or $\theta = 0°$ and a spreading coefficient of zero. Positive spreading coefficients are found experimentally because a substrate in contact with saturated vapor is in a metastable state prior to nucleation and condensation, and so has a higher energy than the same substrate covered with condensate. The metastable condition, in the absence of supersaturation, can persist indefinitely; therefore the spreading coefficient may indeed be positive to the investigator while its theoretical value of zero awaits the Greek kalends. When complete spreading on a macroscopic scale is observed, the system may not be at equilibrium. If at equilibrium, $S = 0$; if not at equilibrium, $S > 0$.

The same reasoning that concludes that at equilibrium the spreading coefficient cannot be larger than zero also leads to the conclusion that at equilibrium the work of adhesion cannot be more than the work of cohesion, and that at equilibrium the free energy of wetting cannot be more than the surface energy of the wetting liquid.

The arguments assume that the liquid occupying the substrate retains the properties that it has when not in contact with the substrate. Should any interaction take place with the substrate, such as an acid–base interaction, the arguments no longer hold. Positive spreading is then likely to occur, even at equilibrium. A work of adhesion greater than the work of cohesion can be obtained under the same conditions, namely, an acid–base interaction between the substrate and the occupying liquid. In formulating composites, the strongest bonding between fiber and matrix is obtained by arranging for an acid–base interaction, that is, electron transfer, between them.

Another possibility exists at equilibrium. The solid substrate may have a surface free energy initially less than $F_{sl} + \sigma_l$, in which case Young's equation predicts a finite contact angle and a negative spreading coefficient. In such a case the spreading coefficient would become more negative, and the contact angle would increase on approaching equilibrium. For substrates of low surface energy, the difference between initial and final spreading coefficients equals π_e, as long as the solid is insoluble in the liquid; and π_e approaches a negligible quantity for systems with high contact angles. Therefore such systems show hardly any difference between initial and final spreading coefficients.

With some systems the energy difference between F_{s0} and F_{sv}, which is π_e of Eq. [A.48], can be quite large. With water on metallic oxides, $\pi_e \sim 300 \text{ mJ/m}^2$; with organic liquids on oxides, $\pi_e \sim 60 \text{ mJ/m}^2$; on the other hand some values of π_e are quite small, as shown in Table A.7.

If the substrate is an insoluble solid, the difference between the initial and final spreading coefficients resides entirely in the reduction of the free energy of the solid by the adsorbed vapor of the liquid, as neither the surface tension of the liquid nor the energy at the solid–liquid interface is affected by prolonged contact; hence,

$$\pi_e = F_{s0} - F_{sv} = S_{b/s} - S'_{b/s'} \qquad [A.61]$$

Fowkes et al.[26] introduced a new laboratory technique for the direct measurement of π_e of a vapor on a low-energy substrate, using advancing contact angles of one liquid to measure the π_e generated by the vapor of a second liquid. Figure A.13 represents the effect of cyclohexane vapor on the contact angle of water on polyethylene. In the presence of water vapor alone,

$$\sigma_{H_2O} \cos \theta_1 = F_{sv(1)} - F_{sl} \qquad [A.62]$$

When cyclohexane is added to the vapor phase, both the contact angle of the water and the free energy of the unwetted substrate are affected, giving

$$\sigma_{H_2O} \cos \theta_2 = F_{sv(2)} - F_{sl} \qquad [A.63]$$

therefore

$$F_{sv(1)} - F_{sv(2)} = \pi_e = \sigma_{H_2O}(\cos \theta_1 - \cos \theta_2) \qquad [A.64]$$

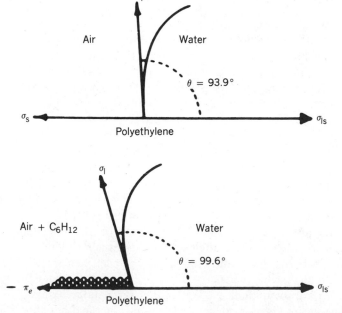

Figure A.13 Effect of cyclohexane vapor on the contact angle of water on polyethylene.[26]

assuming that the π_e due to water vapor on a low-energy substrate is negligible, that is, that $F_{sv(1)} \simeq F_{s0}$. The advancing contact angle of a drop of water on polyethylene was 93.9° in air and 99.6° in air saturated with pure cyclohexane, as illustrated in Fig. A.13. The value of π_e is then 7.19 mJ/m^2.

An alternative way to obtain π_e is by means of the initial spreading coefficient. Since $\pi_e = S_{b/s} - S'_{b/s'}$, where $S_{b/s}$ is the initial and $S'_{b/s'}$ is the final spreading coefficient; and since $S'_{b/s'}$ is usually low (at equilibrium, never greater than zero), then $\pi_e \simeq S_{b/s}$. This equation allows the approximate calculation of π_e for cases where $S_{b/s} > 0$, as follows:

$$\pi_e \simeq S_{b/s} = F_{s0} - \sigma_b - F_{bs}$$

$$= 2(F_s^d \sigma_b^d)^{1/2} - 2\sigma_b \qquad [\text{A.65}]$$

since

$$F_{bs} = F_{s0} + \sigma_b - 2(F_s^d \sigma_b^d)^{1/2} \qquad [\text{A.14}]$$

This equation gives $\pi_e = 7.15$ for cyclohexane on polyethylene ($F_s^d = 33.15 \text{ mJ/m}^2$; $\sigma_b = \sigma_b^d = 25.5 \text{ mJ/m}^2$). The two methods yield answers that agree.

11. DETERMINATION OF SPREADING PRESSURE BY ADSORPTION TECHNIQUES

The methods reported above to determine the equilibrium spreading pressure π_e are limited to planar substrates of large area, such as single crystal faces or polymer substrates. A method suitable for fine solid particles is provided by the technique of gas adsorption. The phenomenon of gas adsorption is due to the spontaneous reduction of the surface free energy of the substrate; thus, for example, since water vapor is not spontaneously adsorbed by paraffin wax at room temperature, π_e is zero for this case.

The equilibrium between a gas phase and gas adsorbed on a substrate at constant temperature is expressed by equating the chemical potentials of the adsorbate in each phase. In the vapor phase

$$dG = V_g \, dp \qquad [\text{A.66}]$$

where V_g is the molar volume of the vapor. In the adsorbed phase

$$dG = V_{ads} \, dp - A \, d\sigma \qquad [\text{A.67}]$$

where V_{ads} is the molar volume of the adsorbed gas; therefore,

$$(V_g - V_{ads}) \, dp = - A \, d\sigma = + A \, d\pi_e \qquad [\text{A.68}]$$

Neglecting V_{ads} by comparison with V_g and assuming the vapor is ideal gives

$$A \, d\pi_e = (RT/p) \, dp = RT \, d\ln p \qquad [\text{A.69}]$$

hence

$$\pi_e = RT \int_0^{p_0} \Gamma \, d\ln p \qquad [\text{A.70}]$$

where p_0 is the saturated vapor pressure and $\Gamma = 1/A$ (moles/unit area). Equation [A.69] is the Gibbs adsorption equation applied to gas adsorption. To obtain π_e, the adsorption isotherm is plotted as the amount adsorbed, Γ, versus the natural logarithm of the equilibrium pressure: The area under the isotherm is then measured from $p = 0$ to $p = p_0$.

The major difficulty with this method is the enormous effect introduced by relatively small concentrations of high-energy sites or patches on the surface. Water is adsorbed by hydrophilic sites but not by lipophilic sites, so the measured adsorption isotherm refers only to the hydrophilic subarea. The heterogeneity of the substrate may also be the result of large differences in the nature of the interaction of a nonaqueous adsorbate with different sites on the surface. The spreading of drops on the other hand is barely affected by occasional high-energy sites. The vapor adsorption method requires great care and is time consuming. It has been used to determine π_e for adsorbed films of nitrogen, argon, and alkanes on graphite, metal oxides, and polypropylene powders (Table A.11).[6] For example, π_e for water on silica is approximately $450 \, \text{mJ/m}^2$ and π_e for water on polyethylene is approximately zero.

12. DYNAMICS OF INITIAL SPREADING

When the initial spreading coefficient is calculated to be positive, and when spreading on a macroscopic scale is observed, the liquid is expected to spread over the whole available area of substrate, and if that is sufficiently extended, to end as a monomolecular layer. Monomolecular layers of lauryl alcohol on water are so obtained. Spreading does not always occur in this way. Hardy observed that a spreading droplet is preceded and surrounded by a sensible though invisible film of liquid, which shows up ahead of the nominal contact line.[27] Certain observations of the spreading of a nonpolar liquid on steel reveal a precursor film (also called a primary film) visible in ellipsometry at the late stages of spreading, with a thickness of only a few tens of nanometers.[28] The precursor film has also been observed by measurements of electrical resistance.[29]

De Gennes[25] pointed out that the molecules in the precursor film have higher potential energies because they are separated farther than in the bulk liquid. Continued spreading is inhibited as it leads to even greater potential energies. The spreading comes to an end therefore before the drop has thinned to monolayer dimensions. What may appear to be a completely spread monolayer may actually be a film a few score nanometers thick.

If volatile impurities of lower surface tension than the spreading liquid are present, they are preferentially lost by evaporation from the precursor film, the surface tension of which then increases. The higher surface tension of the precursor film draws liquid from the rest of the drop and so promotes further spreading (a Marangoni effect). In the lubrication of ball bearings, oil is transported by spreading from the grease supply to the raceway; the spreading is favored by the molecular heterogeneity of ordinary petroleum oils, whose more volatile components have lower surface tensions than the mixture. The tendency to spread can be counteracted by an additive. Undistilled squalane was made nonspreading for several days by adding 5% isopropylbiphenyl, which has a surface tension 7 mN/m higher and a boiling point 50°C lower than squalane. The same nonspreading effect can be produced by adding a liquid of less volatility than the squalane but a lower surface tension, for example, polydimethylsiloxane. Evaporation at the edge of such a film lowers the surface tension relative to the bulk liquid mixture and causes the film to retract. The polydimethylsiloxane liquids by themselves are often troublesome because of excessive spreading, but adding small amounts of a more volatile methyl phenyl silicone with a higher surface tension inhibits spreading for several weeks.[28]

13. THERMODYNAMICS OF CURVED INTERFACES

At any point on a plane surface the components of the tension lie within the surface, so there is none normal to the surface; on a curved surface, however, the components of the tension pull normally (to the concave side) as well as tangentially to the surface. Hence, a curved surface can only be maintained by an equal and opposing force to the normal components, which force may be exerted by gas pressure or by hydrostatic pressure (Fig. A.14). For example, a spherical bubble has normal components of the surface tension pulling inward, and these are counteracted by the higher gas pressure within the bubble. For the same reason the pressure is always larger on the concave side of a curved liquid surface. Another example: The shape of a sessile drop is determined by the

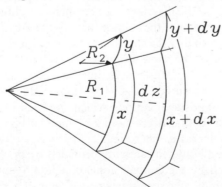

Figure A.14 Balance of pressure drop and curvature across a curved surface.

balance between the forces of surface tension and the counteracting hydrostatic pressure.

Consider a small element of area of a curved liquid surface of length x and width y at equilibrium. If this surface were to expand by a differential amount, dz, normal to the surface, the area expands to $(x + dx)(y + dy)$. The increase in area is, to leading order, $y\,dx + x\,dy$. The corresponding increase in volume is $xy\,dz$. The work required to increase the area against the surface tension is balanced by the work done by the expansion against the pressure, Δp, across the curved surface, that is,

$$\Delta p(xy\,dz) = \sigma(y\,dx + x\,dy) \qquad [A.71]$$

By choosing the x and y axes to lie in the planes of the principal circles of curvature of radii, R_1 and R_2, respectively (the z axis is the line common to both planes), the properties of similar triangles gives

$$\frac{dx}{x} = \frac{dz}{R_1} \text{ and } \frac{dy}{y} = \frac{dz}{R_2} \qquad [A.72]$$

Rearranging and substituting gives the Laplace equation

$$\Delta p = \sigma(1/R_1 + 1/R_2) \qquad [A.73]$$

The pressure is hydrostatic in origin for liquids and is "gasic" for gases; the pressure drop across the surface is directly proportional to the curvature of the surface. The surface tension is the constant of proportionality. This remarkably simple equation describes the shapes of all free-standing liquid surfaces, no matter what their boundary conditions. Some interesting surface shapes are developed by bubbles and soap films. The spherical surface of a bubble has a curvature of $2/R$ where R is the radius of the bubble. The extra pressure of the gas inside a bubble equals $4\sigma/R$; the second factor of 2 arises because a bubble has two liquid surfaces, inside and outside, both of them equally curved (to a close approximation). A cylindrical surface has a principal radius of curvature in one direction equal to the radius of the cylinder and, in the other direction, the radius of curvature is infinite; so that the curvature of a cylindrical surface is $1/R$, that is, half that of a sphere of the same radius. A plane surface has zero curvature because the circles of curvature have infinite radii; therefore the pressure difference across a plane surface is zero. Various saddle-shaped surfaces have radii of curvature in opposite directions, which, in the case of a catenoid are also equal to each other, so that a catenoidal surface has zero curvature and no pressure drop exists across it. These results are listed in Table A.13.

Examples of curved liquid surfaces are provided by bubbles, soap films, and the curvature induced by the contact angle of a liquid surface against a solid substrate, as, for example, the meniscus created against a glass slide positioned vertically to a water surface. The contact angle of water on clean glass is zero

Table A.13   Pressure Drops across Curved Liquid Surfaces

Shape	Curvature	Pressure Drop
Spherical bubble	$2/R$	$4\sigma/R$
Sphere	$2/R$	$2\sigma/R$
Cylindrical lamella	$1/R$	$2\sigma/R$
Cylinder	$1/R$	σ/R
Plane	0	0
Catenoid	0	0
Spiral ramp	0	0

degrees, which requires that the water surface bend upward to meet the vertical surface of the slide; the curvature thus induced creates a lower hydrostatic pressure inside the liquid, which is balanced by the liquid held in the meniscus against the pull of gravity. Another example of curvature induced by contact angle is the approximately hemispherical meniscus of water inside a capillary tube. The pressure difference associated with this curvature is the force that causes the spontaneous rise of liquids in capillaries. The flow of liquid in response to pressure differentials created by curvature of liquid surfaces is known as *capillary flow*.

14. SESSILE AND PENDENT DROPS

A nonspreading drop supported by a solid substrate is sessile or sitting; a drop suspended from a tip is pendent or hanging. Such drops have a shape that is determined by a balance of the forces of gravity and of surface tension: Gravity tends to flatten sessile drops and elongate pendent drops; surface tension tends to confer spherical shape, as the sphere has minimum area for a given volume. The less dense the liquid, the more the shape approaches the sphere; the denser the liquid, the more the shape departs from the sphere. Small drops are less affected by gravity and so are almost perfectly spherical. Large drops of liquid are less affected by surface tension, hence flattened; nevertheless, their ultimate thickness is determined by contact angle and surface tension. A liquid drop in a medium in which it is insoluble forms a perfect sphere if the two liquids have the same density. A good example of a sessile drop is provided by a rain drop on a petal or leaf; another example is the shape frequently used for rural water towers, a concept developed commercially by the Chicago Bridge and Iron Works.

The balance of forces acting on a liquid drop is described mathematically by a differential equation, which gives an expression for the shape in terms of volume, surface tension, and density. Its starting point is Eq. [A.73] written in the form

$$\rho g z + C = \sigma(1/R_1 + 1/R_2) \qquad [A.74]$$

where the pressure difference Δp is replaced by $\rho gz + C$, which is the hydrostatic pressure, ρgz, at a distance z from the apex of the drop plus the pressure difference C across the apex of the drop. Equation [A.74] expressed in Cartesian coordinates is

$$\frac{d^2z/dx^2}{[1+(dz/dx)^2]^{3/2}} + \frac{(1/x)dz/dx}{[1+(dz/dx)^2]^{1/2}} = \frac{\rho gz + C}{\sigma} \qquad \text{[A.75]}$$

This form of the equation is a second-order nonlinear differential equation with singularities at the equatorial extremes. A numerically stable form is the expression in terms of arc lengths s and the angular inclination Φ of the tangent to the horizontal (Fig. A.15) as

$$\sigma\left(\frac{d\Phi}{ds} + \frac{\sin \Phi}{x}\right) = \rho gz + C \qquad \text{[A.76]}$$

Equation [A.76] is a transcendental differential equation. This form of the equation is the most suitable for experimental evaluation. Neither form of the equation has yet received a closed solution. A numerical solution was originally given as hand-calculated tables by Bashforth and Adams in 1883; nowadays more extensive tables giving the shape of sessile drops, pendent drops, and external menisci, are available (Hartland and Hartley, 1976). The inverse problem, to find the surface tension of the liquid given the shape of a sessile or pendent drop and the density of the liquid, is a more common problem, which is solved by a computer program.[30,31]

These equations describe the complete surface of revolution, but actually the drop is truncated by the plane of the substrate at a depth determined by the specific contact angle of the liquid–solid system. A drop of mercury on glass, with a contact angle of 140°, is truncated very little compared to a drop of n-octane on polytetrafluoroethane, with a contact angle of 32°.

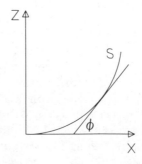

Figure A.15 Notation for axisymmetric surface bounding a sessile or pendent drop.

15.   THE KELVIN EQUATION

The effect of an increase of external pressure on a liquid is to raise its vapor pressure. An equilibrium condition obtains when the differential increase of the molar Gibbs free energy of the liquid, due to the increase in the total pressure dP, is equal to that of its vapor:

$$dG_l = dG_v \qquad\qquad [A.77]$$

At constant T

$$dG_l = V_m dP$$

where V_m is the molar volume of the liquid and

$$dG_v = RT\, d \ln p$$

where p is the vapor pressure. Assuming V_m constant with variation of P, Eq. [A.77] may be integrated between the limits $P = 0(p = p_0)$ and $P = P(p = p)$ to give

$$V_m P = RT \ln (p/p_0) \qquad\qquad [A.78]$$

Surface tension acts as a pressure on a curved surface (Section II.A.13). By combining Eqs. [A.78] and [A.73], written for a sphere, setting P equal to the pressure produced by surface tension, the increase in vapor pressure of a small drop above the normal vapor pressure of the liquid is

$$\ln \frac{p}{p_0} = \frac{2\sigma V_m}{aRT} \qquad\qquad [A.79]$$

where a is the radius of the drop. Equation [A.79] is the Kelvin equation. It states that small drops have higher vapor pressure than flat sheets of liquid, hence small drops evaporate more readily. Conversely, the condensation of small drops from the vapor requires so much supersaturation that the process does not occur spontaneously in the absence of solid particles to act as nucleating sites. Once small drops are nucleated, they grow rapidly as their equilibrium pressure declines. For the same reason, in a mist of polydisperse drops, the larger drops grow at the expense of the smaller ones. This process is an example of what is known as Ostwald ripening.

The Kelvin equation also applies to the concave surface of a liquid meniscus. Where r is the radius of curvature of the meniscus,

$$\ln \frac{p}{p_0} = -\frac{2\sigma V_m}{rRT} \qquad \text{where } p < p_0 \qquad\qquad [A.80]$$

Equation [A.80] states that the vapor pressure of a liquid is decreased at a concave meniscus, such as occurs on a wetting liquid in a capillary tube or in the pores of a solid substrate. Experimental verification of the Kelvin equation is reported by Fisher and Israelachvili[32] for the concave meniscus of cyclohexane between crossed cylinders of molecularly smooth mica. With a clean system, where the establishment of equilibrium and the absence of significant contamination could be demonstrated, the Kelvin equation was found to hold for radii of curvature as low as 2.5 nm.[33]

The fierce debate about the existence of anomalous water or, as it was popularly known, "polywater," arose when investigators found that water, in capillaries of a few microns radius, had a much lower vapor pressure than predicted by the Kelvin equation. The debate ended with the discovery that the "anomalous" behavior is accounted for by the presence of various leachable impurities in solution (Franks, 1981).

Equation [A.79] may be applied to the solubility of small particles, replacing vapor pressures by saturation concentrations:

$$\ln \frac{c}{c_0} = +\frac{2\sigma V_{\mathrm{m}}}{aRT} \qquad [A.81]$$

where c is the solubility of small spherical particles of radius a and c_0 is the solubility of large particles ($r = \infty$). Although solid particles are not usually spherical, the general conclusion that small particles have a higher solubility than large particles is well established, as is the requirement of supersaturation before nucleation occurs in clean systems, and the phenomenon of the Ostwald ripening of precipitates.

16. CAPILLARY FLOW AND MARANGONI FLOW

Capillary flow of a liquid results from differences of hydrostatic pressure within a liquid, created by local differences of curvature of the liquid surface. The pressure is less on the convex side of an interface; liquid at higher pressure then flows to the region of lower pressure. A liquid that makes a low angle of contact in a narrow vertical tube creates a meniscus that is convex to the liquid, consequently liquid flows up the tube. Mercury in a glass tube creates a meniscus that is concave to the liquid, consequently mercury flows down the tube; but since the density of mercury is large, the decline of the mercury level in the tube is small; nevertheless a correction for this capillary effect has to be introduced in precise mercury manometry. The pressure, Δp, required to blow a liquid out of a capillary tube of radius r is also a result of the curvature of the liquid surface.

$$\Delta p = 2\sigma \cos \theta / r \qquad [A.82]$$

The more acute the angle of contact, the greater is the pressure required to blow the liquid out; if the angle is 90°, the effect is zero; if the angle is obtuse, the liquid will not enter the tube spontaneously.

Another example of capillary flow is the motion of a drop of liquid in a narrow conical aperture. The capillary pressure at each end of the liquid drop is different because the radii of curvature of the two surfaces are different. In an aperture narrow enough to neglect deviations from spherical surfaces, the net pressure on the drop is given by (Bikerman, 1970, p. 274)

$$\Delta p = 2\sigma \cos(\theta - \Phi)(1/R_1 - 1/R_2) \qquad \text{[A.83]}$$

where θ is the contact angle, 2Φ is the angle of the cone, and R_1 and R_2 are the radii of curvature at the front and rear surfaces of the drop. For acute contact angles the drop moves toward the narrow end of the aperture; for contact angles sufficiently large, the drop moves toward the wide end.

The importance of capillary forces is measured by the Bond number:

$$B_0 = \frac{\rho g d^2}{\sigma}$$

where ρ is the density difference, g is the acceleration due to gravity, d is a characteristic dimension, and σ is the surface tension. Capillary phenomena are significant when B_0 is less than unity. The Bond number is related to the capillary constant used in the numerical solutions of the Laplace Eq. [A.73].[34] An interesting application of this criterion occurs in the microgravity of space, where the value of g is much reduced. Under this condition capillary flow is significant for much larger dimensions.[35]

Another kind of flow induced by surface tension is *Marangoni flow*. This flow results from local differences of surface tension. At equilibrium, no local differences of surface tension exist; but various factors can lead to non-equilibrium. Some examples of these factors are:

(a) Local thermal differences.
(b) Local differences of composition due to evaporation.
(c) Local compressions and dilatations of adsorbed films at liquid surfaces.

Each of these disturbances makes the surface tension of the liquid depart from equilibrium. Nonequilibrium surface tensions are called "dynamic" and equilibrium surface tensions are called "static." Gradients of tension at an interface are equivalent to shear forces τ at the boundary of the two adjoining phases, α and β:

$$\text{grad}\,\sigma = \tau_\alpha + \tau_\beta$$

These shear forces cause adjoining liquids to flow.

Many processes depend on or are plagued by Marangoni flow, caused by dynamic surface tension. James Thomson, even before Marangoni, explained the phenomenon of brandy tears. Alcohol evaporates faster than water, so that the surface layer of a solution of alcohol in water is more dilute than the bulk solution. In a deep vessel, the loss of alcohol at the surface is restored by diffusion. But a thin layer of brandy adhering to the sides of a glass becomes more dilute by evaporation and remains so, with a consequent rise in its surface tension. The higher surface tension on the walls draws the surface of the brandy in the glass, until the quantity of liquid drawn up becomes so large that it falls back in the form of tears.

A household hint on how to remove grease stains from cloth was explained by Maxwell (Maxwell, 1872). Grease has a higher surface tension than the solvent used to remove it. If the middle of the stain is wetted with solvent, the grease is driven into the clean part of the cloth. The correct method is to apply the solvent in a ring all around the stain, gradually bringing it nearer the center. The grease is thus brought to the center of the stain, from which it may be removed by a paper towel.

Marangoni flow is the explanation of a phenomenon observed by Fowkes.[36] The rate of wetting of a skein of gray unboiled cotton yarn was found to be more rapid in beakers of larger diameter. The solute was adsorbed from the surface of the solution on to the fiber, which depletes the surface locally, and so creates a surface tension gradient that draws the rest of the surface to the cotton. This process is much more rapid than the rate of diffusion to the surface from the bulk solution, which becomes the rate-determining step. The larger the area of the surface, the more molecules are brought into it per unit time, and then rapidly transported to the cotton.

Scriven and Sternling[37] reviewed surface movements due to Marangoni effects. They list various phenomena in which Marangoni effects play an important role, such as in crystal growth, motion of protoplasm, transport of bacteria, surface fractionation, absorption and distillation, brandy tears, foam stability, and the damping of waves by oil. The uneven drying of film coatings, creating thermal gradients, leading to cellular flow patterns at the surface, which were observed and explained correctly in 1855 by James Thomson, and are now known as "Bénard cells," are also examples of Marangoni flow. Thermal gradients create both density and surface tension gradients. In the reduced-gravity environment of spacecraft, density gradients occur but do not lead to convection currents; but surface tension gradients produce flow, and a type of convection called "thermocapillary."[35] The term "thermocapillarity" is an unfortunate choice as it blurs the useful distinction we have made between capillary flow and Marangoni flow. A better term is "thermally induced Marangoni flow," or "TIM flow."

A household example of TIM flow is illustrated by another effectual method to remove a grease spot. A hot iron is applied to one side of the cloth and a paper towel to the other. The thermally induced Marangoni flow drives the grease into the paper towel.

17. METHODS TO MEASURE SURFACE AND INTERFACIAL TENSION

The Laplace equation [A.73] is the basis on which all the methods of measuring surface and interfacial tensions are founded:

$$\Delta p = \sigma(1/R_1 + 1/R_2) \qquad [A.73]$$

In all such methods the pressure difference and radii of curvature are implicated in different ways.

a. Capillary Height

The curved meniscus of a liquid in a narrow glass tube is created by the contact angle of the liquid against glass: If the liquid wets the glass, the meniscus has the form of a hollow sessile drop; if the liquid does not wet the glass, as for example, mercury, the liquid surface assumes the convex form of a sessile drop. Only if the contact angle were exactly 90° would the liquid surface be perfectly plane and coincident with the cross section of the tube. When the tube is sufficiently wide, the meniscuses along the peripheral contact with the liquid do not overlap; a flat spot exists at the center of the tube, and no pressure difference is created there. But in a narrower tube the pressure difference at the center of the tube causes liquid to flow, either up or down the tube depending on whether the curvature is concave or convex. The flow of liquid continues until the hydrostatic pressure, ρgh, just balances the Laplace pressure difference, Δp. Let a liquid of density ρ and surface tension σ make an angle of contact θ against the wall of a glass uniform glass tube of radius r: then

$$\Delta p = \rho gh = 2\sigma \cos \theta/r \qquad [A.84]$$

Equation [A.84] is exact when the meniscus is spherical and its weight is negligible compared to the weight of the column, which is true only for very narrow tubes. The method is not recommended for values of θ other than zero because the angle, if finite, is usually not known and a nonzero angle reduces the capillary rise and hence the precision of the method. When the angle is zero, Eq. [A.84] becomes

$$\rho gh = 2\sigma/r \qquad [A.78]$$

Harkins and Brown[38] have warned that the method has serious limitations. Capillary tubes of uniform bore must be used, readings of high accuracy require a cathetometer, and the results obtained with basic solutions may be 20% or more too low. Viscous liquids and solutions of certain organic compounds also give incorrect results; on the other hand, excellent results are obtained with water, benzene, the lower alcohols, and similar liquids. Refinements of this technique are fully described by Sugden.[39]

b.   Drop Weight

A stream of liquid falling slowly from the tip of a glass tube is nipped off into drops by the surface tension. A general relation is

$$V \rho g = mg = 2\pi r \sigma F \qquad [A.85]$$

where V is the volume of the drop, m is its mass, ρ is the density of the liquid, r is the radius of the tip, and F is a correction factor. The values of the factor and a discussion of the refinements of the method are given by Harkins and Brown[38] and Padday.[40]

The drop weight, mg, is found by weighing a counted number of drops or by counting the number of drops when a measured volume of liquid passes through the tip. A volumetric syringe with a motor-driven plunger is convenient to obtain an accurate volume of liquid. The nature of the method precludes the study of prolonged aging of the surface, which is an important feature of many solutions or of highly viscous liquids.

c.   Wilhelmy Plate

The Wilhelmy method to determine surface tension consists of measuring directly by means of a balance the weight of the meniscus formed on the perimeter of a thin plate, such as a platinum blade, partially immersed in the liquid. The Wilhelmy method has many advantages: It is absolute, simple to set up, and independent of the density of the fluid. To avoid a correction for buoyancy, the blade is positioned with its lower edge parallel to and at the same height as the level liquid surface. The measured weight is that of the meniscus created by the wetting of the blade by the liquid. The surface tension is given by the relation

$$\sigma = \frac{W}{2(l + d)\cos\theta} \qquad [A.86]$$

where W is the weight of the meniscus, $2(l + d)$ is the perimeter of the blade, and θ is the angle of contact. The method is eminently suitable for studying prolonged aging of the surface, as an electronic balance gives continuous readings while maintaining the blade at a constant height. Refinements of the method are described by Harkins and Anderson[41] and by Padday.[40] Gaines has pointed out the advantage of using plates made of filter paper, which are porous enough to give a zero contact angle and are disposable.[42]

d.   The Du Noüy Ring

The adhesion of a liquid to a metal ring is greater than the cohesion of the liquid, if there is no finite contact angle; consequently when a ring is detached from the surface of such a liquid, the force to be overcome is that of cohesion (2σ) rather than adhesion. This is the basis of the ring detachment method of measuring surface tensions. The surface tension is given by

$$\sigma = \frac{mg}{4\pi r} F \qquad\qquad [A.87]$$

where mg is the maximum upward pull applied to the ring of radius r and F is a correction factor given by Harkins and Jordan[43] and by Padday.[40] The correction factor takes account of a small but significant volume of liquid that remains on the ring after detachment, and the discrepancy between the radius r and the actual radius of the meniscus in the plane of rupture. The method is not suited for solutions that attain surface tension equilibrium slowly or for very viscous liquids.

e.   Sessile and Pendent Drop

The sessile and pendent drop methods for measuring surface or interfacial tensions are absolute, require only small volumes of liquid, are readily amenable to temperature control, do not require detachment, and do not depend on the contact angle. They can be brought to the highest level of accuracy by control of external vibration, high-precision measurement of drop dimensions, and computer programs that accept multiple readings of drop shape coordinates. These methods can be used for the study of aging effects and with highly viscous liquids, such as crude oils.

The shape of a sessile (or sitting) drop on a plane surface is determined by the balance between the surface tension of the liquid and hydrostatic pressures within the drop. Surface tension or interfacial tension may be determined by recording the profile of the sessile drop; and contact angle may be obtained by measuring the angle between the tangents to the liquid surface and to the solid substrate at a point of contact. The profile of the drop may be shown by a sharp silhouette thrown on a screen or by measurements of several coordinates along the drop profile. The original tables of Bashforth and Adams are now superseded by a computer program that matches observations of the drop profile to its closest theoretical counterpart.[30] The development of computer techniques has also diminished the importance of the various approximations to the drop shape that have been suggested in the past. The use of the sessile drop to measure surface or interfacial tensions is described by Butler and Bloom[31] and its use to determine contact angle is described below (Section II.A.18).

The principle of the pendent drop method is similar to that of the sessile drop method, inasmuch as the same forces of surface tension and hydrostatic pressure determine its shape. The pendent drop may be formed quickly so that changes of a newly formed surface can be monitored. The pendent drop may be vibrated to study dynamic surface or interfacial tensions, as in the pulsating-bubble technique of Lunkenheimer et al.[44] The use of the pendent drop to measure surface or interfacial tensions is described by Patterson and Ross[45] and Ambwani and Fort.[46]

Sessile and pendent bubbles are possible variants of these techniques that offer advantages under special conditions.

f. Maximum Pull on a Rod

The Wilhelmy plate has the disadvantage that the contact angle often does not remain zero when measuring the surface tension of a monolayer on an aqueous substrate or with some aqueous solutions. A method that is independent of contact angle, but still makes use of essentially the same apparatus as the Wilhelmy plate, was developed by Padday et al.[47,48] The Wilhelmy plate is replaced by a stainless steel rod, 3 cm in length, accurately machined to a diameter of 0.500 cm. The bottom of the rod is cut to form a plane with a sharp circular edge. The plane is roughened so as to be wetted by the contacting liquid, while the cylindrical surface of the rod is highly polished. Highly polished stainless steel is not wetted by aqueous liquids, which therefore do not spread above the meniscus. The rod is hung vertically by a thread from under a standard bottom-loading balance of 1-mg sensitivity, which is supported by a motorized vertical translator that can be raised or lowered at speeds as low as 1.0 mm/min (Fig. A.16.) The axisymmetric shape of the meniscus raised by the rod above the level of the liquid (Fig. A.17) is determined only by its density, its surface tension, and the local gravitational constant, and is independent of the contact angle. In this respect the method resembles that of the sessile drop or of the pendent drop. No liquid should reside on the polished side of the rod, as only the weight of the meniscus is to be measured. When the balance is raised, the reading increases to a maximum, W_{max}, and then falls prior to detachment of the rod.

In place of the general tables given by Padday[47] a linear approximation may

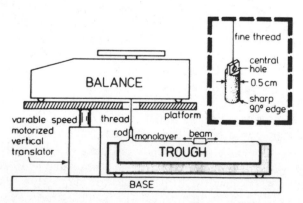

Figure A.16 Schematic drawing of Langmuir trough and balance for measuring surface tensions (D'Arrigo, 1986, p. 111).

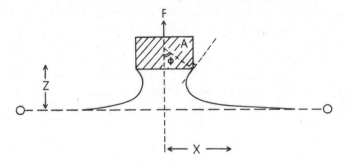

Figure A.17 Axisymmetric meniscus formed between the free surface of a liquid and a circular rod.[48]

be used (D'Arrigo, 1986). For a rod 0.500 cm diameter at 21°C,

$$\sigma_0 - \sigma = 0.54[171 - (W_{max} - W_{rod})]$$ [A.88]

where $\sigma_0 = 72.9 \, \text{mN/m}$ and $(W_{max} - W_{rod})$, the weight of the meniscus, is expressed in milligrams. With a balance that reads ± 1 mg or better, values of σ can be determined to at least $\pm 0.5 \, \text{mN/m}$.

g.   Maximum Bubble Pressure

If a bubble is blown from the bottom of a tube that dips vertically into a liquid, the pressure in the bubble increases at first as the curvature increases or as the radius of curvature decreases. A bubble small enough to be taken as spherical will grow to a hemisphere with a continuous decrease of the radius of curvature, after which further growth increases its radius, thus decreasing its curvature and its pressure. The highest pressure is reached at the point where the bubble is the closest to the hemisphere; at this point the pressure in the bubble is

$$P = \rho g h + 2\sigma/r$$ [A.89]

where $\rho g h$ is the part of the total pressure P required to force the liquid down the tube to the level h, which is the depth of immersion of the tip of the tube, and r is the radius of the tube. If the tube is wetted by the liquid, the radius r is its internal radius, since the liquid covers the lower edge of the tube completely. By determining the maximum pressure that is attained just prior to the detachment of the bubble, the surface tension is evaluated. The method is especially suitable for aqueous solutions.

A newly developed instrument using maximum bubble pressure to monitor surface tension consists of dual capillaries of different radii using a differential pressure transducer to sense the pressure difference. The transducer output is conditioned and sent through an analog interface board to the computer where it is scaled and offset in relation to a previously computer-calculated calibration curve. The resulting value of surface tension has a resolution of 0.1 nM/m. The value is measured a new each time a new surface is formed and the bubble is released from the orifice.[49] An instrument based on this description is the SensaDyne 6000 of the Chem-Dyne Research Corporation.

Recent improvements in the maximum-bubble-pressure method make it particularly suitable to study rates of adsorption at an aqueous surface.[50] Improvements are the use of an inclined capillary with a siliconized bore, but with a hydrophilic face and outside surface.

h.   Spinning Drop

A drop of liquid of about 0.02 cm^3 is placed in a liquid of higher density in a straight precision bore capillary tube, which is spun on its axis (1200–24,000 rpm). The drop is deformed by the imposed and innate forces to a threadlike cylinder, to an extent dependent on its interfacial tension, density difference, and speed of rotation. The lower the interfacial tension, the greater the elongation and the more precisely determined is the result; so the method is well suited for very low interfacial tensions. A traveling microscope is used to measure the length and width of the drop. For example, the interfacial tension between n-octane and water containing 0.2% Petronate TRS 10-80 (Witco Chemical Co.) and 1.0% NaCl at 27°C, measured by this technique, is 10^{-3} mN/m.[51] An analysis of the method is given by Princen et al.[52]

i.   Vibrating Jet

The vibrating jet method, originated by Lord Rayleigh, is used to determine the rate of attaining surface tension equilibrium in solutions of surface-active solutes. A plate orifice is made from a thin sheet of brass (0.005 in. thick) in which an elliptical hole, approximately 1 mm by 0.8 mm, is bored. The solution is ejected through the orifice. The initial elliptical cross section of the jet is unstable because of its unequal curvatures, and oscillates around a cylindrical cross section with a frequency determined by the surface tension. The age of the surface in seconds at the midpoint of successive waves is found from the relation $t = \pi r^2 L/v$, where L is the distance from the orifice, r is the radius of the jet where it has a circular cross section, and v is the flow rate. The technique is used to measure dynamic surface tensions at an age of milliseconds. Addison,[53] Burcik,[54] and Ross and Haak[55] describe the technique and report data.

18.   METHODS TO MEASURE CONTACT ANGLE

The gross geometry of the solid substrate determines the appropriate method to measure the contact angle. On large flat surfaces, the sessile drop allows the angle to be observed directly. The contact angle of a liquid on a fiber is measured by weighing the meniscus, and the contact angle of a liquid on a fine powder is obtained by measuring the pressure required to push the liquid through a porous bed of packed fines. Reviews of these techniques and the results obtained have been published.[56,57]

a.   Goniometer

The contact angle goniometer is a device, mounted on an optical bench, to examine a single liquid drop resting on a plane solid substrate. The drop is illuminated from the rear to form a silhouette and viewed through a telescope equipped with adjustable cross hairs, by means of which the angle of contact can be determined directly by aligning one hairline along the substrate and the other tangential to the drop at its point of contact with the substrate. The angle between the cross hairs, measured through the liquid, is the contact angle. A useful available accessory is an environmental chamber to control conditions of temperature and pressure. Elevated temperatures to 300°C are produced by integral electrical heaters, while subambient temperatures can be produced by circulating coolant through the base of the chamber. The chamber is also suitable for vacuum use and pressure of an inert gas to 0.6 atm. This method was described by Zisman et al.[58] An instrument based on their design is commercially available from Ramé–Hart Inc., which has also published a bibliography of the use of contact angle data in surface science.[20] Another optical device, which also uses the outline of a drop on a plane substrate, enlarges and projects the image onto a calibrated frosted-glass screen (Imass, Inc.).

b.   Interfacial Meniscus

A direct and convenient method to measure contact angles of a liquid–vapor or a liquid–liquid interface against glass or any transparent solid substrate is by means of a photograph of the meniscus inside a cylindrical tube. Two requirements have to be satisfied: to eliminate the distortion of the image of the meniscus introduced by the tube acting as a cylindrical lens and to reduce the deviation of the meniscus from sphericity, so that the contact angle may be derived from a single dimension of the meniscus, namely, its radius. The optics of such systems were analyzed by Bock[59] who showed that an undistorted image can be obtained by immersing the tube in an external fluid of appropriate index of refraction, obtained by computation. A meniscus that departs only negligibly from a sphere is obtained by the use of tubes whose diameter is less than a limiting value that depends on the angle of contact.[60] This method is particularly applicable to measure the angle of contact as a function of temperature.

c.   Fiber Balance

The instrument consists of a recording microbalance, to which the fiber is attached, and an elevator holding the liquid sample. The liquid surface is slowly moved along the filament and the forces exerted on the solid fiber are recorded as a function of position. The elevator is moved slowly enough to avoid effects of dynamic wetting. If the fiber surface is uniform, the liquid–solid line of contact is horizontal. The downward force due to surface tension at any location of the three-phase boundary on the fiber is given by the following closed integral:

$$f(h) = \int \sigma \cos\theta \, ds \qquad \text{[A.90]}$$

where h is the position along the fiber and ds is the element of line contact. Equation [A.90] holds for a fiber of uniform perimeter and contact angle. Under these conditions, advancing and receding angles can be obtained as follows. The liquid is first brought into contact with the fiber; the liquid level is brought up for a short distance; then the motion is reversed; and the liquid is withdrawn. If the advancing and receding angles differ, the plot shows a hysteresis loop. The difference between the force on advancing the liquid and the force on its recession is

$$\Delta f = \sigma s(\cos\theta_2 - \cos\theta_1) \qquad \text{[A.91]}$$

where θ_1 and θ_2 are the advancing and receding contact angles, respectively, and s is the length of the perimeter (Fig. A.18).

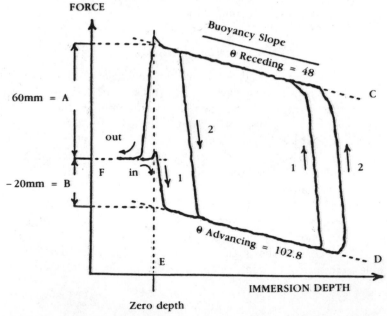

Figure A.18 The force–depth isotherm of the fiber balance, showing a nonwetting advancing angle causing a buoyancy B, and a wetting receding angle, causing an added weight A. The arrows indicate the direction of motion of the fiber. While partially submerged, the fiber is raised and lowered (inner loop) to check for hysteresis (none observed). (Courtesy of Cahn Instruments.)

If the fiber is not of uniform perimeter and contact angle, the balance measures only $f(h)$, which cannot be further analyzed unless the value of the perimeter is known at every height.

d.   The Bartell Cell

The wetting of a powdered solid by a liquid is often practically important. The powder can be compressed to give a flat upper surface upon which the contact angle of a sessile drop can be measured. Although this method is frequently used, the results are unreliable as porosity of the substrate affects the measured angle. A sounder method was developed by Bartell and co-workers.[61-63] The powder is formed into a porous plug by compression and liquid fills the pores by dint of capillary pressure. The pressure to remove the liquid equals the Laplace pressure:

$$\Delta p = 2\sigma \cos \theta / R \qquad [\text{A.92}]$$

where R is the average pore radius, defined as the radius of the equivalent cylinder. The equivalent radius of the porous plug is obtained by measuring Δp for a liquid that completely wets the solid. Such a liquid would have a low surface tension and a zero contact angle. The average pore radius is then calculated by Eq. [A.92] with $\cos \theta = 1$. Care must be taken not to disturb the porous plug in any way that might affect its porosity during the time between the calibration and the test measurements. The apparatus consists of a plug holder, both ends of which are connected to a pressure-measuring device and an indicator tube containing liquid, to observe the liquid position. Usually a simple glass tube with a fritted-glass filter makes a suitable plug holder. Metal apparatus is required only for pressures greater than one atmosphere. The rate of flow of liquid in the plug is measured as a function of the applied pressure, both for the advance and retreat of the contacting liquid. No movement over a range of pressures gives the difference between advancing and receding contact angles.[64] Commercially available mercury porosimeters can be adapted for this purpose.

e.   The Washburn Equation

The flow of liquid in a capillary tube is caused by the Laplace pressure across the curved liquid surface. If the capillary is vertical, the liquid flow continues until the hydrostatic head created by liquid in the capillary is just balanced by the Laplace pressure. If the capillary is horizontal, the flow continues. The rate of flow of liquid in a horizontal capillary is described mathematically by combining the Laplace equation for the pressure difference across a curved surface and the Poiseuille equation for the flow of liquid in a tube.

The volume rate of flow ω of a liquid with viscosity η in a capillary of radius R and length L, driven by a pressure gradient Δp, is given by the Poiseuille equation:

$$\omega = \frac{\pi \Delta p R^4}{8\eta L} \qquad [\text{IB.11}]$$

The velocity of the liquid front, dL/dt, is the volume rate of flow divided by the cross-sectional area

$$\frac{dL}{dt} = \frac{\omega}{\pi R^2}$$

Hence,

$$\frac{dL}{dt} = \frac{\Delta p R^2}{8\eta L} \qquad [\text{A.93}]$$

Equation [A.93] describes the velocity of the liquid as a function of the applied pressure. If that pressure is due to the curvature of the liquid surface, then the Laplace equation [A.82] may be introduced, giving the Washburn equation[65]:

$$\frac{dL}{dt} = \frac{\sigma R \cos\theta}{4\eta L} \qquad [\text{A.94}]$$

The Washburn equation is used to study the penetration of liquid into fiber beds or packed powders since it relates the rate of liquid penetration to the average pore radius, surface tension, and contact angle. By measuring the rate of liquid penetration, the product $R\cos\theta$ is obtained. If the liquid has a nonfinite contact angle, the average pore radius R is obtained. Once the average pore radius for a given fiber bed or packed power is known, the contact angle of liquids with finite contact angles can be determined. Fiber beds or packed powders can be characterized either by the Bartell method or the Washburn method; the former measures the pressure to prevent liquid penetration and the latter measures the rate of liquid penetration with no applied pressure.

19.   CONTACT ANGLES ON IRREGULAR SURFACES

In the experimental determination of contact angles various effects are observed, such as variation of the angle from place to place on the substrate, differences in the final static value of contact angle depending on whether the liquid advances over a dry substrate or recedes from a previously wetted substrate. Thermodynamics requires that the contact angle be a single-valued parameter of the system. Some of these effects are due to slow adsorption of vapor or of a component of the liquid. Other valid reasons for the difference in advancing and receding contact angles are surface roughness and a heterogeneity of surface energy. Surface roughness makes the observed angle appear to differ from the equilibrium angle, depending on whether the edge of the liquid is in contact with an ascending or descending slope of the substrate. If the substrate has patchwise heterogeneity, the advancing angle is affected more by the lower energy patches, which resist wetting and give a greater angle of contact; on receding, the liquid dewets the lower energy patches more readily and the observed angle is now the smaller one characteristic of the wetting of high-energy patches.

The contact angle is given by the Young–Dupré equation:

$$F_{sv} - F_{sl} = \sigma_1 \cos \theta \qquad [A.54]$$

If the surface heterogeneity occurs as small patches on a smooth solid substrate, the effective free energies per unit area of the solid–liquid and solid–vapor interfaces are averages around the three-phase boundary. The cosine of the observed contact angle is therefore an average of the cosines of the microscopic contact angles:

$$\cos \theta = \sum f_i \cos \theta_i \qquad [A.95]$$

where f_i is the fraction of total perimeter with contact angle θ_i. Although this derivation is not rigorous and is probably not accurate, it helps to explain some observed phenomena. A drop of liquid on a felt or on a woven fabric bridges apertures for which the local angle of contact is taken as 180°. Equation [A.95] becomes

$$\cos \theta = f_1 \cos \theta_1 - f_2 \qquad [A.96]$$

where $f_1 + f_2 = 1$ and f_1 is the fraction of the substrate occupied by the textile. Clearly, the presence of the voids has caused the contact angle to increase. That is one reason why the plumage of birds is waterproof. Once water has filled the voids, the waterproofing is destroyed. Anyone who has occupied a tent during a rainstorm is aware that rubbing the fabric brings water in. The influence of such surfaces on the apparent contact angle was analyzed by Cassie and Baxter.[66]

A surface irregularity such as a notch, produced by a scratch, causes a liquid to spread even when its contact angle θ is finite, as long as it is not greater than 90°. Figure A.19 shows that when the angle of the notch is less than 180° − 2θ, the surface of the liquid in the notch is concave; when the angle of the notch equals 180° − 2θ, the surface of the liquid is plane; and when the angle of the notch is larger than 180° − 2θ, the surface of the liquid is convex. The Laplace pressure in the first case causes the liquid to spread along the notch; that is, there is a critical value of the notch angle, 180° − 2θ, below which the liquid will spread, even though it has a negative spreading coefficient. Similarly for a given notch angle there is a critical value of the contact angle below which the liquid will spread along the notch. This is the reason why scratches on the Teflon coating of a frying pan reduces its efficacy in preventing sticking of fried eggs, and why in general a rough surface is more readily coated than a smooth one.

Several theories have been proposed to obtain a relation between the apparent contact angle and the microscopic angle of a liquid drop on a rough solid substrate. A useful model, proposed by Wenzel,[67,68] is that the roughness of the substrate influences the apparent contact angle by providing an interfacial area larger than the apparent area. Using an argument similar to that used to

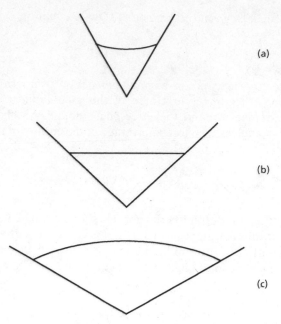

(a)

(b)

(c)

Figure A.19 A liquid with a contact angle of 45° in (*a*) a notch of 60°, (*b*) a notch of 90°, and (*c*) a notch of 135°.

derive the Young equation, the balance of free energies at the three-phase line is

$$r\sigma_{sv} = r\sigma_{sl} + \sigma_1 \cos \varphi \qquad \text{[A.97]}$$

where r is the roughness factor, that is, the ratio of the real area to the apparent area, and φ is the apparent contact angle. Combining Eq. [A.97] with the Young–Dupré equation [A.45] and rearranging gives

$$\cos \varphi = r \cos \theta \qquad \text{[A.98]}$$

Equation [A.98] is the Wenzel equation relating the apparent contact angle to the real, microscopic contact angle and the surface roughness. When the real contact angle is acute, then surface roughness makes the apparent contact angle appear smaller, that is, the liquid appears to wet the substrate more. When the real contact angle is obtuse, surface roughness makes the apparent contact angle appear larger, that is, the liquid appears to wet the substrate less.

The real interfacial area may differ from the apparent area not only by liquid penetrating into microscopic irregularities (Wenzel model) but also by the liquid bridging irregularities leaving entrapped air (Cassie and Baxter model). Thorough mathematical analyses of both these models are given by Johnson and Dettre[69,70] in a book on the topic (Fowkes, 1964).

20.   THE JAMIN EFFECT

In 1860 Jamin[71,72] noticed that a ordinary cylindrical capillary tube filled with a chain of alternate air and water bubbles is able to sustain a finite pressure. If a series of constrictions is present in the tube, the sustained pressure is considerably increased. Each of these two effects, which might be termed cylindrical and noncylindrical, respectively, is referred to by Jamin's name. In the former case, Jamin himself found that substituting alcohol or oil for water eliminated the effect, and subsequently[73] it was found that careful cleaning and avoidance of contamination could render a glass capillary incapable of sustaining pressure even with water. What is attained with each of these liquids is a condition of spreading. When the tube is deliberately contaminated, as by a dilute solution of oleic acid in benzene, the Jamin pressure is reestablished. The Jamin effect for cylindrical capillaries in which the advancing and receding angles differ is analyzed as follows. Let one end of such a tube be exposed to the atmosphere and the other end acted upon by a pressure P. Then, if n is the number of bubbles, each designated by i, r the radius of the tube, σ the surface tension of the liquid, θ_1 and θ_2 the contact angles of the advancing and receding ends of a bubble, respectively, the following equilibrium relation holds for tubes so small that gravity effects are negligible:

$$P = \frac{2}{r} \sum_{i=1}^{n} \sigma(\cos\theta_2 - \cos\theta_1)_i + P_0 \qquad [A.99]$$

where P_0 is the atmospheric pressure. If the advancing and receding angles are the same on each bubble, the expression becomes

$$P = \frac{2n\sigma(\cos\theta_2 - \cos\theta_1)}{r} + P_0 \qquad [A.100]$$

The Jamin effect scales linearly with the number of bubbles, the surface tension, the difference in the cosines of the advancing and receding angles of contact, and inversely with the radius of the capillary.

CHAPTER IIB

The Relation of Capillarity to Phase Diagrams

1. INTRODUCTION

A surface-active solute can only be called so with respect to a particular solvent. Such a solute contains in its molecular structure some moiety that interacts strongly with the solvent, whether it be by solvation, hydrogen bonding, or acid–base interaction (these may be merely different names for the same effect); and another moiety in which interaction with the solvent is insufficient to overcome the cohesional energy of the solute. The former moiety is termed "lypophilic" and the latter "lyophobic"; meaning "affinity for the solvent" and "lack of affinity for the solvent," respectively. The lyophilic moiety confers solubility, and the lyophobic moiety ensures that the solubility is limited. An example of an amphipathic solute for a hydrocarbon solvent is a molecule containing a hydrocarbon moiety, which interacts with the solvent by strong dispersion force attraction, and a hydrogen-bonding moiety, which, because of its cohesional energy has limited solubility in hydrocarbon, but which segregates by adsorption at the oil–water interface. Such a solute would be recognized as a typical oil-soluble emulsifying agent, for example, the Spans of I.C.I. Americas, Inc.

The moieties need not even be chemically linked in one molecule: A combination of two solutes, incompatible when together, may be brought into solution by a third component, as, for example, a mixture of water and benzene may be dissolved by a cosolvent such as ethanol. Such a combination is likely to include partial miscibility at some part of its (ternary) phase diagram. The position of a solution of a given concentration on the phase diagram indicates degrees of heteromolecular interaction: compositions near a phase boundary have a weaker solute–solvent interaction than those farther from a phase

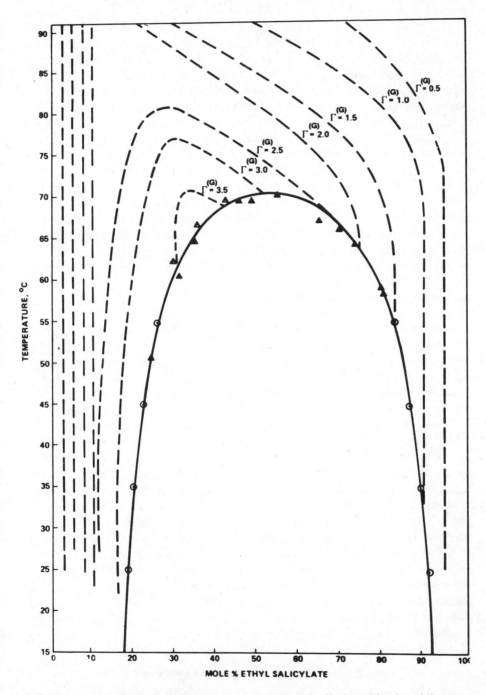

Figure B.1 Phase diagram of diethylene glycol and ethyl salicylate. The dotted lines are the Gibbs excess concentrations of ethyl salicylate.[77]

127

boundary. Adsorption is the precursor of imminent phase separation, as the surface offers a region for partial segregation of molecules prior to their more complete separation as a bulk phase. As compositions approach those of a phase boundary, therefore, adsorption increases and other interfacial phenomena associated with surface activity begin to occur. A propensity toward phase separation is therefore a general guide to surface-active behavior. A prescient but passing and incidental remark to this effect was made by Langmuir 60 years ago. Langmuir wrote: "In mutually saturated liquids, especially near the critical temperature, the conditions are favorable for orientation and segregation of the molecules in the liquid."[74] Adsorption of solute is usually accompanied by micellization, and micelles are well known to occur near the critical temperature.

The surface activity of a solute is not primarily due to its amphipathic molecular structure but to its weak interaction with the solvent. This interaction must of course still be sufficient to dissolve the solute but need be no greater than the least degree required to do so. A tendency toward phase separation is a general indicator of surface activity. This statement is akin to Lundelius's rule[75] that the least soluble materials are the most readily adsorbed. Thus, for example, the foam stability of gelatin solutions is greatest at the isoelectric point where the gelatin is least soluble. Similarly, the foaminess of polymer solutions is maximum in poorer solvents, declining again, however, when insolubility supervenes.[76] See also the Ferguson effect (Section II.E.2). Not surprisingly, therefore, the surface activity displayed by solutions or mixtures can be related to the phase diagram.

The fundamental index of the surface activity of a solute in any solvent is positive adsorption, which is measured quantitatively by the excess surface concentration of solute. For a binary solution, this quantity is given by Γ_2^G in the adsorption isotherm equation of Gibbs (Eq. [G.20]). If Eq. [G.20] is applied to treat experimental data of a binary system at concentrations greater than about $0.01M$, activities rather than concentrations must be used. Nishioka et al.[77] used available data on surface tensions and activity coefficients as functions of composition and temperature to calculate Gibbs excess concentrations of solute for the binary system diethylene glycol and ethyl salicylate. The Gibbs excess concentrations were then plotted as cosorption contours superimposed on the phase diagram for the system (Fig. B.1). The thesis that surface activity is the precursor of phase separation is amply confirmed by these observations of increasing surface activity of the unsaturated solutions as they approach saturation, with maximum surface activity near the critical-solution point. In these solutions ethyl salicylate, the component of lower surface tension, is the surface-active component. The maximum surface excess has an epicenter that does not correspond exactly to the consolute point, but is shifted toward the component of higher surface tension (diethylene glycol).

2.   REGULAR SOLUTION THEORY

Nishioka et al.[77] compared these experimental results with a theoretical model of the same kind of system, using the "two-surface-layer" regular solution model of

Defay and Prigogine.[78] In this model only the top two molecular layers are considered to differ in composition from the bulk phase, arising from molecular coordination numbers less than those of molecules in the bulk phase. Consider a molecule in the bulk phase: it has z nearest neighbors, of which lz are in the same lattice plane, where l is the fraction of nearest neighbors in that plane (e.g., $\frac{6}{12}$ for a close-packed lattice; $\frac{4}{6}$ for a cubic lattice); and mz are in either contiguous plane, where m is the fraction of nearest neighbors in that plane (e.g., $\frac{3}{12}$ for a close-packed lattice; $\frac{1}{6}$ for a cubic lattice). Then

$$l + 2m = 1 \qquad \text{[B.1]}$$

A molecule in the surface layer has a smaller number z' of nearest neighbors given by

$$z' = (l + m)z = (1 - m)z \qquad \text{[B.2]}$$

Using the Bragg–Williams approximation, and assuming that the molecular surface areas of the two components are the same, the compositions in the upper two layers are given by the following two equations:

$$\log\frac{x_1 x_2''}{x_1'' x_2} - \frac{2\alpha(x_2'' - x_2)}{RT} - \frac{2\alpha m(x_2 + x_2' - 2x_2'')}{RT} = 0 \qquad \text{[B.3]}$$

$$\frac{RT}{\alpha}\log\frac{x_1 x_2'}{x_2 x_1'} + \frac{a(\sigma_2 - \sigma_1)}{\alpha} + 2l(x_2 - x_2') + m(x_2 - x_1) + 2m(x_2 - x_2'') = 0$$
$$\text{[B.4]}$$

where x_1, x_2 = mole fraction of component 1 or 2 in the bulk phase
x_1', x_2' = mole fraction of 1 or 2 in the top monolayer
x_1'', x_2'' = mole fraction of 1 or 2 in the intermediate monolayer
α = interaction constant ($2RT_c$ for a regular solution)
σ_1, σ_2 = surface tension of pure component 1 or 2
a = area per molecule

Equations [B.3] and [B.4] were solved numerically for x_2' and x_2'' as a function of the composition of the solution (x_1). A close-packed lattice was assumed and the term $a(\sigma_2 - \sigma_1)/\alpha$ in Eq. [B.4] was assumed to be constant and equal to -0.5. This assumption was based on reasonable values of $a = 0.30\,\text{nm}^2/\text{molecule}$, $T_c = 300\,°\text{K}$, and $(\sigma_2 - \sigma_1) = -6.9\,\text{mN/m}$. The dimensionless surface-excess concentration of component 2 is then

$$\omega\Gamma_2^G = \frac{x_2' - x_2 + x_2'' - x_2}{x_1} \qquad \text{[B.5]}$$

where ω is the surface occupancy in square centimeters per mole.

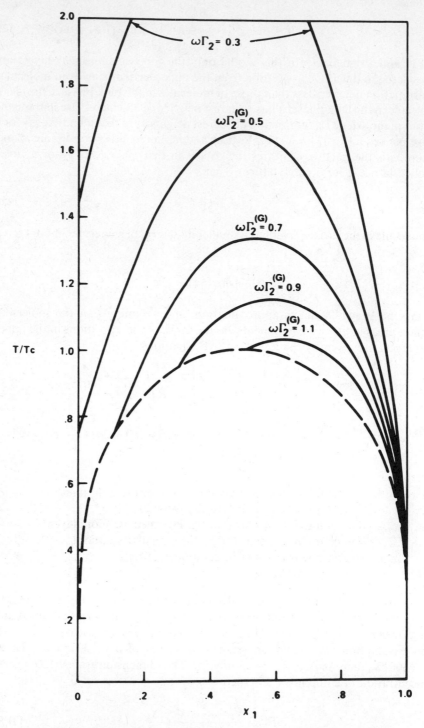

Figure B.2 Dimensionless surface-excess concentration of component 2, $\omega\Gamma_2^G$, calculated for a two-layer regular solution model. Component 2 was assumed to have an appreciably lower surface tension than component 1.[77]

The dimensionless surface excess $\omega\Gamma_2^G$, calculated for a regular solution is shown in Fig. B.2 as a series of contours of equal surface concentration, called cosorption lines. In this hypothetical system the component of lower surface tension is surface active at all points in the phase diagram. The maximum surface activity, which is called the epicenter, occurs at a point on the solubility curve, thus agreeing with Lundelius's rule; but it does not coincide with the critical point, being biased toward a lower concentration of the component of lower surface tension. Calculations based on the theory disclose that the greater the difference in surface tension between the two components, the greater is the bias toward that side of the composition scale; and also that the greater that bias, the greater is the surface activity.

Figure B.2 shows that at higher temperatures a regular solution tends toward having maximum surface activity at a mole fraction of one-half, but experimental data do not confirm this prediction: They show the maximum shifting toward still lower concentrations of the component of lower surface tension. This behavior is also typically observed for solutions at temperatures far above a critical point, such as water–ethanol solutions at room temperature, which have a maximum surface activity at 5–10% alcohol. Nevertheless, Fig. B.2 bears a marked resemblance to published diagrams reporting observed foam stabilities of two-component systems as functions of composition and temperature.

CHAPTER IIC

Surface-Active Solutes

1. INTRODUCTION

The capillary properties of pure liquids, discussed in the previous chapters, are modified by certain solutes called "surface active." Surface activity of a solute is defined as the ability to reduce the surface tension at an interface without requiring concentrations so large that the distinction between solute and solvent is blurred. The surface tension of water or the interfacial tension between water and a hydrocarbon may be reduced by 50mN/m at concentrations of less than 0.1% of a surface-active solute. In nonaqueous solutions the effects are much smaller. The quality of the solute is measured by how little is required for a given effect. The most marked results are obtained from a solute that combines in its molecular structure an element having a high affinity for the solvent with an element having minimal affinity for the solvent. The combination of such disparate elements produces a molecule that has its lowest potential energy at a phase boundary. This subject cannot be discussed further without a specialized vocabulary, given in Table C.1.

Technical terms (neologisms) are formed by combinations of the words given in Table C.1, such as the following adjectives:

amphipathic = combining both natures (oil and water understood)
amphiphilic = with affinity for both (oil and water understood)
hydrophilic = with affinity for water
hydrophobic = lack of affinity for water
lipophilic = with affinity for oil
lyophilic = with affinity for the solvent
lyophobic = lack of affinity for the solvent

132

Table C.1   Glossary of Some Classical Prefixes and Suffixes[a]

English	Greek	Latin
Oil	Lipo-	Oleo-
Water	Hydro-	Aqua-
Solvent	Lyo-	
Both	Amphi-	
Flow	Rheo-	
Affinity	-philic	
Lack of affinity	-phobic	
Nature	-pathic	
Science	-logy	

[a]The English words are not literal translations but interpretations of how the Greek words are understood in the vocabulary of this branch of science.

Corresponding concrete nouns are amphipaths, hydrophiles, hydrophobes, lipophiles, lipophobes, lyophiles, and lyophobes. Corresponding abstract nouns are amphipathy, amphiphilia, hydrophilia, hydrophobia, lipophilia, lipophobia, lyophilia, and lyophobia. Scholars do not consider it good form to combine Latin and Greek roots in the same word; therefore, the following terms are best avoided: hydrophilicity, hydrophobicity, oleophilic. The widely used term "surfactant" is a combination of syllables from the phrase, "surface-active agent." The descriptive expression "surface-active solute" is used in this book where others might use "surfactant."

The molecule of a surface-active solute may not be a combination of hydrophilic and lipophilic elements, as moderate surface activity is found in solutions near a phase separation. Nevertheless, the more powerful surface-active solutes used as detergents, emulsifiers, or wetting agents have a molecular structure composed of hydrophilic and lipophilic elements. These elements are present in different proportions; the balance between them determines whether the solute is oil soluble or water soluble, and various degrees between determine properties useful in different applications. Every such molecule occupies a place on a scale of hydrophile–lipophile balance (HLB). Most industrial surface-active solutes are not pure materials, hence their HLB is based on performance rather than molecular structure.

A useful glossary of terms pertaining to surface-active solutes and their application as detergents is provided by the Soap and Detergent Association.[79]

2.   RANGE OF SOLUTES FROM LIPOPHILIC TO HYDROPHILIC

Materials used as solutes can be classified on a scale ranging from a lipophilic extreme to an hydrophilic extreme. Starting at the lipophilic end of the scale with, for example, a long-chain hydrocarbon, materials that display various degrees of

Table C.2 Graded Series of Solutes in Terms of Hydrophile–Lipophile Balance

Lipophilic End of Scale					Hydrophilic End of Scale	
Stearane	Stearic Acid	Sodium Stearate	Sodium Laurate	Sucrose	Sodium Sulfate
Soluble in oil; insoluble in water	Soluble in oil; insoluble in water	Soluble in oil; and in hot water	Slightly oil-soluble; soluble in water	Insoluble in oil; soluble in water	Insoluble in oil; soluble in water
Nonspreading on water substrate	Spreads on water substrate	Spreads on water substrate	Reduces surface tension of aqueous solution	Does not affect the surface tension in aqueous solution	Increases surface tension in aqueous solution
Does not affect interfacial tension at oil–water interface	Reduces interfacial tension at oil–water interface	Reduces interfacial tension at oil–water interface	Reduces interfacial tension at oil–water interface	Does not affect interfacial tension at oil–water interface	Increases interfacial tension at oil–water interface
Does not stabilize emulsions	Stabilizes water in oil emulsions	Stabilizes either type of emulsion	Stabilizes oil in water emulsions	Does not stabilize emulsions	Decreases the stability of emulsions

1 |⟵————— Useful range of ICI Americas' HLB scale —————⟶| 20
(see Section I.VA.6)

134

hydrophilia can be obtained by adding hydrogen-bonding substituent groups, such as hydroxyl, or amino, or carboxylate. The greatest hydrophilic character is conferred when the substituent group is ionized. In this way we proceed from stearane ($C_{18}H_{38}$,) for example, to stearyl alcohol, stearyl amine, stearic acid and, finally, sodium stearate. Stearane is entirely lipophilic: It is all but completely insoluble in water and will not spread on a water surface but floats thereon as a compact liquid lens. On adding a hydrophilic group, such as hydroxyl or amino, to the stearane molecule, the solubility in water increases, although it is still practically insoluble; but the hydrogen-bonding group now enables the compound to spread spontaneously on a water surface. The spreading would continue all the way to monomolecular thickness given an appropriately small quantity of compound or a sufficiently large surface of water. This type of monomolecular layer (abbreviated as "monolayer") is known as an insoluble film. Let the hydrogen-bonding group be ionic, such as carboxylate, sulfate, sulfonate, or ammonium ions, and the corresponding compounds are much more soluble in water but are still positively adsorbed at the water–air surface or water–oil interface. By virtue of their positive adsorption these materials as solutes reduce surface or interfacial tension, the relation between adsorption and surface tension lowering being a thermodynamic consequence first made quantitatively explicit by J. Willard Gibbs (the Gibbs adsorption theorem). The surface region, that is, that which contains the excess adsorbed solute, is known as a "soluble film" to distinguish it from the insoluble monolayer already mentioned. Soluble films are usually thicker than a single layer of molecules.[80]

As ions are the most hydrophilic groups known, increasingly hydrophilic compounds cannot be made by further substitution of hydrophilic groups; instead, the degree of lipophilia of the organic radical can be diminished by decreasing its molecular weight. Ultimately, when fewer than six carbon atoms are present in the radical, the compound, for example, sodium propionate, hardly differs from an ordinary strong electrolyte comprised of a small anion and a small cation. Small ions are so extremely hydrophilic that their lowest potential energy occurs when they are fully surrounded by water molecules, which is to say, when they are in the bulk solution phase rather than in the surface region. Such a condition leads to negative adsorption, for which experimental evidence can be found in surface tension that increases with higher concentrations of solute. Ordinary strong electrolytes show this effect.

Table C.2 lists examples of solutes from stearane, the most lipophilic, to sodium chloride, the most hydrophilic. Their properties and the properties they confer on the solution vary continuously. A graded series of solutes in terms of hydrophile–lipophile balance is also shown in Table C.2.

3. TYPES: ANIONIC, CATIONIC, AND NONIONIC

A surface-active solute may be synthesized by combining in the same molecule lipophilic and hydrophilic moieties; these must also be sterically separable. For

Table C.3   Hydrophilic and Lipophilic Moieties

Hydrophilic		Lipophilic	
Ionic		Hydrocarbon	
Carboxylate	$-CO_2^-$	Straight-chain alkyl	(C_8-C_{18})
Sulfate	$-OSO_3^-$	Branched-chain alkyl	(C_8-C_{18})
Sulfonate	$-SO_3^-$	Alkylbenzene	(C_8-C_{16})
Quaternary ammonium		Alkylnaphthalene	
	R_4N^+	Perfluoroalkyl	
Nonionic		Polymeric	
Fatty acid	$-CO_2H$	Polypropylene oxide:	
Primary alcohol	$-CH_2OH$	$H-[OCH(CH_3)CH_2]_n-OH$	
Secondary alcohol	$-CRHOH$		
Tertiary alcohol	$-CR_2OH$	Polysiloxane:	
Ether	$-COC-$	$H-[OSi(CH_3)_2]_n-OH$	
Polyethylene oxide:			
$H-[OCH_2CH_2]_n-OH$			

example, a block copolymer of polyethylene oxide and polypropylene oxide makes a surface-active solute, whereas the polymer produced by mixing the monomers is not surface active. Table C.3 lists a number of hydrophilic and lipophilic moieties, which may be combined to give solutes with different hydrophile–lipophile balances.

The terms "anionic," "cationic," and "nonionic" refer in each case to the hydrophilic moiety. Of these the anionic types account for about 75% of consumption in the United States; sodium and potassium salts of the fatty acids (soaps) are historically the first detergents known and still account for about 25% of U.S. consumption. The raw materials, animal fat (tallow) and alkali, are cheap. The chief disadvantages of the alkali-metal soaps are that they form water-insoluble salts with divalent and trivalent cations (particularly calcium, magnesium, and iron), they are insoluble in brine, and they hydrolyze to their insoluble parent fatty acids in acid solutions.

With the advent of the petrochemical industry, stimulated by world war, other types of anionic surface-active solutes appeared, namely, long-chain sulfonates and sulfates. Of these, the sodium salt of the linear alkylbenzene sulfonate is the most widely used as an industrial and domestic detergent. It does not have the disadvantages listed above for the alkali-metal soaps. Examples of commonly used surface–active solutes are listed in Table C.4.

Cationic surface-active agents are adsorbed at negatively charged surfaces by electrostatic attraction. Their first effect on such substrates is to neutralize the charge and, by virtue of their lipophile, to render the surface lipophilic. At greater concentrations the cationic is adsorbed in the usual way, that is, by the lipophile; thus placing a second layer of solute on top of the first, thereby creating a positively charged surface and converting it from lipophilic to hydrophilic. Cationic surface-active solutes have wider bactericidal activity than anionics and find their major use as germicides. Although they are detergents, they are seldom used for that purpose except where sterile conditions are important, as in the laundering of diapers or in rinsing bar glassware. Cationic solutes are incompatible with anionic solutes; and since latexes and dispersions are often mixed together in industrial processes, the use of cationics as emulsifying or dispersing agents is generally avoided. Special cases exist, of course, where the destabilizing of an emulsion or a latex or a foam is desired, and the addition of a cationic solute accomplishes that purpose. For example, asphalt emulsions, used in road-

Table C.4   Commonly Used Emulsifiers, Detergents, Dispersants and Builders

Chemical Class	Application
1. Anionic	
Alkyaryl sulfonates	Detergents, emulsifiers
Fatty alcohol sulfates	Detergents, emulsifiers
Lignosulfonates	Dispersants
Alkali soaps of tall oil	Anionic emulsifiers
Alkali soaps of rosin	Anionic emulsifiers
Sulfosuccinates	Wetting agents
2. Cationic	
Alkyltrimethylammonium chloride	Emulsifier, corrosion inhibitor, textile softener, antibacterial agent, detergent
3. Nonionic	
Alkanolamides	Detergents, foam stabilizers
Glyceryl esters	Emulsifiers
Ethylene-oxide condensates of alkylphenols	Emulsifiers
Ethoxylated alkylphenols	Detergents, wetting agents, emulsifiers, dispersants
Ethoxylated fatty esters	Food emulsifiers (oil in water)
Fatty esters	Food emulsifiers (water in oil)
Polyalkylsuccinimides	Oil-soluble dispersants
Lecithins	Oil-soluble dispersants
Metal soaps	Oil-soluble dispersants
4. Builders	
Polyphosphate	Dispersing agent
Tetrasodium pyrophosphate (TSPP)	Dispersing agent
Trisodium orthophosphate	Dispersing agent
Sodium nitrilotriacetate (NTA)	Sequestering agent
Sodium ethylenediaminetetraacetate (EDTA)	Sequestering agent
Sodium carbonate, borax, silicates, citrates	Alkalis
Sodium oxydiacetate (ODA)	Sequestering agent

From McCutcheon (1985).

making, are designed to "break" and release the asphalt when they are in contact with wetted gravel. If the gravel contains flints or silicates, an emulsion stabilized by a cationic emulsifier breaks readily on contact with the negatively charged surfaces of these rocks. Gravels containing limestones, however, are positively charged on wetting and are better treated with asphalt emulsions stabilized by anionic agents. In paper making, cationics inhibit hydrogen bonding between the cellulose fibers to produce the softer texture of tissue and toilet paper. In the same way cationics act as softeners in laundry use. Quaternary ammonium compounds are the most commonly used cationics.

Amphoteric surface-active solutes may be either anionic or cationic in water depending on pH. These substances have both amino and carboxylate groups: The amino group is charged positively at low pH and the carboxylate is charged negatively at high pH. An example is β-N-alkylaminopropionic acid, $RNHCH_2CH_2COOH$. Near the isoelectric point they have minimum solubility in water and greater surface activity. These are frequently used in shampoos and other personal-care products because of their mildness.

By far the largest quantity of nonionic surface-active solutes have polyethylene oxide as the hydrophile. This hydrophile is approved by the Food and Drug Administration for use in foodstuffs. It finds applications as a wetting agent in dehydrated milk and eggs, and in cocoa, flour, and other poorly wetted powders. The lipophile can be attached to fatty-acid esters of glycerol, sorbitol, or propylene glycol. Any solute containing polyethylene oxide as an adduct becomes insoluble in water at higher temperatures, as the hydrogen bonding of water to the ether oxygens is diminished by heat. What was previously a hydrophile then becomes a lipophile. The larger the fraction of polyethylene oxide or other hydrophiles in the molecule, the higher is the cloud point.

Builders are certain components in formulated detergents that remove metal ions either in a soluble form by sequestration or in an insoluble form by precipitation. They also maintain alkalinity, which promotes detergency. Examples of builders are listed in Table C.4.

CHAPTER IID

Physical Properties of Insoluble Monolayers

1. OBSERVATIONS

The first reported influence of oil films on water was the observation of the damping of waves by oils such as olive oil and whale oil. This effect was known to the ancients; it is mentioned in Pliny's *Natural History* (A.D. 77); and it received widespread recognition in the eighteenth century from experiments by Benjamin Franklin. A surprising result of Franklin's experiments was the extent of coverage attained by a small volume of oil. In 1891 Lord Rayleigh requested space in the British journal *Nature*, "for the accompanying translation of an interesting letter which I have received from a German lady, who with very homely [that is, domestic] appliances has arrived at valuable results respecting the behavior of contaminated water surfaces."[81] Miss Pockels' experiments demonstrated that the surface tension of a strongly contaminated water surface declines as the area of the surface is reduced. She measured the surface tension of the film by the force required to separate a small disk (6 mm diameter) from the surface. Rayleigh, repeating some of Miss Pockels' experiments, found that castor oil spreads on water to a minimum thickness of 1.3 nm, from which he concluded that the oil film attenuates to the thickness of a single molecule. Langmuir developed the fundamental methodology to measure the properties of insoluble monolayers as a function of their surface concentration, leading to the determination of molecular dimensions. A typical calculation goes as follows: A solution of palmitic acid (M.W. = 256) in benzene contains 4.24 g/L. When 0.0239 mL of this solution is placed on a water surface of 500 cm², the benzene evaporates and the palmitic acid forms a monomolecular film, which can be calculated to give an area of 0.21 nm² per molecule. The crystal structure of alkanes determined by X-ray diffraction gives a cross-sectional area of $0.186 \, nm^2$ per molecule. The close agreement between these two independent results indicates that under these conditions the palmitic acid molecule forms a close-packed monolayer on the water substrate oriented with the hydrophilic acid group in the water.

Langmuir's original apparatus has been refined (Fig. D.1) and automated. See Appendix J for a list of commercial manufacturers.

Figure D.2 is a schematic diagram of the possible monolayer states at three temperatures, T_1, T_2, T_3, where $T_1 < T_2 < T_3$, as typically obtained from an

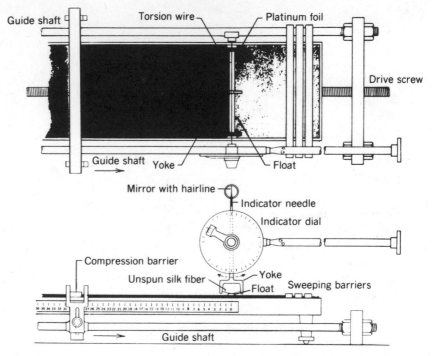

Figure D.1 The Langmuir trough.[86]

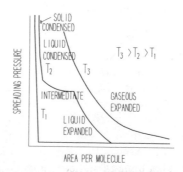

Figure D.2 Possible monolayer states at three temperatures.

automated Langmuir film balance.[82] At large areas per molecule (and low spreading pressures) the film behaves as a two-dimensional analog of the ideal gaseous state, and is described by

$$\pi A = nkT \qquad\qquad [\text{D}.1]$$

where π, the lowering of the surface tension in millinewtons per meter, is the "spreading pressure," A is the area of the water surface, and n is the number of molecules of monolayer substance. At lower temperatures and at smaller molecular areas, the monolayer undergoes a series of phase changes analogous to those in bulk matter. Using the traditional phase names, the compressed monolayer undergoes a first-order phase transition from gaseous to liquid expanded and further second-order transitions to intermediate-condensed and liquid-condensed phases (T_2) or to a solid-condensed phase (T_1). The precise behavior depends on the temperature and on the nature of both monolayer and substrate. In general, compression of the monolayer to smaller areas results in an increase of spreading pressure and changes the molecules from random to more organized structures, tending toward an orientation similar to that of the crystal. Ultimately, the monolayer collapses under sufficient compression. The so-called insoluble monolayer is actually a two-dimensional solution of the adsorbate in water; the spreading pressure is more closely analogous to osmotic pressure than to the pressure of a three-dimensional gas; and the monolayer states are analogous to the demixing of a two-component system.

Ries and Cook[83] compared the π–A isotherms of monolayers of stearic acid (I) and isostearic acid (II), which differ in molecular structure only by branching at the end of the hydrocarbon chain (Fig. D.3). The methyl side chain of isostearic acid should have little effect on molecular packing and film strength because it is small and at the remote end of the long molecule. Furthermore, the melting points of the isostearic and stearic acids are almost identical. Although the difference in structure is so slight, the behavior of the two monolayers is markedly different (Fig. D.4). The limiting area per molecule, extrapolated to zero pressure from the steepest part of the π–A isotherm, is 0.20nm^2 for stearic acid but 0.32nm^2 for isostearic acid. Evidently the tiny branching at the end of the chain in II is enough to prevent the close packing of chains achieved in I. When the chains are fully extended and oriented parallel to one another, considerable cohesion exists between molecules, just as in crystals of paraffin. Such a monolayer is very rigid and resistant to collapse: For example, the collapse pressure of I is 42mJ/m; but the same cohesion is not reached in II: its monolayer collapses at the relatively low pressure of 15mJ/m. A third substance, tri-p-cresyl phosphate (III) spreads out like a three-leafed clover on the surface of water. The π–A curve of the III monolayer shows great compressibility, but it collapses at 9mJ/m and has the very extensive limiting area per molecule of 0.96nm^2. These comparisons show the effects of differences of molecular structure.

Ries and co-workers[84–86] obtained gold-shadowed electron photomicrographs of some features of monolayers. In the discontinuous region of the first-

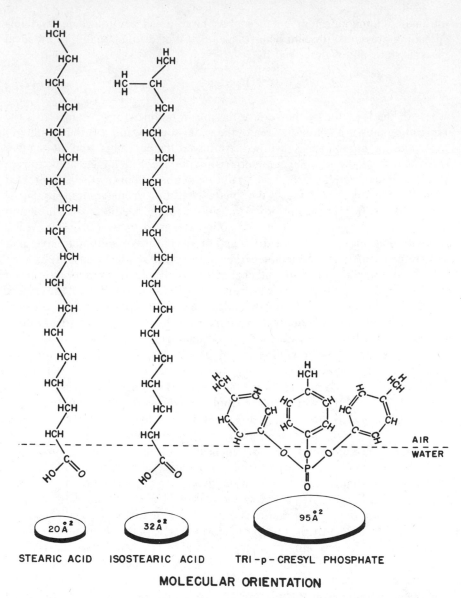

STEARIC ACID   ISOSTEARIC ACID   TRI−p−CRESYL PHOSPHATE

MOLECULAR ORIENTATION

Figure D.3 Structures of stearic acid, isostearic acid, and tri-*p*-cresyl phosphate.[86]

order phase transition of *n*-hexatriacontanoic acid, $CH_3(CH_2)_{34}COOH$, from two-dimensional gaseous to two-dimensional liquid, at which the adsorbate should exist in two phases, namely, islands of two-dimensional liquid disks surrounded by a matrix of gaseous phase, the electron photomicrograph clearly shows the circular outline of the separated disklets. Successive photomicro-

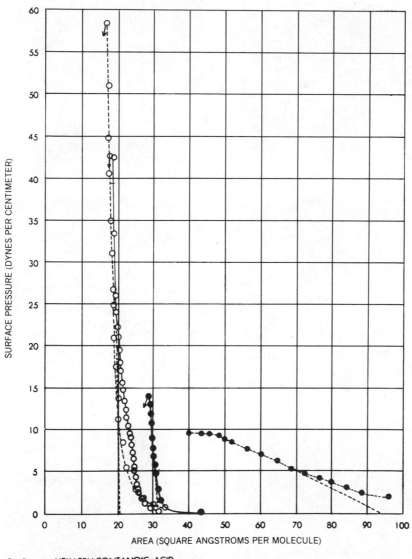

Figure D.4 π–A isotherms of *n*-hexatriacontanoic acid, stearic acid, isostearic acid, and tri-*p*-cresyl phosphate.[86]

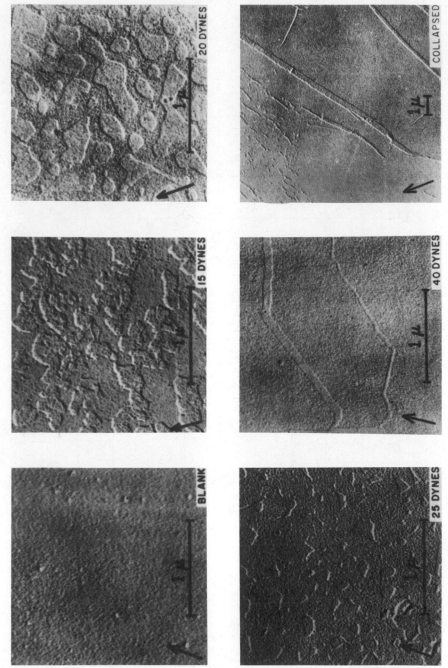

Figure D.5 Electron photomicrographs of monolayers of *n*-hexatriacontanoic acid.[86]

graphs, taken as the available area is reduced, show the inversion of the two-phase system to a state where the two-dimensional liquid is the continuous matrix, and finally to the state of the collapsed film, where bimolecular leaflets have been raised and then dropped on top of the underlying monolayer (Fig. D.5).

Multicomponent monolayers of biological origin constitute the lipid bilayers that are the basic structural framework of cell membranes. The inner and outer monolayers of the bilayer are oriented with their respective hydrophilic extremities toward the internal and external aqueous phases. Studies of membrane components and model compounds on aqueous substrates are intended to probe the structure and properties of lipid bilayers. Another type of biogenetic monolayer is "lung surfactant," a substance in the mammalian lung whose surface tension is greatly reduced by compression on every exhaled breath, and so regulates capillary pressure.[87] Biogenetic monolayers are usually extremely strong, with maximum spreading pressures at the collapse point of 68–72mN/m, which is remarkable when compared with that of stearic acid, 40 mN/m. Ries and co-workers have published electron micrographs of collapsed films of such monolayers, revealing structures such as flat ribbons, platelets, and ridges.[88,89] Once formed these structures relax very slowly, giving rise to large hysteresis loops of the surface tension when the area is expanded and contracted. Microlayers of biological origin also occur on the surfaces of oceans and lakes, where they are immediately detected by their effect on the reflection of light from the water surface.[90]

The sensitivity of the π–A curve to minute changes of molecular structure of the adsorbate makes the film balance technique suitable for deciding fine points of difference between competing versions of structural formulas, as happened, for example, with those suggested for sterols (Adam, 1941, pp. 79–82).

2.   VISCOELASTICITY AND INSOLUBLE MONOLAYERS

A characteristic of insoluble monolayers is that their surface tension varies when the film is dilatated or compressed. Surface tension gradients lead to Marangoni flow, by which an applied strain is spontaneously opposed. This property is an elasticity caused by dilatation; liquid surfaces that are incapable of creating surface tension gradients would show no dilatational elasticity. The viscoelastic properties of insoluble monolayers can be detected and measured by means of a periodic input of stress and the recording of the resulting strain. Dilatational elasticity and viscosity of a monolayer can be determined quantitatively by linear systems analysis if the response of the system to a periodic excitation has the same frequency as the input, and if the phase difference between excitation and response is independent of the amplitude of the input signal. These conditions can be met by appropriate control of the external stimuli applied to the surface.

A technique suitable for use with the Langmuir film balance is to apply a periodic strain to the monolayer by oscillating a barrier between two points. The frequency applied can be varied by adjusting the speed or amplitude of the oscillation. The output is measured as the variation of the spreading pressure by means of a Wilhelmy plate or a pressure transducer.[91]

3. BUILT-UP FILMS: LANGMUIR–BLODGETT FILMS

Langmuir[92] observed in 1919 that an insoluble monolayer could be transferred from the water substrate to a hydrophilic solid surface, such as a glass slide, by raising the slide through the liquid–gas interface. In 1934 Katharine Blodgett announced the discovery that built-up films could be formed by sequential monolayer transfer—the structures now universally referred to as Langmuir–Blodgett films.[93] On hydrophilic substrates no film is deposited during the first immersion; the first monolayer is deposited only during the first withdrawal. Further layers are deposited by repeated dippings of the solid through the interface. The layers are deposited hydrophile to hydrophile and lipophile to lipophile, or head to head and tail to tail, by the successive folding back and forth of the monolayer, as in Fig. D.6. The slide, in air, always has an odd number of

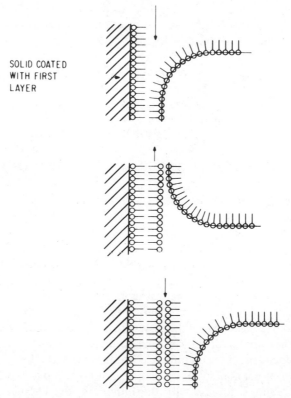

SOLID COATED
WITH FIRST
LAYER

Figure D.6 The building up of a Langmuir–Blodgett film (Gaines, 1966, p. 337).

layers on it: one on the first withdrawal, three on the second withdrawal, and so forth.

The outside surface is composed of the methyl groups that terminate the fatty-acid chain, and these surfaces exhibit high contact angles for water and organic liquids in accordance with the low surface energy of the methyl groups. Optical measurements by Langmuir and Blodgett[94,95] show that each double layer produces an increment of thickness nearly equal to twice the length of the fatty-acid chain.

Built-up multilayers are important materials in thin-film technology, with applications to electron beam microlithography, integrated optics, electro-optics, electronic displays, photovoltaic cells, two-dimensional magnetic arrays, field-effect devices, and biological membranes. Langmuir–Blodgett films are also used to build well-defined structures for fundamental studies of their electrical and optical properties. (See *Physics in Technology*, **12**, 1981: Technological Applications of Langmuir–Blodgett Films; *Thin Solid Films*, **99**, 1983, which is devoted entirely to papers presented at the first international conference on Langmuir–Blodgett films, held at the University of Durham in September 1982; and *Thin Solid Films*, **132–134**, 1985, for the second international conference on Langmuir–Blodgett films, held at Schenectady, NY, in July, 1985).

CHAPTER IIE

Aqueous Solutions of Surface-Active Solutes

1. ADSORPTION

In aqueous solutions the adsorption process that retains amphipathic molecules at liquid–vapor or liquid–liquid or solid—liquid interfaces is traced to an effect in the solution phase, namely, the reordering of the water molecules in the vicinity of an alkyl hydrocarbon chain, and not to a lack of sufficient attractive forces between the hydrocarbon and water. This statement derives from the following argument: The work of adhesion between an alkyl chain and water, due to dispersion forces, is

$$W^{\text{adh}} = 2(\sigma_{\text{hc}}\sigma_{\text{w}}^{\text{d}})^{1/2} \simeq 50 \, \text{mJ/m}^2 \qquad [\text{E.1}]$$

The work of adhesion therefore is about the same as the work of cohesion $(= 2\sigma_{\text{hc}})$ of an alkyl hydrocarbon. One cannot claim, therefore, that the alkyl chain is pushed out of the water by being repelled by water molecules, although adsorption phenomena appear to correspond to such an action. To take the term "hydrophobic" literally is therefore wrong.

The cohesion of water, $W^{\text{coh}} = 146 \, \text{mJ/m}^2$, is much larger than the adhesion between water and an alkyl chain, which accounts for the insolubility of the paraffin hydrocarbons in water. An amphipathic solute, however, because of its solubility in water introduces discrete molecules containing a hydrocarbon chain into an aqueous environment. Water molecules in the vicinity of the alkyl chain are subjected to a stronger cohesional force pulling them away from that vicinity than the adhesional force that would keep them there. As the water molecules are labile, they respond to the stronger force, thus creating an aqueous region of

148

lower density around the hydrocarbon. Around each hydrocarbon moiety of the dissolved molecule, therefore, the condition of the surrounding water is the same as was described for water at the liquid–vapor surface (Chapter IIA.1), namely, a region of lower density, which at (dynamic) equilibrium produces a tension between two bulk phases. While it is inappropriate to invoke macroscopic effects such as interfacial tension to describe the condition at a molecular interface, nevertheless water molecules in the vicinity of a hydrocarbon chain have greater potential energy than those farther away. In the case of water, the lower density in this region reduces the number of nearest neighbors. The water molecules rearrange to conserve the number of hydrogen bonds between them but now have fewer configurational states with which to do so than they had previously, that is, than water has in its normal liquid state. A decrease in the number of possible states implies a decrease in entropy of the water, and hence an increase of free energy of the system. A similar behavior of water occurs also at the liquid–vapor surface, at an oil–water interface, and at the interface between a lipophilic solid and water. Water contiguous to each of these interfaces has a greater potential energy than normal water, and manifests this condition by large surface and interfacial tensions at the liquid interfaces and by spontaneous flocculation of dispersed lipophilic particles.

The process at the water–vapor surface, while similar, is not exactly the same as at a water–hydrocarbon interface. The entropy of water at the surface is likewise less than that of water in the bulk phase; but the corresponding enthalpy change is large rather than small because at the one-component liquid–vapor surface water–hydrocarbon dispersion forces are not present to compensate for the loss of water–water dispersion forces that occurs when water molecules transfer from the bulk phase to the surface (Tanford, 1980).

The lipophilic interface induces ordering effects, which correspond to a loss of entropy within the vicinal water. The formation of ice from liquid water is a similar process, hence the structure of the water around a hydrocarbon chain is sometimes called "icelike." The interfacial free energy of water has, therefore, a large contribution from the $T \Delta S$ term.[96,97]

We have traced the adsorption behavior of amphipathic solutes in aqueous solution to entropy changes within the water, rather than as a specific property of the lipophile. When the solute molecule happens to arrive at a surface or an interface, it is partially withdrawn from the water; its potential energy is thereby reduced; and it lingers at that location, as it takes work to bring it back into the bulk phase. Adsorption is a state of dynamic equilibrium, which is why the adsorbed molecules are said to linger than to remain static at the interface.

2. LUNDELIUS' RULE AND THE FERGUSON EFFECT

The lipophilic moiety limits the solubility in water of the molecule to which it is attached. The function of the hydrophile is to provide enough interaction with water to bring the insoluble lipophile into solution. A negative or a positive ion can bring a 16-carbon chain into aqueous solution at room temperature and an 18-carbon atom chain into solution in hot water. Solubility decreases with increasing chain length, and surface activity becomes more pronounced. Lundelius's rule is an explicit statement of a direct connection between the two. It states: Any factor, such as a change of composition, concentration, or temperature, tending to decrease solubility promotes surface activity. We have already traced the effects of this rule in the relation of capillarity to phase diagrams. It is also to be seen in an homologous series as an effect depending on the molecular weight of the lipophile.

Ferguson[98] discovered an optimum effect at a certain carbon chain length in an homologous series of surface-active solutes, when applied for various purposes, such as detergency, antibiotic action, hemolytic action, and emulsification. Lipophilic character, and therefore surface activity in an aqueous medium, is continuously increased by lengthening the hydrocarbon chain; the trend is accompanied by decreasing solubility in water, so that ultimately too few molecules are present in solution to be effective. The balance between the two opposite trends gives a maximum in the function that relates effectiveness in a given application to the number of carbon atoms in the lipophile. The behavior is sufficiently general to be named "the Ferguson effect" (Fig. E.1).

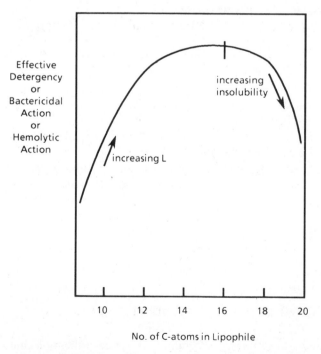

Figure E.1 The variation of effectiveness of a surface-active solute with the number of carbon atoms in the lipophile.

3. MICELLIZATION AND SOLUBILIZATION

a. Spontaneous Association

The same mechanism that causes adsorption of an amphipathic molecule also leads to spontaneous association of such molecules. The greater potential energy of water molecules in the vicinity of a hydrocarbon chain implies that any way to withdraw hydrocarbon from the surrounding water reduces the potential energy and so occurs spontaneously. Adsorption of hydrocarbon at surfaces and interfaces is one such way; another is the spontaneous association of the hydrocarbon chains of amphipathic molecules to form colloidal aggregates, varying in size from dimers to complexes of 50 or more molecules. These aggregates of molecules are called "micelles" (*micellae*, crumbs), and the mechanism of their association is referred to as "hydrophobic bonding." This infelicitous term should be construed as a shorthand reminder of a rather involved series of events: that primarily the entropic drive of the contiguous water layers, obscurely underlying the adjective *hydrophobic*, rather than the *bonding* between hydrocarbon and hydrocarbon, is responsible for the phenomenon.

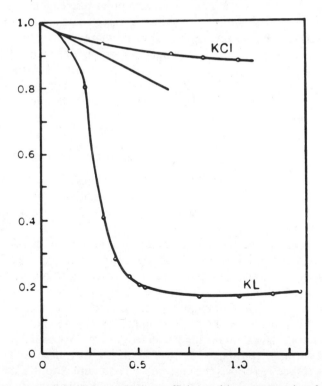

Figure E.2 The variation of the osmotic coefficient with concentration for potassium chloride and for potassium laurate. The straight line is the Debye–Hückel–Onsager slope (McBain, 1950, p. 248).

Experimental evidence for the presence of micelles in aqueous solutions of amphipathic solutes is obtained from measurements of the colligative properties of such solutions, that is, the osmotic pressure, the lowering of the vapor pressure, the elevation of the boiling point, and the depression of the freezing point. These properties are linked, as the name *colligative* suggests: Each of them depends only on the concentration of osmotic units in the solution, whether these units be ions, molecules, macromolecules, or micelles. As dissolved ions or molecules of an amphipathic solute associate in solution, the concentration of osmotic units loses its proportionality to the total concentration of solute. The osmotic coefficient is defined as the ratio of the observed colligative property per mole of solute to the calculated colligative property based on the concentration of ions or molecules given by Avogadro's number and the theory of electrolytic dissociation. The osmotic coefficient is equal to unity at infinite dilution and is always less than that at any finite concentration, being reduced by intermolecular or by interionic attraction as the concentration increases. The variation of the osmotic coefficient with concentration is shown in Fig. E.2 for potassium chloride and for potassium laurate, based on measurements of the lowering of the freezing point. Data for any of the other colligative properties could be used to construct a similar diagram. The lowering of the freezing point restricts the investigation to temperatures near 0°C. Modern techniques, using matched thermistors, make measurements of the lowering of the vapor pressure equally precise, and these measurements are not restricted to a single temperature. The abrupt decline of the osmotic coefficient at what appears to be a critical concentration of solute is clearly marked on the diagram, and certainly indicates that molecular aggregates begin to exist in the solution at and beyond a well-defined concentration, called the *critical micelle concentration*, or CMC.

b.   Methods to Determine Critical Micelle Concentration

Methods to determine the CMC of amphipathic solutes can be divided into those that require no additive and those in which an additive is present in the bulk solution. The former type of method is preferable as the presence of an additive can affect the CMC.

The CMC is well defined experimentally by a number of physical properties besides the variation of the osmotic coefficient with concentration. Ionic amphipaths show a discontinuity at the same concentration in the variation of specific electrical conductivity with concentration. For such amphipaths, locating this discontinuity is the simplest experimental technique to determine CMC. Another electrical property of ionic amphipath solutions is the Hittorf transport number, which shows an interesting peculiarity at the CMC and above. When the anode and cathode compartments are analyzed for anion and cation concentrations, respectively, as is the customary procedure with this determination, the anode compartment is found to contain more equivalents of anion than the electrical equivalents of current that were passed through the solution; and the cathode compartment is found to have lost rather than gained cations. The explanation is that the micelle, an aggregate of anions, must also contain cations in the form of ion pairs, which are carried to the anode with the micelle instead of to the cathode. Let us suppose, following Gonick,[99] that the micelle of the amphipath potassium laurate, KL, has a composition corresponding to the formula $K_{24}L_{27}^{3-}$, that is, an aggregate of 27 anions of which 24 are paired with cations, leaving a net ionic charge of -3; and let the transport number of the micelle be 0.227, which would give for the K^+ ion a transport number of $1 - 0.227 = 0.773$. On passing one equivalent of electricity, that is, 96,500 C, 0.227 equivalents of micelles move to the anode. Now, one equivalent of micelles is one-third of a mole of micelles, and in terms of laurate ion contains one-third of the laurate content of the micelle, that is, one-third of 27 mol or 9 mol. The quantity of laurate ion in the anode compartment is therefore 0.227×9 or 2.04 mol. By analysis, therefore, the transport number of the anion appears to be 2.04, which is tantamount to saying that 204% of the current is carried by the anion. At the same time in the cation compartment, per equivalent of electricity passed through the solution, 0.773 equivalents of K^+ has passed in, but one-third of 24 mol of K^+ has moved out with the micelle, with a transport number of 0.227; that is, a net change of $0.773 - 8 \times 0.227 = -1.043$, or -104.3% of the current is transported by the cations. These transport numbers, of course, are not real, as they are calculated on the wrong assumption, namely, that the ions are not associated; they add algebraically to 1.000, in agreement with the Hittorf mechanism, but the true transport numbers are the assumed values of $t_- = 0.227$ and $t_+ = 0.773$. The Hittorf method cannot obtain the true values; instead it yields manifestly absurd values. Below the CMC the method is valid, and the presence of micelles in the solution is marked by the deviation from reasonable to absurd values of the transport numbers.

The presence of a new species in solution can be detected by observing changes in any physical property that can be measured with high precision. Three such properties are turbidity, density, and refractive index. Solutions of amphipathic solutes (amphipaths) show a discontinuity at the CMC when these properties are plotted as a function of concentration. The most popular of these methods is the measurement of the intensity of scattered light. The intensity of the scattered light increases sharply at the CMC because the micelles scatter more light than the medium[100] (Section I.D.3b).

Other properties by which CMC can be measured are surface and interfacial tensions. Single molecules of an amphipath are readily adsorbed, and positive adsorption is related to the lowering of surface or interfacial tension by the Gibbs equation, which, for a two-component system, is

$$\Gamma_2^G = \frac{-1}{RT}\frac{d\sigma}{d\ln a_2} \qquad \text{[E.2]}$$

where a_2 is the activity of the solute. Adsorption and micellization are two effects promoted by the same cause, but they are independent of each other; each has its own equilibrium constant; but once micelles form in significant quantities, the concentration of single molecules does not increase, or increases only slightly, with increase of total concentration, since all or most of the solute dissolved above the CMC goes into micelles. The adsorption equilibrium, therefore, receives no, or only slight, further contribution; and the surface or interfacial tension does not reduce any further, or reduces only slightly, below the value it reaches at CMC. The surface tension isotherm, plotted with the surface tension as ordinate versus the logarithm of concentration, shows an abrupt change of slope at the CMC. A minimum in the isotherm, should it occur, is now recognized as the consequence of contamination or the presence of hydrolysis products. The hydrolysis products are more surface active than the prime component but are removed from the surface at higher concentrations by solubilization in micelles (Section II.E.3e.)

c.   Equilibrium or Phase Separation?

The major thermodynamic process taking place in micelle-forming solutions is the formation of molecular aggregates from molecularly dispersed solute. This process may be described thermodynamically either as a kinetic equilibrium between the single ions and the micellar species[101] or as a phase separation.[102] Both approaches explain major features of micelle formation.

Micellization and adsorption are separate and independent processes, taking different paths, with different equilibrium constants and rates of reaction. Micelles are brought into the solution at a concentration of solute determined by the micellization equilibrium constant, independently of the surface concentration of adsorbate, which is determined by a different equilibrium constant. But when micelles appear at the CMC, they virtually bring adsorption to an end at whatever surface concentration has been reached, because they greatly inhibit further increase of concentration of the amphipathic ion or molecule. The adsorbed layer may not be fully saturated at the CMC. If the soluble amphipath is converted into an insoluble monolayer by placing it on a strong brine solution, it can be compressed farther, by means of a barrier, to a greater surface concentration than it ever could reach by adsorption from solution (Section II.G.6).

Let us consider micellization as an equilibrium process. Take as a model of a micelle the formula $K_{24}L_{27}^{-3}$ that we used before. Then

$$K = \frac{[K_{24}L_{27}^{-3}]}{\{[K^+]^{24}[L^-]^{27}\}} = \frac{[m]}{(c - 24m)^{24}(c - 27m)^{27}} \qquad [\text{E.3}]$$

where $[K_{24}L_{27}^{-3}] = [m]$ and c is the total concentration of solute. The variation of the concentration of micelles $[m]$ with the total concentration of solute $[c]$ is calculated by Eq. [E.3] with an arbitrary value of $K = 1.000$ to illustrate the effect of large exponents in the denominator of such an expression.

Figure E.3 shows how the concentration of micelles $[m]$ grows with an increase of total concentration $[c]$. The CMC is the point on the curve at which

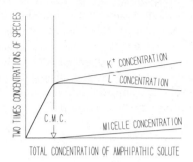

Figure E.3 Micellization as an equilibrium process. Micelle taken as $K_{24}L_{27}^{-3}$.

$[m]$ shows a sudden increase; the larger the values of the exponents in Eq. [E.3], the more abruptly does this happen and the more pronounced is the "criticality" of the "transition." It therefore becomes of interest to explore the behavior expected from a genuine transition and a genuine critical point. The model to consider is analogous to the separation of a phase when maximum solubility is reached. If micelles separate as a phase at a critical concentration, the concentration of unassociated solute does not increase beyond its saturation value reached at the CMC. Although more solute goes into solution beyond the CMC, it is taken up entirely by micelles; the concentration of micelles, therefore, continues to increase while the concentration of unassociated amphiphiles remains constant at its CMC value. If the average number of amphiphiles in a micelle is n, the molar concentration of micelles grows at a rate that is only $1/n$ times that of the moles of solute put into the solution (Fig. E.4).

Comparison of Figs. E.3 and E.4 shows how nearly alike are the results obtained from these two models: namely, concentrations based on micellar equilibrium and concentrations based on a separation of micelles as a distinct phase. The result the two models have in common is that micelles appear in the solution abruptly enough to define a true critical or a pseudocritical micelle

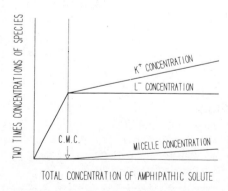

Figure E.4 Micellization as a pseudophase separation. Micelle taken as $K_{24}L_{27}^{-3}$.

concentration. Experimental measurements of extreme precision on scrupulously purified solutes would be required to differentiate between the two models. Mitchell and Ninham[103] have shown that the law of mass action is an appropriate description of the process of micellization; the alternative description as a phase separation does provide a simpler basis for subsequent development of a theory of self-assembly of amphipathic molecules into micelles, but is designated "the pseudophase approximation."

d.   Effects of Varying Molecular Structure

A continuous variation of the hydrophile–lipophile balance of solutes is provided by an homologous series of the lipophile. The free energy of micellization, whichever mechanism of micelle formation is considered, is given by $-RT\ln$ [CMC]. The logarithms of the critical micelle concentrations of an homologous series of sodium soaps (taken from Table E.1) are plotted in Fig. E.5 against the number of carbon atoms in the alkyl chain. The straight line obtained on this plot shows that each additional carbon atom makes the same contribution to the chainge of free energy on micellization. The slope of the straight line gives a value of 1.64 kJ/mol per carbon atom. This value may be compared with the free energy of adsorption of an homologous series of fatty acids at a water–air surface, which is 2.72 kJ/mol. (Section II.G.5). The two values are of the same order, indicating the common source of the two effects; but complete agreement is not to be expected because of the different hydrophiles and the different structures of micelles and adsorbed films.

The same relation that holds for the variation of the CMC with the chain length of an homologous series also holds for the concentrations of successive members of an homologous series of sodium soaps required to give the same

Table E.1   Critical Micelle Concentrations of Aqueous Solutions of Sodium Soaps without Added Electrolyte

Name	Temperature (°C)	CMC (M)
Sodium pentanoate	20	2.35
Sodium hexanoate	20	1.6
Sodium heptanoate	20	9.5×10^{-1}
Sodium octanoate	25	3.45×10^{-1}
Sodium nonanoate	20	2×10^{-1}
Sodium decanoate	25	9.5×10^{-2}
Sodium dodecanoate (laurate)	25	2.4×10^{-2}
Sodium tetradecanoate (myristate)	25	6.9×10^{-3}
Sodium hexadecanoate (palmitate)	50	3.2×10^{-3}
Sodium octadecanoate (stearate)	50	1.8×10^{-3}

From Mukerjee and Mysels (1971).

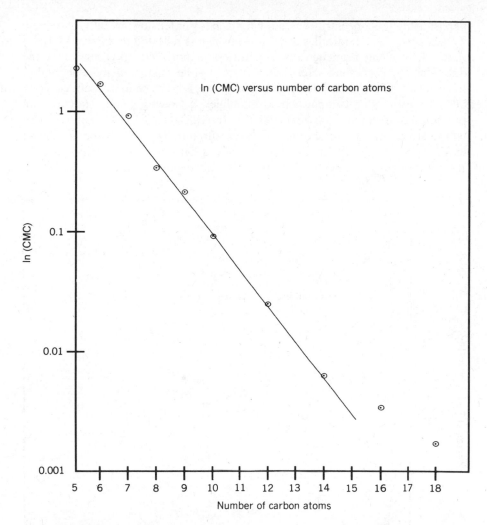

Figure E.5 Critical micelle concentrations of aqueous solutions of sodium soaps without added electrolyte as a function of chain length.

lowering of surface tension, where it is known as Traube's rule. This rule can be interpreted, by the same thermodynamic reasoning, as an indication of the increase in the work done when a molecule passes from the bulk phase to the surface layer for each additional CH_2 group.

Surface tension and interfacial tension isotherms show that the initial steep decline of the tensions is arrested at the CMC. Only the single ions are surface active; the micelles are surface inactive as they are symmetrically hydrophilic. Micelle formation is an alternative way to remove the lipophile from the water; it is a parallel property to adsorption and competitive with it. If micelle formation can be inhibited, therefore, the lowering of surface and interfacial tension might persist to higher concentrations, so that much lower tensions would be attained. The ease with which micelles are formed (which is reflected in a lower CMC) is reduced by using branched, asymmetric, or subdivided alkyl chains as the lipophile. The potassium salts of the sulfonate of hexadecyl phenyl ether and of the sulfonate of resorcinol dioctyl ether are powerful surface-active solutes. Their behavior provides a comparison, as lipophiles, between a single straight-chain paraffin and the same chain divided in two. Interfacial tension isotherms at room temperature, against cyclohexane, were measured by Hartley[104] and are shown in Fig. E.6. At low concentrations the hexadecyl ether is the more effective surface-active agent as shown by its greater adsorption at the interface; but it

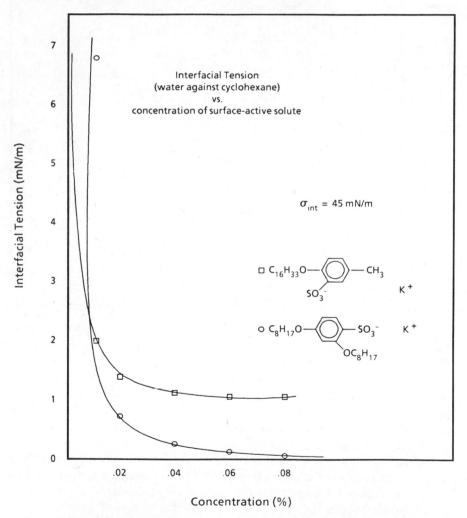

Figure E.6 Interfacial tensions of water against cyclohexane as a function of the concentration of the sulfonates of hexadecyl phenyl ether and resorcinol dioctyl ether.[104]

forms micelles at a lower concentration than does the dioctyl ether. Its ability to lower interfacial tension further essentially terminates at its CMC. The dioctyl compound continues to lower the interfacial tension to values well below those obtainable with the hexadecyl analogue. A disadvantage of the double-chain or branched-chain compounds is their comparatively low solubility, which is also related to their difficulty in forming micelles.

$$C_{16}H_{33}-O-Ph-CH_3 \qquad C_8H_{17}-O-Ph-SO_3^-K^+$$
$$\quad\quad\quad\quad |\qquad\qquad\qquad\qquad\qquad |$$
$$\quad\quad\quad SO_3^-K^+ \qquad\qquad\qquad\qquad OC_8H_{17}$$

Potassium hexadecyl phenyl ether sulfonate   Potassium resorcinol dioctyl ether sulfonate

Hartley observed that the dioctyl solution at a concentration of 0.12%, the most concentrated clear solution of this compound obtainable at room temperature, emulsified various oils easily. Gentle handshaking in an open test tube produced emulsions stable for at least several weeks. In terms of Rosen's "efficiency" of surface-active solutes (the lower the concentration at which a given surface tension is reached), and "effectiveness" (the lower the surface tension at the CMC), the hexadecyl ether is more efficient and less effective (Rosen, 1978, pp. 153–159).

Oil–water interfaces with ultralow tensions are used to penetrate porous systems, as in the removal of oil from oily soil, and in promoting spontaneous emulsification. Interfacial tensions as low as 10^{-5} mN/m have been measured in connection with tertiary oil recovery. Compounds in which the hydrophile is located near the middle of the carbon chain rather than at the end are among the most powerful wetting agents and penetrants. A well-known example is sodium dioctylsulfosuccinate (Aerosol OT, American Cyanamid Chemical Corporation):

$$O{=}C-O-(CH_2)_7CH_3$$
$$\quad\quad |$$
$$\quad CH_2$$
$$\quad\quad |$$
$$HC-SO_3^-Na^+$$
$$\quad\quad |$$
$$O{=}C-O-(CH_2)_7CH_3$$

Sodium dioctylsulfosuccinate

Traube's rule for an homologous series, originally found for a series of soaps, also holds for the behavior of the salts of the secondary alcohol sulfates, but does not hold for salts of primary alcohol sulfates.[105] All such attempts to obtain simple relations for molecular and atomic contributions to surface activity share the same defect of limited application. Davies and Rideal have worked out a quantitative evaluation of the HLB of a surface-active solute by means of molecular and atomic contributions (Section IV.A.6), but the HLB scale as a measure of surface activity also has many exceptions. While certain variations in molecular structure are clearly influential, constitutive effects overshadow any simple rule. Since the CMC of a surface-active solute, even of commercial agents

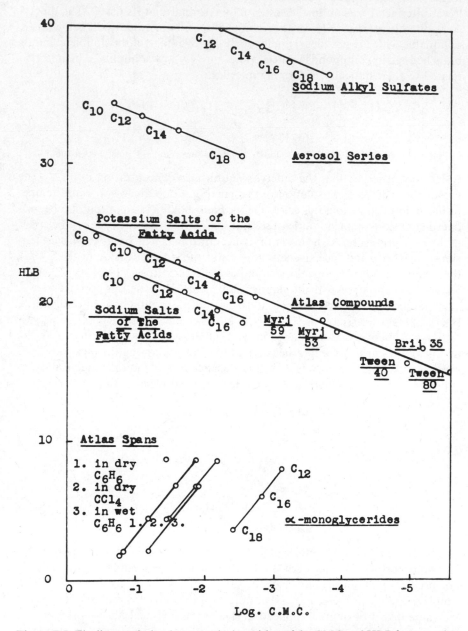

Figure E.7 The linear relation between the logarithm of the CMC and HLB for a number of homologous series. Aqueous solutions have negative slopes and nonaqueous solutions have positive slopes.[106]

that are far from being single-component systems, is readily measured by any of a number of techniques, the relation between the CMC and HLB can be established. A linear relation obtains between the logarithm of the CMC of members of a homologous series and their HLB number (Fig. E.7).

e.   Solubilization by Micelles (Microemulsions)

Solubilization was discovered when it was found that a concentrated aqueous soap solution absorbs liquid benzene to form a clear isotropic liquid. Other organic compounds, normally insoluble in water, also dissolve spontaneously in aqueous solutions of surface-active solutes at concentrations above the CMC. The solubility of 2-nitrodiphenylamine in dilute potassium laurate solutions is shown in Fig. E.8. The solubility remains constant and is the same as in pure water until the concentration reaches CMC; the solubility of the amine then increases with concentration of the soap, due to solubilization in the micelles. That micelles are responsible for the solubilizing action is also shown clearly by

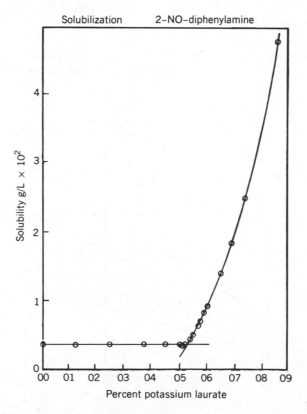

Figure E.8 Solubility of 2-nitrodiphenylamine in aqueous solutions of potassium laurate (Harkins, 1952, p. 323).

the variation of the Henry's law constant for the solubility of a gas as a function of concentration of surface-active solute. The absorption isotherms at 25.0°C of butadiene in water and in aqueous solutions of a cationic surface-active solute, *p*-diisobutylphenoxyethoxy-ethyl dimethylbenzyl ammonium chloride (Hyamine 1622, Rohm & Haas Co.) are shown in Fig. E.9. The regular decrease of the Henry's law constant is linear with the logarithm of the concentration of Hyamine 1622, as shown in Fig. E.10. The Henry's law constant of butadiene in pure water is 2.96×10^6 torr/mol gas/mol solution. Extrapolating the experimental straight line to the concentration of solute corresponding to the value of the Henry's law constant of 2.96×10^6 would presumably yield a minimum concentration of solute below which no additional solubility of butadiene occurs. The value for the minimum concentration obtained from Fig. E.10 is 0.19% Hyamine 1622. The CMC of Hyamine 1622 from conductivity measurements is 0.16%. The role of micelles in solubilization is thus demonstrated.

X-ray diffraction of micellar solutions shows that the hydrocarbon chains of the micelle interior are essentially liquid like. The photographs have a broad, diffuse halo corresponding to a spacing of 0.45 nm, similar to that given by liquid alkanes.[108] The application of NMR techniques also indicate that the hydrocarbon is in a liquid like state.[109]

As more solubilizate is brought into the micelles, they change gradually to swollen micelles (microemulsion) and then to miniemulsion droplets. At this stage the system certainly contains two phases. The interior of the micelle is essentially hydrocarbon in a liquid state; the solubilizate is a second component in the micelle and accordingly has all the thermodynamic properties of a solute,

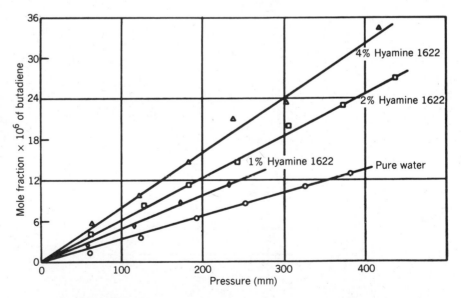

Figure E.9 Mole fraction of butadiene dissolved in pure water and in solutions of Hyamine 1622 as a function of the equilibrium partial pressure of butadiene at 25.0°C.[107]

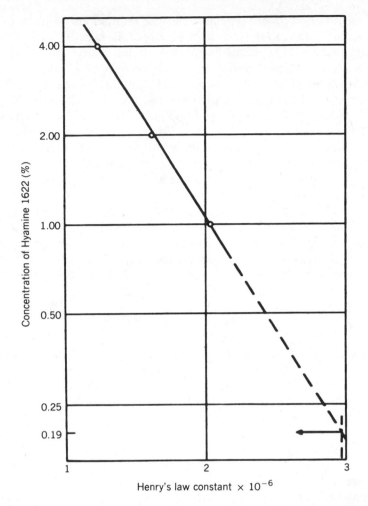

Figure E.10 Variation of the Henry's law constant for the solubility of butadiene in aqueous solutions of Hyamine 1622 with the concentration of solute. The vertical dotted line is the Henry's law constant for pure water. The left arrow is the concentration of Hyamine 1622 below which the solubility is not enhanced.[107]

including osmotic pressure. These droplets have also capillary (or Laplace) pressure due to their small radius of curvature. The capillary pressure causes solubilizate to diffuse from the smaller droplets to the larger ones (Ostwald ripening); at the same time the loss of solubilizate reduces the osmotic pressure in the smaller droplets with respect to the larger ones. These two opposing effects balance at equilibrium, which occurs at a particular droplet size. Uniform microemulsion droplets in the range 10–100 nm may be stabilized in this way at an equilibrium point.[110]

The same principle is applied in emulsion polymerization, by including a water-insoluble non volatile paraffin along with the monomer in the micelles. The paraffin acts similarly to the hydrocarbon core of the micelle. Capillary pressure promotes diffusion of monomer from smaller to larger droplets; but in doing so the osmotic pressure inside the smaller droplets is reduced, which puts a terminus to the Ostwald ripening. When, for example, styrene is homogenized with water containing emulsifier, the resulting emulsion is unstable due to Ostwald ripening. A small amount of water-insoluble component present in the droplet strongly inhibits the Ostwald ripening because, by dissolving styrene, it creates an osmotic pressure. Since only the styrene diffuses, the osmotic pressure in the larger droplets increases, compared to that in the smaller droplets. The gradient of osmotic pressure tends to send the styrene back to the smaller droplets. Thus the Ostwald ripening sets up a counteraction that brings about a state of equilibrium. Ugelstad et al.[111] prepared finely dispersed styrene microemulsions stabilized by the presence of a co-emulsifier that had a low solubility in water and a low molecular weight. Examples of such a co-emulsifier are long-chain fatty alcohols and alkanes, like hexadecanol and hexadecane. When polymerization takes place, the resulting latex has uniform particles of a controlled size.

4.   THE KRAFFT POINT

In 1895 Krafft and Wiglow noticed[112] that the solubility of an homologous series of sodium soaps in water shows an abrupt increase with temperature at what has since been called the "Krafft point." They ascribed the effect to the melting of the hydrocarbon chains of the soaps, based on the coincidence of the Krafft points of the soaps with the melting points of their parent fatty acids. The modern interpretation is that the Krafft point marks a transition between the dissolution of the soap to form ions and its dissolution to form micelles. Krafft's notion of partial melting is supported, however, by modern evidence that the hydrocarbon core of the micelle is in the liquid state. The solubility of soap increases sharply with temperature once micelles have appeared, by a process that can only be described as self-solubilization, which packs more solute into micelles. Mixed micelles are also capable of being formed with numerous additives above the Krafft point—further evidence of the liquid character of the micellar core.

The solubility relations of a surface-active solute are shown in Fig. E.11 in the form of a pseudo phase diagram. The CMC does not vary much with temperature, as micellization is chiefly an entropic effect. Solubility increases with temperature to a greater degree than does micellization: Consequently at some temperature the solubility curve intersects the CMC curve. The Krafft point can be seen from the diagram to be the temperature at which the CMC and the solubility coincide. At temperatures below the Krafft point, the CMC is at a higher concentration than the solubility, that is, micelles are formed only in supersaturated solutions. The CMC can still be determined quite readily, however, as these solutions can be supercooled for many days without crystallizing. At temperatures above the Krafft point, the solute forms micelles in solutions that are

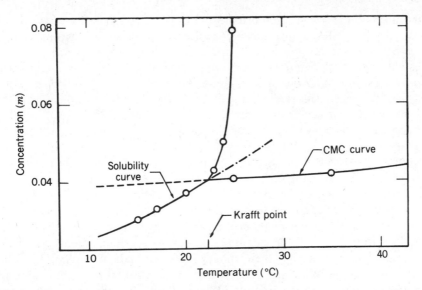

Figure E.11 The pseudo phase diagram of sodium dodecyl sulfonate and water (Shinoda, et al., 1963, p. 7).

unsaturated. Surface-active solutes do not function well as detergents or emulsifying agents below their Krafft point, either because of too low a solubility or because micelles are lacking to act as solubilizers.

At 60°C the detergent power of the sodium alkyl sulfates is in the order C_{16} > C_{14} > C_{18} > C_{12}. Measurements of foam number at 60°C by the Stiepel method show the compounds to lie in the same order.[113] But at 40°C the C_{14} compound gives maximum foam, while at 20°C the C_{12} is the most powerful foamer.[114] These results confirm the rule that surface activity at higher temperatures is promoted by using a longer alkyl chain in the lipophile, that is, an amphipath with a higher Krafft point.

Krafft points of some surface-active solutes in aqueous solution are given by Rosen (Rosen, 1978, Table 5-4). These are for purified compounds. Mixtures generally have Krafft points considerably lower than those of single components.

5. DYNAMIC SURFACE TENSIONS

The surface tension of a pure liquid is not established immediately on creation of the surface. To reach equilibrium the molecules have to rearrange to a preferred orientation at the created surface, with an increase in their intermolecular distance. The equilibrium state in a pure liquid is attained rapidly, within a few milliseconds. In solutions of surface-active solutes the rate at which equilibrium is reached is much slower, of the order of several seconds, as it depends on the rate of diffusion of solute molecules to the surface. The rate of diffusion, in turn, is determined by the size of the molecule and its concentration. For conventional surface-active solutes, diffusion brings molecules to the surface within seconds; yet experience shows that some solutions take hours or even days to reach constant (static) surface tension. Such long periods of aging are ascribed to traces of polyvalent ion impurities, if the water had not been quartz distilled or treated with ion exchange resins (Davies and Rideal, 1963, p 167), or to highly surface-active impurities present in trace amounts.[115] These impurities, because of their low concentration, may require hours to reach the surface and produce their full effect.

Dynamic surface tensions are defined as any nonequilibrium values of surface tension that arise when the surface of a solution is extended or contracted. As the surface moves toward equilibrium, either by adsorption or desorption of solute, the surface tension changes toward its static value. If the time scale is in milliseconds, a convenient method of measurement is the vibrating jet (Section II.A.17). Another method for the same time scale is the falling meniscus.[116] For time scales larger than 30 s, the ordinary static methods of measuring surface tension, such as the Wilhelmy plate, are applicable. All these methods give the variation of surface tension with time. Other methods give indirect evidence of the presence of a surface-active solute by the effect of dynamic surface tension on the damping of capillary waves or the reduction of the velocity of a rising bubble. These methods are useful in oil solutions where the surface tension lowering is small, and measurements of surface tension are not sufficiently precise to detect dynamic surface tensions directly.

If the solution is very dilute, less than about $10^{-6} M$, the adsorbed excess concentration is so dilute that its concentration, and hence the surface tension of the solution, is not much affected by extension or contraction of the surface; hence no change of surface tension is observed. The rate of solute adsorbed at a newly created surface in the absence of stirring or of an energy barrier to adsorption, increases with concentration, according to an equation given by Ward and Tordai[117]:

$$n = 2(D/\pi)^{1/2}ct^{1/2}(N_0/1000) \qquad [E.4]$$

where n = number of molecules/cm^2 after time t in seconds
   D = bulk diffusion constant, cm^2/s
   c = bulk concentration, mol/L
   N_0 = Avogadro's number

The time needed to replace solute at a new surface becomes progressively less with increasing concentration; whereas the time required to change the surface area remains about the same. As a consequence the extent of the dynamic surface tension, and related phenomena such as the Marangoni effect, the Marangoni elasticity (Section II.E.6), and the foaminess of the solution, all tend to show greatest effects at a concentration near that of the CMC.

A major difference between aqueous and nonaqueous solutions lies in the magnitude of the effects produced by surface-active solutes. The surface tension of water is reduced from 73 to 25 mN/m quite readily by amphipathic organic solutes; but the surface tension of most organic solvents is already in the low range of 25–30 mN/m, so that only a small reduction can be achieved by an organic solute. Thus, although Marangoni effects may arise in nonaqueous solutions, they are usually much less pronounced than in aqueous solutions of soaps or detergents. Oil lamellae, therefore, have a relatively low resistance to mechanical shock; consequently oil foams are transient or evanescent, resembling the foam produced from very dilute aqueous solutions of detergents or more concentrated solutions of weakly surface-active solutes. Of course, special solutes have been developed to stabilize nonaqueous foams for application in the field of cellular plastics. These solutes incorporate polyalkylsiloxane or perfluoroalkyl moieties in the molecule, which are able to reduce the surface tension of organic liquid monomers by 12–15 mN/m to boost the Marangoni effects; and the result is obvious in an increased stability of the foam.

6. ELASTICITY OF SURFACE FILMS

The elasticity of a surface film under any given conditions is defined as the ratio of any small increase in spreading pressure π to the areal compression. The areal compression is the relative contraction of the area, that is, the ratio of the change of area to the area:

$$E = \frac{-d\pi}{dA/A} = \frac{-d\pi}{d \ln A}$$ [E.5]

Since the areal compression is dimensionless, surface elasticity has the same units as spreading pressure. The elasticity at the point P on the π–A curve in Fig. E.12 is obtained by extending the tangent at point P to intersect the π axis at point G. Let

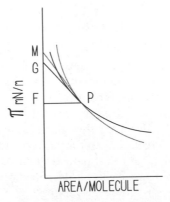

Figure E.12 How to obtain elasticity of a surface film from a π–A curve. The solid line represents the equilibrium isotherm; the dotted line represents a nonequilibrium curve. The equilibrium elasticity is FG; the dynamic elasticity is FM.

F be the value of π at P; then

$$E = \frac{-A\,d\pi}{dA} = \frac{FP \cdot FG}{FP} = FG \qquad\qquad [\text{E.6}]$$

Hence, if the relation between the spreading pressure and the area per molecule under certain conditions, as for instance at a given temperature, is represented by the locus of P, the elasticity of the film when in the state represented by P may be found by drawing PG a tangent to the curve at P, and PF a horizontal line. The portion FG of the π axis represents, on the scale of π, the elasticity of the film.

In most bodies compression introduces an increase of temperature, the effect of which is to increase the elasticity. Elasticity can be measured isothermally or adiabatically. The former elasticity pertains when stresses and strains occur slowly enough to remain at thermal equilibrium. The latter pertains in the case of rapidly changing forces, where there is not time for the temperature to be equalized. The adiabatic elasticity is larger than the isothermal elasticity.

$\pi{-}A$ curves can be drawn that do not represent equilibrium. If the surface of the solution has recently undergone rapid expansion or contraction, a non-equilibrium state of dynamic surface tension exists. On compression the film becomes more concentrated and the value of π increases; on expansion the film is more dilute and the value of π decreases. In Fig. E.12 the dynamic $\pi{-}A$ curve at point P lies above the equilibrium curve on compression and below it on expansion. If enough time is allowed, these changes in π disappear, but if measurements are made before equilibrium is reached, the condition is as shown in the diagram. The tangent to the dynamic curve at point P is represented by PM, and the dynamic elasticity is represented by FM.

If the surface of a solution containing a surface-active solute is expanded suddenly, then, before adsorption from the bulk phase to the newly created surface has had time to take place, the same number of adsorbed molecules remain on the surface, so that the differential increase in area per molecule is

$$da = d\left(\frac{A}{n_s}\right) = \frac{dA}{n_s} \qquad\qquad [\text{E.7}]$$

where a is the area per adsorbed molecule, A is the total area, and n_s is the number of molecules of adsorbed solute on the total area. The suddenly expanded surface is not a stable state, but subsequent changes are so slow on a molecular scale that one can assume it to be at thermal equilibrium. Thermal equilibrium implies that thermodynamics is applicable even though the states of the system are not in mechanical equilibrium. We may, therefore, use equilibrium equations of state for the surface film at each stage of its path to final equilibrium. Let us suppose initially that the adsorbed surface film is described by the two-dimensional ideal equation of state:

$$\pi a = (\sigma_0 - \sigma)a = kT \qquad\qquad [\text{E.8}]$$

or

$$(\sigma_0 - \sigma) = RT\Gamma_2^G \tag{E.9}$$

where $\Gamma_2^G = n_s/N_0 A \,\text{mol/cm}^2$.

Elasticity is defined as the ratio of the increase in the surface tension resulting from an infinitesimal increase in area and the relative increment of the area. For a lamella with adsorbed films on both sides, the elasticity E is given by

$$E = \frac{2\,d\sigma}{d\ln A} = \frac{-2\,d\sigma}{d\ln \Gamma_2^G} \tag{E.10}$$

Differentiating Eq. [E.9] and substituting in Eq. [E.10] gives

$$E = 2RT\Gamma_2^G \tag{E.11}$$

Equation [E.11] says that at a given temperature the elasticity is proportional to the Gibbs excess surface concentration in the case of a two-dimensional ideal equation of state. For more compressed films the elasticity is given by a few terms of a power series in Γ_2^G:

$$E = 2RT\Gamma_2^G - b(\Gamma_2^G)^2 + c(\Gamma_2^G)^3 \tag{E.12}$$

Elasticity determined isothermally from the equilibrium π–A curve is called the Gibbs elasticity; and nonequilibrium elasticities measured by any dynamic method are known as Marangoni elasticities.[118] Marangoni elasticity is larger in value than the Gibbs elasticity that could be obtained in the same system. While Gibbs elasticity is a thermodynamic property of a surface film, Marangoni elasticities depend on the nature of the stresses and strains applied to the surface, as do the dynamic surface tensions. Neither Marangoni elasticities nor dynamic surface tensions have characteristic values. Some investigators measure dynamic surface tensions occurring at rather low frequencies of dilatation–compression cycles, from 1/min to 1 every 30 mins; others[119] have used frequencies as high as 15–135 Hz (cycles/second,) although such disturbances are far from corresponding to the extension–contraction cycles occurring in foam. The measurement must be made coincidentally with measurements of changes of surface area.

The modulus of surface elasticity of a solution can be determined by measuring the damping of transverse ripples as a function of their frequency. The surface of a wave is contracted at the crest and extended at the trough. In the absence of any surface-active material, this contraction and extension does not change the surface tension; but if an adsorbed layer is present, it is compressed on the contraction and dilatated on the extension, causing the surface tension to decline at the crest and to increase at the trough. These local differences of tension alter the pattern of subsurface flow, giving rise to a greater rate of energy dissipation by viscous friction, and consequently to a greater damping of the waves than would otherwise occur. The distance-damping coefficient β may be measured by the logarithmic decrease of the wave amplitude, A, with distance x from the source:

$$\beta = \frac{-d \ln A}{dx} \qquad\qquad [E.13]$$

The damping coefficient β is constant for Newtonian liquid surfaces: The surface of a solution contains adsorbed solute if the solute is surface active, and such a surface may not be Newtonian. The plot of $\ln A$ versus x would then not be linear. The testing of Eq. [E.13] therefore is informative about the presence of a non-Newtonian shear viscosity at the surface of a solution. If, however, the surface should be Newtonian, the data have a further use: The relative change of the damping coefficient compared to pure solvent ($\beta = \beta_0$), as a function of wave frequency, is directly related to the surface elasticity,[120] assuming that diffusional interchange between bulk and surface is negligible during dilatation and compression of the adsorbed layer.[121]

7. ULTRALOW DYNAMIC SURFACE TENSION

Some processes that use solutions of surface-active solutes depend on the creation of regions where the interfacial tension is lower by a few orders of magnitude than its equilibrium value. Contraction of a surface that contains adsorbed solute compresses the layer and causes the surface tension to decline. If subsequent relaxation is rapid, the ultralow tension is fleeting, but slow relaxation processes are known that allow such ultralow dynamic surface tensions to be retained long enough to be useful. A prime example is a monolayer of dipalmitoyl lecithin, which has an equilibrium surface tension of more than 70 mN/m but can be compressed to a long-persisting tension of 1 mN/m. This compound is physiologically important in preventing the collapse of lung alveoli. No synthetic surface-active solute has yet been found to compete in effectiveness with this product of millenia of evolution (or Divine creation!).

8. THEORIES OF RATE OF BUBBLE RISE IN VISCOUS MEDIA

The rate of rise of single air bubbles in a solution is determined by the behavior of newly formed surfaces and how they respond to forces of dilatation and compression. In a pure liquid, under conditions of laminar flow, a rising bubble moves faster than predicted by Stokes' law, since the mobility of its interface allows lower velocity gradients in the liquid than those that develop at an immobile, or rigid, interface. When a surface-active solute is adsorbed at the interface, however, the movement of the interface and of air inside the bubble is restricted; the velocity gradient in the outer fluid is increased; until at the limit the bubble acquires the property of a rigid sphere, and its rate of rise is reduced to that given by Stokes' law.

a.   Rise in Pure Liquids—Hadamard Regime

Stokes's theory for the terminal velocity of a solid sphere in a viscous medium was extended by Rybczynski[122] and Hadamard[123] to fluid spheres. For a liquid drop or a gas bubble of radius a, density ρ_1, and viscosity η_1, moving through an infinite volume of a medium of density ρ_2, and viscosity η_2, the terminal velocity is given by

$$v_s = \frac{2(\rho_1 - \rho_2)ga^2}{9K\eta_2} \qquad \text{[E.14]}$$

where κ, the Rybczynski–Hadamard correction factor, has the value

$$K = \frac{3\eta_1 + 2\eta_2}{3\eta_1 + 3\eta_2} \qquad \text{[E.15]}$$

The derivation of Eq. [E.15] postulates that the medium exerts a viscous drag on the surface of the bubble or liquid drop, and so sets up a circulation of the fluid contained inside, whether gas or liquid. According to the theory, a bubble containing a circulating gas, with $\eta_1 \ll \eta_2$, would move 50% faster ($K = \frac{2}{3}$) than one in which the gas, for any reason, does not circulate; for in the latter case the Rybczynski–Hadamard factor is unity, and the velocity of the bubble is given by the unmodified form of the Stokes law.

Garner et al.[124,125] demonstrated experimentally the existence of the circulation inside air bubbles and examined the effects of bubble size and shape. They showed that the validity of the above equations is limited to the range of the Reynolds number less than one, and to the same conditions as for Stokes' law to hold, including the requirement that the gas bubbles be spheres.

b.   Effect of Surface-Active Solutes—Stokes Regime

Levich (1962, pp 395–452) was the first to provide a satisfactory explanation of the retardation of the velocity of a rising bubble caused by surface-active solutes in the medium. He postulated that adsorbed solute is not uniformly distributed on the surface of a moving bubble. The surface concentration on the upstream part of the bubble is less than the equilibrium concentration, while that on the downstream part is greater than equilibrium. This disequilibration of the concentrations is brought about by the viscous drag of the medium acting on the interface, which in turn creates a disequilibrium of surface tensions, with the lower tension where the concentration of adsorbate is greater. The liquid interface then flows (Marangoni flow) from the region of lower tension to that of higher tension, and the direction of this flow offsets the flow induced by the shear stress in the outer fluid acting on the interface. The increasing rigidity of the interface inhibits the circulation of internal gas to a greater or lesser degree; when completely inhibited, the interface is effectively completely rigid and the terminal velocity of rise is reduced to that given by the Stokes law.

Figures E.13–E.15 show that the ratio of the observed velocity of ascent of a bubble to the calculated Stokes velocity varies between the limits $0.99 < K < 1.52$ for the solutions in mineral oil, and the limits $0.95 < K < 1.47$ for the solutions in trimethylolpropane–heptanoate. The value of $K = 1.52$ obtained for the mineral oil and 1.47 for the trimethylolpropane–heptanoate demonstrates the virtual absence of any surface-active contaminant in these solvents. No significance is attached to the difference of 0.05 units from the theoretical value at the lower limit, $K = 1.00$, as fluctuations of that magnitude at that limit are within the experimental error. Figures E.13 and E.14 also show that the range in which K is concentration dependent is $0.1\,\text{ppm} < c < 20\,\text{ppm}$ for solutions of polydimeth-

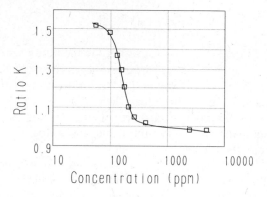

Figure E.13 Ratio of the observed velocity of ascent of a bubble to the calculated Stokes velocity in solutions of various concentrations of polydimethylsiloxane in mineral oil.[144]

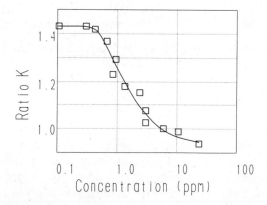

Figure E.14 Ratio of the observed velocity of ascent of a bubble to the calculated Stokes velocity in solutions of various concentrations of polydimethylsiloxane in trimethylolpropane–heptanoate.[144]

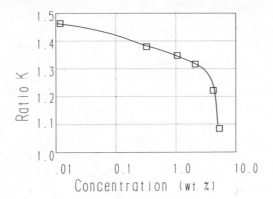

Figure E.15 Ratio of the observed velocity of ascent of a bubble to the calculated Stokes velocity in solutions of various concentrations of N-phenyl-1-napthylamine in trimethylolpropane–heptanoate.[144]

ylsiloxane in mineral oil and in trimethylolpropane–heptanoate. For solutions of Span 20 in mineral oil, the lowest concentration for any measurable variation in the value of K is 500 times greater; but the range of concentration through which the variation of K takes place is relatively narrower, that is, 50 ppm $< c$ < 500 ppm. The solute N-phenyl-1-naphthylamine in trimethylolpropane–heptanoate requires a still greater concentration before its effect is found (Fig. E.15), showing that it is even less surface active in this solvent.

The inhibition of movement of the surface and of air circulation inside the rising bubble is accepted as the explanation of the remarkable slowing of its rate of ascent. Okazaki et al.[126] found that a concentration of sodium dodecyl sulfate (SDS) in water as low as $10^{-5}\,M$ retards the rate of ascent of a bubble in that solution and also confers stability on a single bubble at its surface, even though the static properties of the liquid, that is, the density, the bulk compressibility, the viscosity, and the surface tension, all remain unchanged from those of pure water. A time dependence of the surface tension is also absent in $10^{-5}\,M$ SDS in water; and no measurable increase of the shear viscosity of the surface of a solution so dilute was observed.

Oil solutions behave in many respects as do the very dilute aqueous solutions of SDS in water at concentrations of 10^{-6}–$10^{-4}\,M$, investigated by Okazaki et al. In oil systems, as in these aqueous systems, certain static and dynamic properties of the solution that might appear to be pertinent to surface activity show no change from those of the solvent, yet the rate of bubble rise is greatly affected by the presence of the solute.

9.   PHASE BEHAVIOR OF SURFACE-ACTIVE SOLUTES

Micellar solutions are only a part of the total phase behavior of binary systems containing water and a surface-active solute. Many such solutes form elaborate structures at higher concentrations. The milkiness of concentrated soap solutions is not due to micelles (they are too small to scatter a significant quantity of light) but is due to larger drops of liquid crystals. The structures of liquid crystals are determined by putting them in small (approximately 1 mm) capillaries and by examining them through a microscope with crossed polarizers. Figure E.16 represents idealized structures that may occur as concentration of solute increases in a binary aqueous system. Mesomorphic phases, or mesophases, are structures in which one or two dimensions are highly extended. Such a structure cannot but be anisotropic. A mesophase, that is, an in-between or intermediate phase, has some properties characteristic of crystals, such as birefringence, and yet flows like a liquid. Lehmann called it a liquid crystal (LC) as it combines long-range order with fluidity. Their application in liquid-crystal displays depends on the ability to change orientation with changes in electric field: an ability that requires both structure and fluidity.

Phase diagrams are used to specify the temperature and concentrations at which various structures exist at equilibrium. Figure E.17 is a binary diagram of condensed phases of an alkyl polyoxyethylated nonionic solute in water. The right-hand side of the diagram shows definite hydrates with congruent melting points formed in very concentrated solutions. The middle and neat phases occur at high concentrations and are more extensive on the phase diagram the longer the alkyl chain.

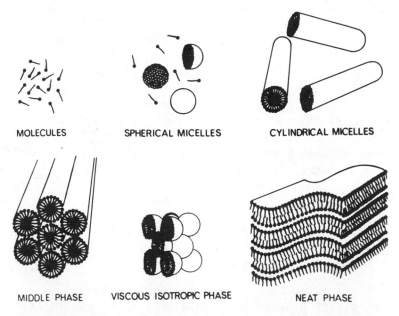

MOLECULES   SPHERICAL MICELLES   CYLINDRICAL MICELLES

MIDDLE PHASE   VISCOUS ISOTROPIC PHASE   NEAT PHASE

Figure E.16 Schematic representation of idealized structures formed by water and a surface-active solute as concentration of solute increases. The terms "middle" and "neat" are derived from nineteenth-century soap-boilers' practice.[127]

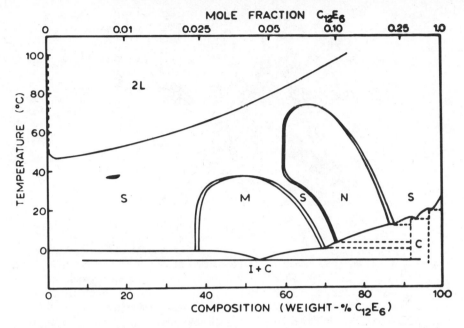

Figure E.17 Binary phase diagram of condensed phases of dodecylhexaoxyethylene ($C_{12}E_6$) and water showing: 2L, two isotropic liquids, S, one isotropic liquid; M, middle phase; I + M, ice + middle phase; I + C, ice + crystals; and N, neat phase. Full lines: experimental boundaries; dotted lines: interpolated boundaries. The estimated extent of the two-phase coexistence region between phases S and M and S and N is shown by the thickness of the boundary line.[127]

In ternary systems in which the third component is an alkane or fatty derivative, such as an acid, alcohol, or amide, the same associated structures can be identified. Figure E.18 is based on the ternary system of water, sodium octyl sulfonate, and decanol. The surface-active solute has limited solubility both in water and in decanol, but the alcohol is not at all soluble in water. The aqueous solution phase, L_1, contains hydrophilic micelles, and the alcohol solution phase, L_2, contains inverse, lipophilic micelles. The phases N and M are both liquid crystals; N represents the neat phase and M the middle phase. This diagram contains features that occur as general behavior in all such ternary systems of surface-active solutes, water, and fatty-acid derivatives. Changes of temperature, salinity, fatty-acid derivative, and structure of the surface-active solute cause systematic shifts in the relative positions of the one-phase regions.

Some compounds (as a rule amphiphilic compounds) with moderately strong hydrophilic groups, such as alcohols, may act both as stabilizers and destabilizers for foams. In fact, identical concentrations may fill either role depending on how the compound is added to the foaming mixture. If added prior to foam formation, the action is stabilizing; if added directly to the foam, it destabilizes. These results are readily interpreted using Fig. E.18. When the alcohol is added prior to foam

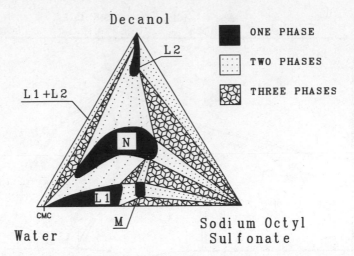

Figure IIE.18 Ternary phase diagram of condensed phases of sodium octyl sulfonate, water, and decanol, at 20°C, showing: L_1, homogeneous isotropic aqueous solution (the L_1 phase at concentrations below cmc contains no decanol); L_2, homogeneous isotropic alcoholic solution; N, homogeneous mesomorphic phase displaying lamellar structure (neat-phase type); M, homogeneous mesomorphic phase displaying normal two-dimensional hexagonal structure (middle-phase type).[129]

formation, the total composition of the system is changed from the aqueous solution toward the liquid crystalline phase, and stabilization of foam occurs. In fact, large amounts may be incorporated without destabilization taking place. The only condition is for the ratio of alcohol to surfactant to be less than the one characterizing the liquid crystal. The liquid crystal by itself does not give a stable

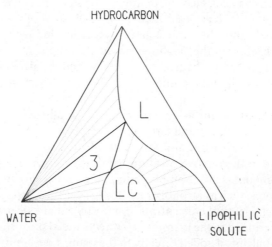

Figure E.19 Ternary phase diagram of hydrocarbon, water, and surface-active solute.[130]

foam. The destabilizing action when the alcohol is sprayed onto the foam can be understood from the conditions of local concentration at the site of the mixing. The alcohol droplet changes the *local* ratio of alcohol to solute, giving a pronounced excess of alcohol. This means that the local composition at the site of mixing corresponds to a composition of the liquid crystal, N, which gives unstable foams.

To obtain a stable foam from an oil solution, a combination of the oil solution and liquid crystal is required. Figure E.19 is the generalized phase diagram for the ternary system of hydrocarbon, water, and surface-active solute showing an isotropic oil solution, L, and a liquid-crystal phase, LC. Water is not miscible with the hydrocarbon; the surface-active solute is soluble in oil but not in water. The solution phase alone, or the liquid crystal phase alone, does not foam. Only compositions in the two-phase area between the liquid crystal and the oil solution produce stable foams, which have a lifetime of several hours. Compositions outside this two-phase area have a foam life of seconds only.

CHAPTER IIF

Surface Activity in Nonaqueous Media

Surface activity in nonaqueous media is not usually observed at the liquid–air surface as the surface tension of the medium is already low, so that neither adsorption of solute nor decrease of surface free energy occurs appreciably, unless the solute is a material of exceptionally low surface energy, such as a siloxane polymer or a compound that contains perfluoroalkyl substituents. But given an opportunity to react with either a Lewis acid or base, the hydrophilic group of the oil-soluble amphiphile can be brought to a liquid–liquid interface or to the interior of an inverse micelle. Such groups interact readily with water, which accounts for the formation of inverse micelles and for the low interfacial tension at the oil–water interface. Indeed an oil-soluble amphiphile in an organic medium will extract dissolved water from the solvent and transfer it into the center of a micelle.

Most amphiphilic substances in apolar organic solvents have groups of ionic character, such as carboxylates or sulfonates, which, with a certain amount of water, bond together to form micelles. In hydrocarbon solutions neither long-chain fatty acids nor the sulfonic acids of alkylated aromatics form micelles, but the more ionic sodium, ammonium, or calcium salts of these acids form micelles readily. Values of CMC of various solutes in benzene and in carbon tetrachloride are listed by Rosen (1978, p 109). In general, the CMCs in these solvents are lower than in aqueous solutions.

The interior of the nonaqueous micelle is an environment of exceptionally high ionic strength. Fowkes[131] points out that the concentration of water can be as low as one tenth of a molecule per ion, corresponding to a $550m$ solution! This is like the ionic environment of crystals, but the small size of the micelle and its dynamic character allow ready access to reactants and solvents. The micellar core

178

can also be strongly acidic or basic, and hence a medium for catalysis for organic reactions (Fendler and Fendler, 1975).

The nonaqueous micelle is capable of carrying a net charge conferred by an excess ion in the core. This property is important in the application of such structures to provide conductivity to carry away electrostatic charges and to prevent buildup of electric fields to dangerous levels in the handling of low-conductivity fluids, as in the filling of fuel tanks and the emptying of supertankers (Klinkenberg and van der Minne, 1958).

Thermodynamics of Adsorption from Solution

The stability of interfaces depends on adsorption, inasmuch as the presence of the spontaneously adsorbed species confers an electrical charge or a steric barrier, or both, to retard coalescence of particles, drops, or bubbles. A gas adsorbed at a solid surface constitutes (or may be treated as) a one-component system, with applications to catalysis; a single solute adsorbed at a solvent surface constitutes a two-component system, with applications to foam, wetting, and spreading; a single solute distributed between two immiscible liquids and adsorbed at the liquid–liquid interface constitutes a three-component system, with applications to emulsions, and solvent extraction. In what follows we discuss two- and three-component systems, the discussion of which can be simplified for one-component systems.

1. THE GIBBS ADSORPTION ISOTHERM

Consider the interfacial phase σ between two bulk liquid phases α and β (Fig. G.1). The σ phase, however, is not an autonomous phase, as it is subject to an external force field, measured by the interfacial tension, caused by the difference in nature of the two adjacent bulk phases. The external field creates heterogeneity of composition, just as the presence of a gravitational or electric field might act analogously. The σ phase, consequently, has a continuously varying composition and density.

By the first and second laws of thermodynamics, the energy of the entire three-phase system is (see Table A.6)

$$dU = T\,dS - p\,dV + \sigma\,dA + \sum \mu_i\,dn_i \qquad \text{[G.1]}$$

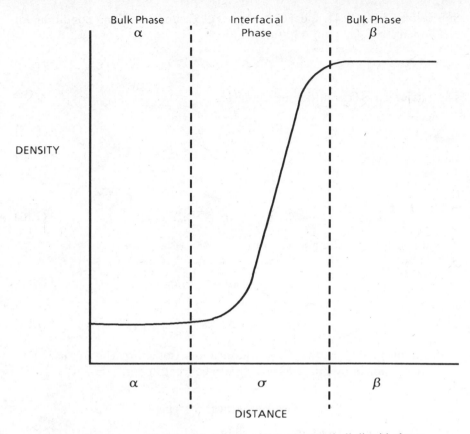

Figure G.1 Schematic of interfacial phase between two bulk liquid phases.

where U, S, V, A, n_i are the extensive variables and T, p, σ, and μ_i are the intensive variables. The two work terms, $p\,dV$ and $\sigma\,dA$ have opposite signs because the spontaneous tendency of a gas is to expand $(dV > 0)$ and that of a liquid surface is to contract $(dA < 0)$.

Integrating Eq. [G.1] gives

$$U = TS - pV + \sigma A + \sum \mu_i n_i \qquad [\text{G.2}]$$

Taking the total differential of Eq. [G.2] and comparing with Eq. [G.1] gives the Gibbs–Duhem equation

$$S\,dT - V\,dp + A\,d\sigma + \sum n_i\,d\mu_i = 0 \qquad [\text{G.3}]$$

Equation [G.3] can be written for each of the bulk phases separately:

$$S^\alpha\,dT - V^\alpha\,dp + \sum n_i^\alpha\,d\mu_i^\alpha = 0 \qquad [\text{G.4}]$$

$$S^\beta\,dT - V^\beta\,dp + \sum n_i^\beta\,d\mu_i^\beta = 0 \qquad [\text{G.5}]$$

Subtracting Eqs. [G.4] and [G.5] from [G.3], and using the condition for equilibrium that $d\mu_i = d\mu_i^\alpha = d\mu_i^\beta$ gives

$$(S - S^\alpha - S^\beta)\,dT - (V - V^\alpha - V^\beta)\,dp + A\,d\sigma + \sum(n_i - n_i^\alpha - n_i^\beta)\,d\mu_i = 0 \quad [G.6]$$

The surface excess quantities are defined as

$$S^\sigma = S - S^\alpha - S^\beta \qquad\qquad [G.7]$$

$$V^\sigma = V - V^\alpha - V^\beta \qquad\qquad [G.8]$$

$$n_i^\sigma = n_i - n_i^\alpha - n_i^\beta \qquad\qquad [G.9]$$

Hence

$$A\,d\sigma + S^\sigma\,dT - V^\sigma\,dp + \sum n_i^\sigma\,d\mu_i = 0 \qquad\qquad [G.10]$$

At constant temperature and pressure

$$-d\sigma = \sum \frac{n_i^\sigma}{A}\,d\mu_i = \sum \Gamma_i\,d\mu_i \qquad\qquad [G.11]$$

where

$$\Gamma_i = n_i^\sigma/A \quad \{\text{moles } m^{-2}\} \qquad\qquad [G.12]$$

Equation [G.11] is the Gibbs adsorption isotherm. The surface concentrations defined by Γ_i are *excess* surface concentrations.

a.   Two-Component Systems

For a two-component system Eq. [G.11] is

$$-d\sigma = \Gamma_1\,d\mu_1 + \Gamma_2\,d\mu_2 \qquad\qquad [G.13]$$

The Gibbs–Duhem equation applied to the bulk phase at constant temperature and pressure is

$$X_1\,d\mu_1 + X_2\,d\mu_2 = 0 \qquad\qquad [G.14]$$

where X_1 and X_2 are the mole fractions of the two components. Therefore,

$$d\mu_1 = -(X_2/X_1)\,d\mu_2 \qquad\qquad [G.15]$$

Substituting in Eq. [G.13] gives

$$-d\sigma = (\Gamma_2 - \Gamma_1 X_2/X_1)\,d\mu_2 \qquad\qquad [G.16]$$

The whole coefficient of the term in $d\mu_2$ has the units of mole fraction per unit area; it is therefore some kind of surface concentration of the solute, which is convenient to designate by a separate symbol, Γ_2^G; or

$$\Gamma_2^G = \Gamma_2 - \Gamma_1 X_2 / X_1 \qquad\qquad [\text{G.17}]$$

Although variations in the conventions for fixing the plane boundary between the bulk and surface phases cause variations in the separate values of Γ_1 and Γ_2, they cannot cause any variation in $d\sigma$. Therefore the expression for Γ_2^G given in Eq. [G.17] is invariant regardless of how one regulates the convention. The invariant quantity given by Eq. [G.17] is the same as the quantity defined by Gibbs, denoted by him with the symbol $\Gamma_{2(1)}$, and usually called the Gibbs excess concentration.

Using the definition of Γ_2^G for a two-component system in Eq. [G.16] and using activities instead of chemical potentials gives

$$-d\sigma = RT\Gamma_2^G \, d\ln a_2 \qquad\qquad [\text{G.18}]$$

where Γ_2^G is the surface excess in moles per square centimeter and a_2 is the activity of the surface-active solute in the bulk solution. Since we are using the Gibbs convention to define the position of the boundary between bulk and surface phases, only a term for the solute appears in Eq. [G.18].

For a nonionizing single solute, in dilute aqueous solution where it may be taken as ideal in behavior

$$d\mu_2 = RT \, d\ln X_2 \qquad\qquad [\text{G.19}]$$

where X_2 is the mole fraction of the surface-active solute in the bulk solution. The Gibbs equation then becomes

$$-d\sigma = RT\Gamma_2^G \, d\ln X_2 \qquad\qquad [\text{G.20}]$$

An ionic solute introduces the possibility of having both ions in the surface phase, as well as H^+ if surface hydrolysis occurs. Taking sodium lauryl sulfate, NaLS, as an example,

$$-d\sigma = RT\Gamma_{Na^+} \, d\ln X_{Na^+} + RT\Gamma_{LS^-} \, d\ln X_{LS^-} + RT\Gamma_{H^+} \, d\ln X_{H^+} \qquad [\text{G.21}]$$

The pH of the bulk solution does not change appreciably as the result of adsorption, even if surface hydrolysis does occur, hence $d\ln X_{H^+} = 0$. Furthermore, in the absence of other electrolytes that might contribute Na^+ to the solution,

$$d\ln X_{Na^+} = d\ln X_{LS^-}$$

and the Gibbs equation is

$$-d\sigma = 2RT\Gamma_2^G \, d\ln X_2 \qquad\qquad [\text{G.22}]$$

In general, for dilute aqueous solutions of ionic surface-active solutes, in the absence of other electrolytes, the Gibbs equation takes the form

$$-d\sigma = vRT\Gamma_2^G \alpha\, d\ln X_2 \qquad [G.23]$$

where v is the number of ions per surface-active solute and α is the degree of dissociation. For a 1:1 strong electrolyte, v is 2 and α is 1.

For dilute solutions of a completely dissociated 1:1 electrolyte in the presence of a swamping amount of an electrolyte containing a common (but not surface-active) ion to the surface-active solute, the dissociation is suppressed, and the Gibbs equation has the same form as that for a nonionic surface-active solute, namely, Eq. [G.20]. That situation exists, for example, if H^+ is adsorbed rather than Na^+, which is the result of surface hydrolysis. The literature contains many examples where surface hydrolysis is not assumed and still Eq. [G.20] is used, whereas Eq. [G.23] would be the correct form.

b.   Three-Component Systems

We shall use subscripts 1, 2, and 3 to designate oil, solute, and water, respectively; and L and W to designate the oil (lipoid) and water phases. In applying Eq. [G.11], two geometrical plane surfaces have to be placed to separate the interfacial phase from the bulk oil and water phases. More than one convention is possible in placing these surfaces because Eq. [G.11] remains true no matter where the imaginary surfaces are drawn, although the individual values of Γ_i are affected. A convenient and natural choice would be so to select the limits that they enclose only the solute and solvent molecules in the interface, which, at low concentrations, is likely to be no more than a monolayer. This convention corresponds to the U-convention of Hutchinson.[132-134] The argument that follows is not, however, restricted to that case. Eq. [G.11] for three components reads

$$-d\sigma = \Gamma_1\, d\mu_1 + \Gamma_2\, d\mu_2 + \Gamma_3\, d\mu_3 \qquad [G.24]$$

The Gibbs–Duhem equation applied to the oil phase (L) and to the water phase (W) gives us two more relations, which, if we suppose the oil and water phases to have negligible mutual solubility, are as follows:

$$X_1^L\, d\mu_1 + X_2^L\, d\mu_2 = 0 \qquad [G.25]$$

$$X_2^W\, d\mu_2 + X_3^W\, d\mu_3 = 0 \qquad [G.26]$$

We can use Eqs. [G.25] and [G.26] to obtain the following:

$$d\mu_1 = -(X_2^L/X_1^L)\, d\mu_2 \qquad [G.27]$$

$$d\mu_3 = -(X_2^W/X_3^W)\, d\mu_2 \qquad [G.28]$$

Substituting Eqs. [G.27] and [G.28] in Eq. [G.24] gives

$$-d\sigma = [\Gamma_2 - (X_2^L/X_1^L)\Gamma_1 - (X_2^W/X_3^W)\Gamma_3]\,d\mu_2 \qquad [G.29]$$

The whole coefficient of the term in $d\mu_2$ in Eq. [G.29] has the units of moles of component 2 per unit area; it is, therefore, some kind of surface concentration of the solute, which it is convenient to designate by a separate symbol, Γ_2^G, or

$$\Gamma_2^G = [\Gamma_2 - (X_2^L/X_1^L)\Gamma_1 - (X_2^W/X_3^W)\Gamma_3] \qquad [G.30]$$

Although different conventions for fixing the plane boundaries betweeh the bulk and the interfacial phases cause variations in the separate values of Γ_1, Γ_2, and Γ_3, they cannot cause any variation in $d\sigma$. Therefore, the expression for Γ_2^G given by Eq. [G.30] is invariant regardless of what convention is adopted. The invariant quantity given by Eq. [G.30] is essentially the same as the quantity defined by Gibbs, denoted by him by the symbol $\Gamma_{2(1,3)}$ (Eq. [G.35]). When both liquid phases contribute the same solute to the interface, the quantity adsorbed is the sum of the two quantities from each immiscible phase.

A clearer understanding of the meaning of the U-convention (or any other convention for defining Γ_i) can be gained by using small whole numbers to designate relative concentrations in the bulk and in the interfacial phases. Suppose the oil and aqueous phases are in equilibrium with the solute distributed in each phase, having a mole fraction of $\frac{1}{5}$ and $\frac{1}{7}$, respectively. Let the solute be positively adsorbed at the interface from each bulk phase according to the numbers in the diagram (Fig. G.2). Applying Eq. [G.30] gives

$$\Gamma_2^G = 51 - \tfrac{1}{4}(84) - \tfrac{1}{6}(102) = 13$$

The variation of concentration of component 2 across the interface for the numerical example shown in Fig. G.2 is represented in Fig. G.3, showing the

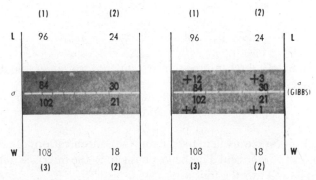

Figure G.2 A numerical example of the variation of concentration of component 2 across an interface.[143]

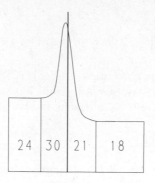

Figure G.3 The variation of concentration of component 2 in the σ phase and in the homogeneous L and W phases.

heterogeneity of composition in the σ phase and the homogeneity of the L and W phases.

An equivalent expression for Eq. [G.30] can be obtained by multiplying Eq. [G.25] by a multiplier x and [G.26] by a multiplier y and subtracting from Eq. [G.11], giving

$$-d\sigma = (\Gamma_1 - x\Gamma_1^L)\,d\mu_1 + (\Gamma_2 - x\Gamma_2^L - y\Gamma_2^W)\,d\mu_2 + (\Gamma_3 - y\Gamma_3^W)\,d\mu_3$$
$$= \sum \Gamma_i(x, y)\,d\mu_i \qquad\qquad\qquad [G.31]$$

where $\Gamma_i(x, y)$ is a surface concentration dependent on the values selected for x and y. Equation [G.31] is valid for any numerical values assigned to x and y; in particular, it is convenient to choose the multipliers in such a way as to make vanish two of the terms in Eq. [G.31]. Let $\Gamma_{1(2,3)}$ represent $\Gamma_1(x, y)$ when $\Gamma_2(x, y)$ and $\Gamma_3(x, y)$ are made to equal zero by an appropriate choice of x and y (say, x_1, y_1); similarly, let $\Gamma_{2(1,3)}$ represent $\Gamma_2(x, y)$ when $\Gamma_1(x, y)$ and $\Gamma_3(x, y)$ are made to equal zero by a second appropriate choice (say, x_2, y_2); and let $\Gamma_{3(1,2)}$ represent $\Gamma_3(x, y)$ when $\Gamma_1(x, y)$ and $\Gamma_2(x, y)$ are made to equal zero by a third appropriate choice of x and y (say, x_3, y_3). Then by Eq. [G.31],

$$-d\sigma = \Gamma_{1(2,3)}d\mu_1 = \Gamma_{2(1,3)}d\mu_2 = \Gamma_{3(1,2)}d\mu_3 \qquad\qquad [G.32]$$

For our example of a three-component system, the terms in μ_1 and μ_3 are selected to vanish, that is,

$$(\Gamma_1 - x\Gamma_1^L) = 0 \quad \text{and} \quad (\Gamma_3 - y\Gamma_3^W) = 0 \qquad\qquad [G.33]$$

These operations are equivalent to bringing the concentrations of component 1 in the L phase and component 3 in the W phase up to the interface unchanged; and so bring Eq. [G.31] to the following:

$$-d\sigma = (\Gamma_2 - x\Gamma_2^L - y\Gamma_2^W)d\mu_2 = \Gamma_{2(1,3)}d\mu_2 \qquad\qquad [G.34]$$

Comparing Eqs. [G.29] and [G.34] yields

$$\Gamma_2^G = \Gamma_{2(1,3)} \qquad\qquad [G.35]$$

Thus the two derivations (of Eq. [G.30] and [G.34]) are equivalent. In Eq. [G.34], $\Gamma_{2(1,3)}$ is expressed as the surface concentration of component 2 reduced by two terms that are functions of how much of component 2 is present in the lipoid and the water phase; that is the reason for describing $\Gamma_{2(1,3)}$ as an *excess* concentration.

A numerical representation of $\Gamma_{2(1,3)}$ is shown in Fig. G.2, with the planes of separation drawn so as to make the surface excess of solvent in each phase equal zero. To do so 12 units of solvent 1 is added to the σ phase to bring its total up to 96 and also 6 units of solvent 3 is added to the σ phase to bring its total up to 108. These additions are required to meet conditions expressed by Eq. [G.30]. The number of solute molecules in the σ phase has, meanwhile, been increased proportionately by 3 units from the L phase and 1 unit from the W phase. The total excess surface concentration of solute is $33 - 24 = 9$ units with respect to the L phase plus $22 - 18 = 4$ units with respect to the W phase, or 13 units in all. This is the same answer already obtained by means of the U-convention.

The numerical values quoted in Fig. G.2 correspond to a weakly surface-active solute, such as a lower alcohol. Soaps and detergents are much more strongly adsorbed, so that the terms other than Γ_2 in Eq. [G.29] may be neglected. The Gibbs excess concentration Γ_2^G, may then be taken as equal to the actual surface concentration without significant error. The Gibbs concentration Γ_2^G, can be evaluated from a plot of the surface tension against the logarithm of the *concentration* (for a dilute solution) or the logarithm of the *activity* for a more concentrated solution. The reciprocal of Γ_2^G is the area per mole of component 2 in the surface, that is, $A = 1/\Gamma_2^G$, where A is the area per mole.

2. THE SURFACE TENSION ISOTHERM

At sufficiently low concentrations of a surface-active solute, the surface tension isotherm is linear at constant temperature:

$$\sigma_0 - \sigma = mX_2 \qquad\qquad [G.36]$$

where m is the slope of the isotherm and X_2 is the mole fraction of solute. The lowering of the surface tension of solvent by a solute, $\sigma_0 - \sigma$, is called the spreading pressure and is designated by π, that is,

$$\pi = \sigma_0 - \sigma \qquad\qquad [G.37]$$

Differentiating Eq. [G.37] gives

$$d\pi = - d\sigma$$

Substituting Eq. [G.37] into Eq. [G.36], taking logarithms, and differentiating, gives

$$d \ln \pi = d \ln X_2 \qquad \text{[G.38]}$$

Substituting Eqs. [G.37] and [G.38] into Eq. [G.20] gives

$$d\pi = RT\Gamma_2^G \, d \ln \pi$$

or

$$\pi = \Gamma_2^G RT \qquad \text{[G.39]}$$

or

$$\pi A = RT \qquad \text{[G.40]}$$

Equation [G.40] is the ideal equation of state of an adsorbed film; it appears therefore that a linear surface tension isotherm signifies an ideal solution in the interface.

The linear relation between surface tension and concentration, which is the behavior by which we recognize an ideal interfacial solution, is observed only with dilute solutions. With solutes of pronounced surface activity, such as soaps or other detergents, the dilution required to get into the range of ideality of the interfacial solution is so extreme that such minute concentrations are rarely the subject of measurements. Solutes of low or moderate surface activity do, however, produce ideal solutions in the interfacial phase at equilibrium bulk concentrations that are several orders of magnitude greater.

At concentrations greater than the range of application of Eq. [G.36], the surface tension isotherm is no longer linear. An empirical equation due to von Szyszkowski[135] is often used:

$$\pi = \sigma_0 - \sigma = RT\Gamma_m \ln(X_2/\mathbf{a} + 1) \qquad \text{[G.41]}$$

where Γ_m = moles of component 2 per unit area at saturation in the interface
$\mathbf{a}$ = an empirical constant

3.   ADSORPTION ISOTHERM: LANGMUIR

The Szyszkowski equation, when combined with Eq. [G.20], gives the adsorption isotherm equation, as follows. Differentiating Eq. [G.41] gives

$$-\frac{d\sigma}{dX_2} = \frac{\Gamma_m RT}{X_2 + \mathbf{a}} \qquad \text{[G.42]}$$

Rewriting the Gibbs equation [G.20] as

$$-\frac{d\sigma}{dX_2} = \frac{RT\Gamma_2^G}{X_2} \qquad \text{[G.43]}$$

Equating Eqs. [G.42] and [G.43], and rearranging, gives

$$\Gamma_2^G = \frac{\Gamma_m X_2}{a + X_2} \quad \text{or} \quad \theta = \frac{X_2}{a + X_2} \tag{G.44}$$

Equation [G.44] is the adsorption isotherm equation of Langmuir.

4. EQUATION OF STATE: FRUMKIN

By eliminating the concentration of solute X_2 between Eqs. [G.41] and [G.44], the equation of state of Frumkin is obtained[136]:

$$\sigma_0 - \sigma = -RT\Gamma_m \ln(1 - \Gamma_2^G/\Gamma_m) \tag{G.45}$$

The derivations given demonstrate that the surface tension isotherm of von Szyszkowski, the adsorption isotherm of Langmuir, and the equation of state of Frumkin are equivalent descriptions. Unfortunately, these equations do not apply exactly to most experimental systems although they fit data well in the region of practical interest, that of close-to-maximum surface tension lowering for nearly surface-saturated solutes in solution. The danger lies in extrapolating the fit found in the region of large surface tension lowering to the

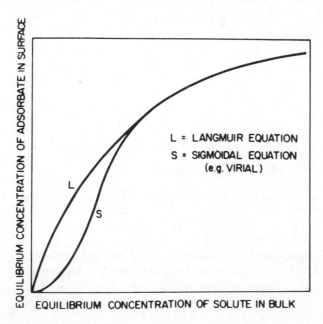

Figure G.4 The extrapolation to low concentrations of the description of data at high concentrations by a Langmuir equation.[138]

region of small surface tension lowering at very low concentrations of solute. The region of low concentration is the region of ideal behavior. When data at these low concentrations are considered, the surface tension isotherm often has a sigmoidal shape, which these equations entirely overlook. Figure G.4 shows the nature of the error that may be introduced by assuming that the Langmuir equation, or its congeners, when extrapolated to zero concentration, gives a good description of behavior in situations where a sigmoidal representation would be a more accurate description.

5.   STANDARD-STATE FREE ENERGIES

The constants found for the fit to experimental data of these congeneric equations at higher concentrations may not be extrapolated to calculate changes in standard-state thermodynamic functions where the standard states are defined by the Henry's law slope. The limiting initial slope of the surface tension isotherm, $-d\sigma/dX_2$, if determined by measurements in the Henry's law region, may have the form $RT\Gamma_m/a$; but if determined by extrapolation from data at higher concentrations fitted to Eq. [G.41] equals $RT\Gamma_m/\mathbf{a}$ where $\mathbf{a}$ may or may not equal a.

The Henry's law region is the linear portion of the adsorption isotherm, corresponding to the linear surface tension isotherm, Eq. [G.36], and refers to an ideal surface solution described by $\pi A = RT$. If the surface solution is ideal, the concentration of the bulk solution in equilibrium with it is also low enough to be ideal. When both solutions are ideal their activity coefficients are unity, and the ratio of their concentrations is equal to the ratio of their activities. Let K be the equilibrium constant for the process of adsorption; therefore,

$$K = \frac{\gamma_2^\sigma \theta}{\gamma_2^\alpha X_2} = \frac{\theta}{X_2} = \frac{1}{a} \quad \text{(by Eq. [G.44])} \qquad [G.46]$$

where

$$\theta = \Gamma_2^G/\Gamma_m \qquad [G.47]$$

γ_2^σ = activity coefficient of the solute in the interfacial or σ phase

γ_2^α = activity coefficient of the solute in the bulk or α phase

From the expression for the standard change of the Gibbs function on adsorption

$$\Delta G^\circ = -RT \ln K = RT \ln a \qquad [G.48]$$

An alternative and more convenient standard state can be chosen. The slopes of Eqs. [G.41], [G.44], and [G.45] applied to data at high concentrations, when extrapolated to infinite dilution, may be substituted into the equation

$$\Delta G^\circ = RT \ln \mathbf{a} \qquad [G.49]$$

This purely mathematical standard is not the real behavior of the solute but rather an imaginary Henry's law based on its behavior near saturation. This standard state of the solute (pure solute) still derives from a Henry's law slope but on an extrapolated Henry's law slope based on **a** rather than the actual Henry's law slope based on a. The practical consequence of using this new standard state is that the new standard change of the Gibbs function, given by Eq. [G.49], can be calculated directly from the constants of the best fit of the Szyszkowski or Langmuir equations to experimental data.

The usual forms of the Szyszkowski and Langmuir equations are

$$\Delta = \frac{\sigma_0 - \sigma}{\sigma_0} = \mathbf{b} \ln \frac{c_2}{\mathbf{c} + 1} \qquad \text{[G.50]}$$

$$\Gamma_2 = \Gamma_m c_2 / (c_2 + \mathbf{c}) \qquad \text{[G.51]}$$

where c_2 is the molality and **b** and **c** are constants. The empirical constants have the following significance:

$$\mathbf{b} = \frac{RT\Gamma_m}{\sigma_0} \quad \text{and} \quad \mathbf{c} = \frac{1000\mathbf{a}}{18(1 - \mathbf{a})} \qquad \text{[G.52]}$$

from which Γ_m and **a** can be evaluated from data. The standard free energy of adsorption is then calculated by means of Eq. [G.49].

The meaning of this new standard state is that the solute is considered to behave at infinite dilution as it would in the nearly saturated surface layer. This is thermodynamically equivalent to the standard state suggested by Rosen and Aronson[137] and Ross and Morrison,[138] who selected an imaginary condition based on the extrapolation of the Szyszkowski–Langmuir equations, valid at high concentrations, to establish an apparent Henry's law constant. It is evaluated by the extrapolation, $\pi \rightarrow 0$, of the portion of the surface tension isotherm in which $\ln X_2$ is linear with π. At the high concentrations where this linearity is found, Eq. [G.41], the Szyszkowski equation, reduces to

$$\pi = RT\Gamma_m \ln (X_2/\mathbf{a}) \qquad \text{[G.53]}$$

so that the intercept $X_2 = \mathbf{a}$ occurs at $\pi = 0$. This method provides a ready manner to evaluate **a** without having to determine the Szyszkowski constants for the surface tension isotherm.

The surface tension isotherms for an homologous series of carboxylic acids were described by the Szyszkowski equation by Freundlich (1926). From the values of these constants, Γ_m and ΔG^0 were calculated by means of Eqs. [G.41] and [G.49] (Table G.1). The values of ΔG° of fatty acids in aqueous solutions bear a continuous linear relation to the number of carbon atoms in the chain, as would be expected from Traube's empirical rule[139,140] as well as from the theoretical concept that the free energy of adsorption is due to a decrease in entropy of the aqueous medium caused by the presence of

Table G.1   Values of the Constants, Γ_m and $-\Delta G^0$, for an Homologous Series of Carboxylic Acids from Surface Tension Isotherms

Acid	$T(°C)$	$RT\Gamma_m/\sigma_0$	c (mol/L)	$1/\Gamma_m$ (Å²/molecule)	$-\Delta G°$ (kJ/mol)
Formic	15	0.1252	1.38	43.3	8.85
Acetic	15	0.1252	0.352	43.3	12.12
Propionic	15	0.1319	0.112	41.1	14.86
Butyric	18.5	0.1792	0.052	30.2	16.95
Valeric	17.5	0.1792	0.0146	30.2	19.91
Hexanoic	19	0.1792	0.0043	30.2	22.98
Heptanoic	18	0.2575	0.0018	21.0	25.01
Octanoic	18	0.3489	0.00045	15.5	28.36
Nonanoic	18	0.2389	0.00014	22.7	31.19

From ref. 138.

hydrocarbon in the bulk phase (Tanford, 1980). The slope of the straight line (Fig. G.5) gives the change of the standard Gibbs free energy of adsorption per carbon atom as -2.72 kJ/mol (-649 cal/mol), which is close to the values for the standard free energy of adsorption per carbon atom for homologous series of organic alcohols and ethers. For complete transfer of a hydrocarbon chain from water to a hydrocarbon medium, the $\Delta G°$ per carbon atom also remains constant with a value of -3.70 kJ/mol (-884 ± 13 cal/mol), for chains from 3 to 8 carbons. Comparison of these two values suggests that the adsorbed state of the solute at the surface of an aqueous solution is not wholly a hydrocarbon environment. More significant, perhaps, is the further deduction that the hydrophobic character of a chain is hardly likely to vary linearly from 3 to 8

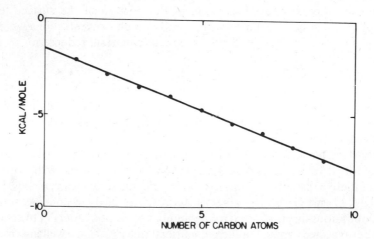

Figure G.5 The variation of $-\Delta G°$ for an homologous series of fatty acids as a function of the number of carbon atoms.[138]

carbons if the chain is curled up in water. Mukherjee[141] suggests that coiling of a hydrocarbon chain in water barely commences at 16 carbon atoms and that shorter chains are fully extended.

6. TRAUBE'S RULE

By comparing curves of σ versus $\log c_2$ for members of an homologous series, Traube[139,140] saw that the surface activity of each member of the series increased regularly with the number of carbon atoms in its molecule. In the series of normal fatty acids in aqueous solution, the surface activity, as measured by the concentration required to reduce the surface tension of water by a constant amount, say 20 mN/m, approximately triples for each additional $-CH_2-$ in the molecule: that is, the surface activity increases geometrically as the number of carbon atoms in the solute molecule increases arithmetically. A similar relation obtains between the geometrically decreasing solubility of a hydrocarbon in water and the arithmetically increasing number of carbon atoms in the chain.

The lowering of surface tension produced by an homologous series of normal aliphatic alcohols, from 6 to 10 carbon atoms, in aqueous solution at 20° C are reported in Fig. G.6 as a function of the logarithm of the concentration. The slopes of these curves at the lower end are almost linear and are the same for each solute, showing, by Eq. [G.20], that the same number of molecules of each is adsorbed per unit area of surface. This result implies that hydrocarbon chains in the adsorbed film are normal to the surface. The quasi-linearity shows, again by

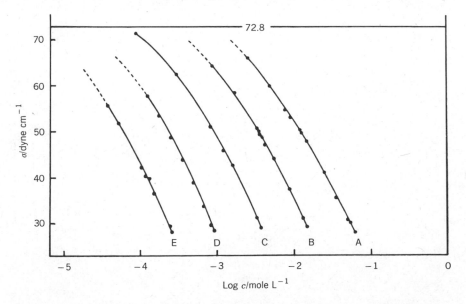

Figure G.6 The lowering of surface tension produced by an homologous series of normal aliphatic alcohols as a function of the number of carbon atoms (Defay et al., 1966, p 95).

the same equation, that Γ_2^G barely increases as the solution concentration is increased, from which we deduce that the adsorbed molecules are quite closely packed together. The area per molecule, calculated by Eq. [G.12], lies between 0.27 and 0.28 nm^2 for each alcohol. This molecular area is 25–30% larger than the limiting area per molecule of a hydrocarbon chain that is measured with insoluble monolayers of long-chain alcohols and acids on a water substrate, which implies that involuntary adsorption of molecules out of solution does not create as great a limiting compression in the soluble monolayer as can be produced by mechanically pushing together the molecules in an insoluble monolayer.

Traube's rule follows from Eq. [G.53] if Γ_m is independent of the chain length of the members of a homologous series. If Γ_m is such a constant, the same ratio of $X_2/\mathbf{a}$ for each member gives the same lowering of the surface tension. For an homologous series the values of $\mathbf{a}$ are found to decrease in a geometric progression (Table G.1), and therefore the values of X_2 do so as well. These conditions appear to hold.

References for Part II

1. Drost-Hansen, W. Aqueous interfaces: Method of study and structural properties, *Ind. Eng. Chem.* **1965**, *57*(3), 38–44; also *Chemistry and Physics of Interfaces*, Ross, S., Ed.; American Chemical Society: Washington, D.C., 1965; pp. 13–20.

2. Brown, R. C. The fundamental concepts concerning surface tension and capillarity, *Proc. Phys. Soc., London* **1947**, *59*, 429–448.

3. Girifalco, L. A.; Good, R. J. A theory for the estimation of surface and interfacial energies. I. Derivation and application to interfacial tension, *J. Phys. Chem.* **1957**, *61*, 904–909.

4. Berthelot, D. Sur le mélange des gaz, *C. R. Hebd. Seances Acad. Sci.* **1898**, *126*, 1703–1706.

5. Berthelot, D. Sur le mélange des gaz, *C. R. Hebd. Seances Acad. Sci.* **1898**, *126*, 1857–1858.

6. Fowkes, F. M. Attractive forces at interfaces, *Ind. Eng. Chem.* **1964**, *56*(12), 40–52; also *Chemistry and Physics of Interfaces*; Ross, S., Ed.; American Chemical Society: Washington, D.C., 1965; pp. 1–12.

7. Fowkes, F. M. Surface effects of anisotropic London dispersion forces in *n*-alkanes, *J. Phys. Chem.* **1980**, *84*, 510–512.

8. Wu, S. Surface tension of solids: Generalization and reinterpretation of critical surface tension, *Polym. Sci. Technol.* **1980**, *12A*, 53–65.

9. Cratin, P. D.; Murray, Jr., J. M. A quantitative surface chemical characterization of pitch, *Tappi* **1970**, *53*, 1960–1963.

10. Drago, R. S.; Vogel, G. C.; Needham, T. E. A four-parameter equation for predicting enthalpies of adduct formation, *J. Am. Chem. Soc.* **1971**, *93*, 6014–6026.

11. Drago, R. S. Quantitative evaluation and prediction of donor-acceptor interactions, *Struct. Bonding (Berlin)* **1973**, *15*, 73–139.

12. Drago, R. S. The interpretation of reactivity in chemical and biological systems with the E and C model, *Coord. Chem. Rev.* **1980**, *33*, 251–277.

13. Fowkes, F. M.; Tischler, D. O.; Wolfe, J. A.; Lannigan, L. A.; Ademu-John, C. M.; Halliwell, M. J. Acid-base complexes of polymers, *J. Polym. Sci.* **1984**, *22*, 547–566.

14. Fowkes, F. M. Donor-acceptor interactions at interfaces, *Polym. Sci. Technol.* **1980**, *12A*, 43–52.

15. Derjaguin, B. V.; Toporov, Yu. P. "Role of the molecular and the electrostatic forces in the adhesion of polymers," in *Physicochemical Aspects of Polymer Surfaces*; Mittal, K. L., Ed.; Plenum: New York, 1983; Vol. 2, pp. 605–612.

16. Derjaguin, B. V.; Smilga, V. P. The present state of our knowledge about adhesion of polymers and semiconductors, *Int. Congr. Surf. Act., 3rd, 1960* **1961**, *2*, 349–367.

17. Young, T. An essay on the cohesion of fluids, *Phil. Trans. Roy. Soc. London* **1805**, 65–87.

18. Melrose, J. C. Evidence for solid-fluid interfacial tensions from contact angles, *Adv. Chem. Ser.* **1964**, *43*, 158–179.

19. Buff, F. P. "The theory of capillarity," in *Encyclopedia of Physics*; Flügge, S., Ed.; Springer-Verlag: Berlin, 1960; pp. 281–304.

20. A bibliography of contact angle use in surface science, *Technical Bulletin TB-100*; Ramé-Hart, Inc.: 43 Bloomfield Ave., Mountain Lakes, NJ 07046.

21. Zisman, W. A. Relation of equilibrium contact angle to liquid and solid composition, *Adv. Chem. Ser.* **1964**, *43*, 1–51.

22. Lee, L.-H. Theory of the effect of phase transition on liquid surface tension, *J. Colloid Interface Sci.* **1971**, *37*, 653–658.

23. Edser, E. "The concentration of minerals by flotation," in *Fourth Report on Colloid Chemistry and Its General and Industrial Applications*; HMSO: London, 1922; pp. 263–297.

24. Ross, S.; Patterson, R. E. Innate inhibition of foaming and related capillary effects in partially miscible ternary systems, *J. Phys. Chem.* **1979**, *83*, 2226–2232.

25. de Gennes, P. G. Wetting: statics and dynamics, *Rev. Mod. Phys.* **1985**, *57*, 827–863.

26. Fowkes, F. M.; McCarthy, D. C.; Mostafa, M. A. Contact angles and the equilibrium spreading pressures of liquids on hydrophobic solids, *J. Colloid Interface Sci.* **1980**, *78*, 200–206.

27. Hardy, W. B. The spreading of fluids on glass, *Phil. Mag.* **1919** (6), *38*, 49–55.

28. Bascom, W. D.; Cottington, R. L.; Singleterry, C. R. Dynamic surface phenomena in the spontaneous spreading of oils on solids, *Adv. Chem. Ser.* **1964**, *43*, 355–380.

29. Ghiradella, H.; Radigan, W.; Frisch, H. L. Electrical resistivity changes in spreading liquid films, *J. Colloid Interface Sci.* **1975**, *51*, 522–526.

30. Rotenberg, Y.; Boruvka, L.; Neumann, A. W. Determination of surface tension and contact angle from the shapes of axisymmetric fluid interfaces, *J. Colloid Interface Sci.* **1983**, *93*, 169–183.

31. Butler, J. N.; Bloom, B. H. A curve-fitting method for calculating interfacial tension from the shape of a sessile drop, *Surf. Sci.* **1966**, *4*, 1–17.

32. Fisher, L. R.; Israelachvili, J. N. Direct experimental verification of the Kelvin equation for capillary condensation, *Nature (London)* **1979**, *277*, 548–549.

33. Fisher, L. R. Forces due to capillary-condensed liquids: limits of calculations from thermodynamics, *Adv. Colloid Interface Sci.* **1982**, *16*, 117–125.

34. Boucher, E. A. Capillary phenomena: properties of systems with fluid/fluid interfaces, *Rep. Prog. Phys.* **1980**, *43*, 497–546.

35. Ostrach, S. "Motion induced by capillarity," in *Physicochemical Hydrodynamics*; Spalding, D. B., Ed.; Advance: London, 1977; Vol. 2, pp. 571–589.

36. Fowkes, F. M. Role of surface active agents in wetting, *J. Phys. Chem.* **1953**, *57*, 98–103.

37. Scriven, L. E.; Sternling, C. V. The Marangoni effects, *Nature (London)* **1960**, *187*, 186–188.

38. Harkins, W. D.; Brown, F. E. The determination of surface tension (free surface energy), and the weight of falling drops: The surface tension of water and benzene by the capillary-height method, *J. Am. Chem. Soc.* **1919**, *41*, 499–524.

39. Sugden, S. The determination of surface tension from the rise in capillary tubes, *J. Chem. Soc.* **1921**, *119*, 1483–1492.

40. Padday, J. F. Surface tension: Part 1. Theory of surface tension, Part 2. The measurement of surface tension, *Surf. Colloid Sci.* **1969**, *1*, 39–251.

41. Harkins, W. D.; Anderson, T. F. I. A simple accurate film balance of the vertical type for biological and chemical work, and a theoretical and experimental comparison with the horizontal type. II. Tight packing of a monolayer by ions, *J. Am. Chem. Soc.* **1937**, *59*, 2189–2197.

42. Gaines, Jr., G. L. On the use of filter paper Wilhelmy plates with insoluble monolayers, *J. Colloid Interface Sci.* **1977**, *62*, 191–192.

43. Harkins, W. D.; Jordan, H. F. A method for the determination of surface and interfacial tension from the maximum pull on a ring, *J. Am. Chem. Soc.* **1930**, *52*, 1751–1772.

44. Lunkenheimer, K.; Hartenstein, C.; Miller, R.; Wantke, K.-D. Investigations on the method of the radially oscillating bubble, *Colloids Surfaces* **1984**, *8*, 271–288.

45. Patterson, R. E.; Ross, S. The pendent-drop method to determine surface or interfacial tensions, *Surf. Sci.* **1979**, *81*, 451–463.

46. Ambwani, D. S.; Fort, Jr., T. Pendant drop technique for measuring liquid boundary tensions, *Surf. Colloid Sci.* **1979**, *11*, 93–119.

47. Padday, J. F. Tables of the profiles of axisymmetric menisci, *J. Electroanal. Chem. Interfacial Electrochem.* **1972**, *37*, 313–316.

48. Padday, J. F.; Pitt, A. R.; Pashley, R. M. Menisci at a free liquid surface: surface tension from the maximum pull on a rod, *J. Chem. Soc., Faraday Trans. 1* **1975**, *71*, 1919–1931.

49. Janule, V. P. Process analysis and control using surface tension measurement, Presented at 1983 Pittsburgh Conference, March 9, 1983.

50. Mysels, K. J. Improvements in the maximum-bubble-pressure method of measuring surface tension, *Langmuir* **1986**, *2*, 428–432.

51. Cayias, J. L.; Schechter, R. S.; Wade, W. H. The measurement of low interfacial tension *via* the spinning-drop technique, *A.C.S. Symp. Ser.* **1975**, *8*, 234–247.

52. Princen, H. M.; Zia, I. Y. Z.; Mason, S. G. Measurement of interfacial tension from the shape of a rotating drop, *J. Colloid Interface Sci.* **1967**, *23*, 99–107.

53. Addison, C. C. The properties of freshly formed surfaces. Part I. The application of the vibrating-jet technique to surface tension measurements of mobile liquids, *J. Chem. Soc.* **1943**, 535–541.

54. Burcik, E. J. The rate of surface tension lowering and its role in foaming, *J. Colloid Sci.* **1950**, *5*, 421–436.

55. Ross, S.; Haak, R. M. Inhibition of foaming. IX. Changes in the rate of attaining surface tension equilibrium in solutions of surface-active agents on addition of foam inhibitors and foam stabilizers, *J. Phys. Chem.* **1958**, *62*, 1260–1264.

56. Neumann, A. W.; Good, R. J. Techniques of measuring contact angles, *Surf. Colloid Sci.* **1979**, *11*, 31–91.

57. Johnson, Jr., R. E.; Dettre, R. H. Wettability and contact angles, *Surf. Colloid Sci.* **1969**, *2*, 85–153.

58. Bigelow, W. C.; Pickett, D. L.; Zisman, W. A. Oleophobic monolayers. I. Films adsorbed from solution in nonpolar liquids, *J. Colloid Sci.* **1946**, *1*, 513–538.

59. Bock, E. J. Correction of astigmatic optical systems consisting of an object inside a cylinder, *J. Colloid Interface Sci.* **1984**, *99*, 399–403.

60. Ross, S.; Kornbrekke, R. E. The wetting of the container wall as a critical-point phenomenon. I. Measurement of contact angles in cylindrical tubes: validation of a method, *J. Colloid Interface Sci.* **1984**, *98*, 223–228.

61. Bartell, F. E.; Whitney, C. E. Adhesion tension, *J. Phys. Chem.* **1932**, *36*, 3115–3126.

62. Bartell, F. E.; Walton, Jr., C. W. Alteration of the surface properties of stibnite as revealed by adhesion tension studies, *J. Phys. Chem.* **1934**, *38*, 503–511.

63. Bartell, F. E.; Osterhof, H. J. The measurement of adhesion tension solid against liquid, *Colloid Symp. Monogr., 5th Nat. Symp., 1927* **1928**, *5*, 113–134.

64. Lucassen-Reynders, E. H. Contact angle and adsorption on solids, *J. Phys. Chem.* **1963**, *67*, 969–972.

65. Washburn, E. W. The dynamics of capillary flow, *Phys. Rev.* **1921**(2), *17*, 273–283.

66. Cassie, A. B. D.; Baxter, S. Wettability of porous surfaces, *Trans. Faraday Soc.* **1944**, *40*, 546–551.

67. Wenzel, R. N. Resistance of solid surfaces to wetting by water, *Ind. Eng. Chem.* **1936**, *28*, 988–994.

68. Wenzel, R. N. Surface roughness and contact angle, *J. Phys. Colloid Chem.* **1949**, *53*, 1466–1467.

69. Johnson, Jr., R. E.; Dettre, R. H. Contact angle hysteresis. I. Study of an idealized rough surface, *Adv. Chem. Ser.* **1964**, *43*, 112–135.

70. Dettre, R. H.; Johnson, Jr., R. E. Contact angle hysteresis. II. Contact angle measurements on rough surfaces, *Adv. Chem. Ser.* **1964**, *43*, 136–144.

71. Jamin, J. C. Mémoire sur l'équilibre et le mouvement des liquides dans les corps poreux, *C. R. Hebd. Seances Acad. Sci.* **1860**, *50*, 172–176; 311–314; 385–389.

72. Jamin, J. C. On the equilibrium and motion of liquids in porous bodies, *Phil. Mag.* **1860**(4), *19*, 204–207.

73. Smith, W. O.; Crane, M. D. The Jamin effect in cylindrical tubes, *J. Am. Chem. Soc.* **1930**, *52*, 1345–1349.

74. Langmuir, I. The distribution and orientation of molecules, *Colloid Symp. Monogr., 3rd Nat. Symp., 1925* **1925**, *3*, 48–75.

75. Lundelius, E. F. Adsorption und löslichkeit, *Kolloid Z.* **1920**, *26*, 145–151.

76. Ross, S.; Nishioka, G. The relation of foam behavior to phase separations in polymer solutions, *Colloid Polym. Sci.* **1977**, *255*, 560–565.

77. Nishioka, G. M.; Lacy, L. L.; Facemire, B. R. The Gibbs surface excess in binary miscibility-gap systems, *J. Colloid Interface Sci.* **1981**, *80*, 197–207.

78. Defay, R.; Prigogine, I. Surface tension of regular solutions, *Trans. Faraday Soc.* **1950**, *46*, 199–204.

79. Layman, P. L. Detergent report, *Chem. Eng. News* **1984**, *62*(4), January 23, 1984, 17–49.

80. Dixon, J. K.; Judson, C. M.; Salley, D. J. Study of adsorption at a solution-air interface by radiotracers, *Monomol. Layers, Symp., 1951* **1954**, 63–106.

81. Pockels, A. Surface tension (A translation by Lord Rayleigh of a letter he received), *Nature (London)* **1891**, *43*, 437–439.

82. Cadenhead, D. A. Monomolecular films at the air-water interface: Some practical applications, *Ind. Eng. Chem.* **1969**, *61*(4), 22–28; also *Chemistry and Physics of Interfaces-II*; Ross, S., Ed.; American Chemical Society: Washington, D.C., 1971; pp. 27–34.

83. Ries, Jr., H. E.; Cook, H. D. Monomolecular films of mixtures I. Stearic acid with isostearic acid and with tri-*p*-cresyl phosphate. Comparison of components with octadecylphosphonic acid and with tri-*o*-xenyl phosphate, *J. Colloid Sci.* **1954**, *9*, 535–546.

84. Ries, Jr., H. E.; Kimball, W. A. Electron micrographs of monolayers of stearic acid, *Nature (London)* **1958**, *181*, 901.

85. Ries, Jr., H. E.; Kimball, W. A. Structure of fatty-acid monolayers and a mechanism for collapse, *Proc. Int. Congr. Surf. Act., 2nd, 1957* **1957**, *1*, 75–84.

86. Ries, Jr., H. E. Monomolecular films, *Sci. Am.* **1961**, *204*(3), 152–164.

87. Exerowa, D.; Lalchev, Z.; Marinov, B.; Ognyanov, K. Method for assessment of fetal lung maturity, *Langmuir* **1986**, *2*, 664–668.

88. Ries, Jr., H. E.; Swift, H. Monolayers of valinomycin and its equimolar mixtures with cholesterol and with stearic acid, *J. Colloid Interface Sci.* **1978**, *64*, 111–119.

89. Ries, Jr., H. E. Interaction of cholesterol, cerebronic acid, valinomycin, and related compounds in monolayers of binary mixtures, *Colloids Surfaces* **1984**, *10*, 283–300.

90. Cini, R.; Lombardini, P. P.; Hühnerfuss, H. Remote sensing of marine slicks utilizing their influence on wave spectra, *Int. J. Remote Sensing* **1983**, *4*, 101–110.

91. O'Brien, K. C.; Lando, J. B. Mechanical testing of monolayers 1. Fourier transform analysis, *Langmuir* **1985**, *1*, 301–305.

92. Langmuir, I. The mechanism of the surface phenomena of flotation, *Trans. Faraday Soc.* **1919**, *15*, Part III, 62–74.

93. Blodgett, K. B. Monomolecular films of fatty acids on glass, *J. Am. Chem. Soc.* **1934**, *56*, 495.

94. Blodgett, K. B. Films built by depositing successive monomolecular layers on a solid surface, *J. Am. Chem. Soc.* **1935**, *57*, 1007–1022.

95. Blodgett, K. B.; Langmuir, I. Built-up films of barium sterate and their optical properties, *Phys. Rev.* **1937**(2), *51*, 964–982.

96. Hühnerfuss, H.; Alpers, W. Molecular aspects of the system water/monomolecular surface film and the occurrence of a new anomalous dispersion regime at 1.43 GHz, *J. Phys. Chem.* **1983**, *87*, 5251–5258.

97. Drost-Hansen, W.; Thorhaug, A. Temperature effects in membrane phenomena, *Nature (London)* **1967**, *215*, 506–508.

98. Ferguson, J. The use of chemical potentials as indices of toxicity, *Proc. Roy. Soc. London* **1939**, *127B*, 387–404.

99. Gonick, E. Stokes' law and the limiting conductance of organic ions, *J. Phys. Chem.* **1946**, *50*, 291–300.

100. Zimm, B. H. The scattering of light and the radial distribution function of high-polymer solutions, *J. Chem. Phys.* **1948**, *16*, 1093–1099.

101. Jones, E. R.; Bury, C. R. The freezing-points of concentrated solutions. Part II. Solutions of formic, acetic, propionic, and butyric acids, *Phil. Mag.* **1927**(7), *4*, 841–848.

102. Stainsby, G.; Alexander, A. E. Studies of soap solutions. Part II. Factors influencing aggregation in soap solutions, *Trans. Faraday Soc.* **1950**, *46*, 587–597.

103. Mitchell, D. J.; Ninham, B. W. Micelles, vesicles and microemulsions, *J. Chem. Soc., Faraday Trans. 2* **1981**, *77*, 601–629.

104. Hartley, G. S. Interfacial activity of branched-paraffin-chain salts, *Trans. Faraday Soc.* **1941**, *37*, 130–133.

105. Dreger, E. E.; Keim, G. I.; Miles, G. D.; Shedlovsky, L.; Ross, J. Sodium alcohol sulfates—properties involving surface activity, *Ind. Eng. Chem.* **1944**, *36*, 610–617.

106. Little, R. C. The physical chemistry of non-ionic surface-active agents, Ph.D. Dissertation, Rensselaer Polytechnic Institute, 1960; University Microfilm Abstract, no. 60–2689.

107. Ross, S.; Hudson, J. B. Henry's law constants of butadiene in aqueous solutions of a cationic surfactant, *J. Colloid Sci.* **1957**, *12*, 523–525.

108. Luzzati, V.; Mustacchi, H.; Skoulios, A. The structure of the liquid-crystal phases of some soap + water systems, *Discuss. Faraday Soc.* **1958**, *25*, 43–50.

109. Lawson, K. D.; Flautt, T. J. Nuclear magnetic resonance studies of surfactant mesophases, *Mol. Cryst.* **1966**, *1*, 241–262.

110. Adamson, A. W. A model for micellar emulsions, *J. Colloid Interface Sci.* **1969**, *29*, 261–267.

111. Ugelstad, J.; El-Aasser, M. S.; Vanderhoff, J. W. Emulsion polymerization: Initiation of polymerization in monomer droplets, *J. Polym. Sci., Polym. Lett. Ed.* **1973**, *11*, 503–513.

112. Krafft, F.; Wiglow, H. Über das verhalten der fettsauren alkalien und der seifen in gegenwart von wasser III. Die seifen als kristallöide (On the behavior of soaps in the presence of water III. Soaps as crystalloids), *Ber. Dtsch. Chem. Ges.* **1895**, *28*, 2566–2573.

113. Stiepel, C. Die schaumzahl der seifen (The foaminess of soaps), *Seifensieder Ztg.* **1914**, *41*, 347.

114. Götte, E. Ein beitrag zur kenntnis der waschwirkung (A contribution to detergency), *Kolloid Z.* **1933**, *64*, 222–227.

115. Mysels, K. J. Surface tension of solutions of pure sodium dodecyl sulfate, *Langmuir* **1986**, *2*, 423–428.

116. Defay, R.; Pétré, G. Dynamic surface tension, *Surf. Colloid Sci.* **1971**, *3*, 27–81.

117. Ward, A. F. H.; Tordai, L. Time-dependence of boundary tensions of solutions I. The role of diffusion in time-effects, *J. Chem. Phys.* **1946**, *14*, 453–461.

118. Rusanov, A. I.; Krotov, V. V. Gibbs elasticity of liquid films, threads, and foams, *Prog. Surf. Membr. Sci.* **1979**, *13*, 415–524.

119. Małysa, K.; Lunkenheimer, K.; Miller, R.; Hartenstein, C. Surface elasticity and frothability of *n*-octanol and *n*-octanoic acid solutions, *Colloids Surfaces* **1981**, *3*, 329–338.

120. Cini, R.; Lombardini, P. P. Experimental evidence of a maximum in the frequency domain of the ratio of ripple attenuation in monolayered water to that in pure water, *J. Colloid Interface Sci.* **1981**, *81*, 125–131.

121. Lucassen-Reynders, E. H.; Lucassen, J. Properties of capillary waves, *Adv. Colloid Interface Sci.* **1969**, *2*, 347–395.

122. Rybczynski, W. Über die fortschreitende bewegung einer flüssigen kugel in einem zähen medium, *Bull. Int. Acad. Pol. Sci. Lett., Cl. Sci. Math. Nat., Ser. A*, **1911**, 40–46.

123. Hadamard, J. Mouvement permanent lent d'une sphère liquide et visqueuse dans un liquide visqueux, *C. R. Hebd. Seances Acad. Sci.* **1911**, *152*, 1735–1738.

124. Garner, F. H.; Hale, A. R. The effect of surface active agents in liquid extraction processes, *Chem. Eng. Sci.* **1953**, *2*, 157–163.

125. Garner, F. H.; Hammerton, D. Circulation inside gas bubbles, *Chem. Eng. Sci.* **1954**, *3*, 1–11.

126. Okazaki, S.; Hayashi, K.; Sasaki, T. Mechanism of antifoaming according to the classification of antifoamers, *Chem., Phys. Appl. Surf. Act. Subst., Proc. Int. Congr., 4th, 1964*, **1967**, *3*, 67–73.

127. Corkill, J. M.; Goodman, J. F. The interaction of non-ionic surface-active agents with water, *Adv. Colloid Interface Sci.* **1969**, *2*, 297–330.

128. Clunie, J. S.; Corkill, J. M.; Goodman, J. F.; Symons, P. C.; Tate, J. R. Thermodynamics of non-ionic surface-active agent + water systems, *Trans. Faraday Soc.* **1967**, *63*, 2839–2845.

129. Ekwall, P.; Mandell, L.; Fontell, K. Solubilization in micelles and mesophases and the transition from normal to reversed structures, *Mol. Cryst. Liq. Cryst.* **1969**, *8*, 157–213.

130. Friberg, S. Liquid crystals and foams, *Adv. Liq. Cryst.* **1978**, *3*, 149–165.

131. Fowkes, F. M. "The interactions of polar molecules, micelles, and polymers in nonaqueous media," in *Solvent Properties of Surfactant Solutions*; Shinoda, K., Ed.; Dekker: New York, 1967, pp. 65–115.

132. Hutchinson, E. Films at oil-water interfaces. I, *J. Colloid Sci.* **1948**, *3*, 219–234.

133. Hutchinson, E. Films at oil-water interfaces. II, *J. Colloid Sci.* **1948**, *3*, 235–250.

134. Hutchinson, E. Films at oil-water interfaces. III, *J. Colloid Sci.* **1948**, *3*, 531–537.

135. von Szyszkowski, B. Experimental studies of the capillary properties of aqueous solutions of fatty acids, *Z. Phys. Chem., Stoechiom. Verwandschaftsl.* **1908**, *64*, 385–414.

136. Frumkin, A. Die kapillarkurve der höheren fettsäuren und die zustandsgleichung der oberflächenschicht, *Z. Phys. Chem., Stoechiom. Verwandschaftsl.* **1925**, *116*, 466–484.

137. Rosen, M. J.; Aronson, S. Standard free energies of adsorption of surfactants at the aqueous solution/air interface from surface tension data in the vicinity of the critical micelle concentration, *Colloids Surfaces* **1981**, *3*, 201–208.

138. Ross, S.; Morrison, I. D. Thermodynamics of adsorbed solutes, *Colloids Surfaces* **1983**, *7*, 121–134.

139. Traube, J. Capillaritätserscheinungen in beziehung zur constitution und zum molekulargewicht, *Ber. Dtsch. Chem. Ges.* **1884**, *17*, 2294–2316.

140. Traube, J. Über die capillaritätsconstanten organischer stoffe in wässerigen lösungen, *Justus Liebig's Ann. Chem.* **1891**, *265*, 27–55.
141. Mukerjee, P. The nature of the association equilibria and hydrophobic bonding in aqueous solutions of association colloids, *Adv. Colloid Interface Sci.* **1967**, *1*, 241–275.
142. Bernett, M. K.; Zisman, W. A. Wetting properties of tetrafluoroethylene and hexafluoropropylene copolymers, *J. Phys. Chem.* **1960**, *64*, 1292–1294.
143. Ross, S.; Chen. E. S. Adsorption and thermodynamics at the liquid–liquid interface, *Ind. Eng. Chem.* **1965**, *57*(7), 40–52.
144. Suzin, Y.; Ross, S. Retardation of the ascent of gas bubbles by surface-active solutes in nonaqueous solutions, *J. Colloid Interface Sci.* **1985**, *103*, 578–585.

PART III
STABILITY OF DISPERSIONS

CHAPTER IIIA

Forces of Attraction

1. FORCES OF ATTRACTION BETWEEN MOLECULES

The flocculation of dispersions is a consequence of attractive forces holding particles together when they collide. If the particles repel each other sufficiently and if they are small enough, they bounce apart on collision and the dispersion is stable. The forces between dispersed particles can be altered drastically: The addition of small concentrations of stabilizing agents can produce large repulsive interactions; the addition of electrolyte to a stable dispersion can eliminate the repulsive interactions and cause the dispersion to coagulate rapidly.

Numerous products and processes depend on controlling the balance between attractive and repulsive interactions: products such as paints and inks, aerosols, jellies and gels, and rubber latexes; processes such as filtration and clarification, emulsification and demulsification, and paper making. Examples of reversals of interparticle forces are seen in the mining and processing of kaolin clay. The clay as it is mined cannot be dispersed in water no matter how vigorous the agitation. With the addition of a few tenths of a percent of a pyrophosphate salt, however, the clay is dispersed readily as a free-flowing suspension. The adsorption of the pyrophosphate anions overcomes the attractive forces between the clay platelets and makes them mutually repellent. The addition of an acid to this suspension immediately coagulates it, showing that the acid eliminates the repulsive forces between the platelets, which it does by combining with the adsorbed phosphate anions to form neutral molecules of phosphoric acid.

An outstanding contribution to colloid science was the development of a quantitative theory to explain the mechanisms of stability in many of these systems. This theory was independently formulated by B.V. Derjaguin and L.

Landau in the U.S.S.R. and E.J.W. Verwey and J.Th.G. Overbeek in the Netherlands and is now denoted by the acronym "DLVO" theory. The basic idea of the theory is that the stability of a dispersion is determined by the sum of attractive and repulsive forces between individual particles. The mutual attraction of particles is a consequence of dispersion forces, often called London–van der Waals forces, and the mutual repulsion of particles is a consequence of the interaction of the electrical double layers surrounding each particle. The DLVO theory, originally developed to describe charge-stabilized systems, has been extended to include steric-stabilized systems, that is, dispersions stabilized by the mutual "repulsion" of the overlap of adsorbed polymer layers.* A description of the DLVO theory includes (1) the magnitudes of the dispersion force interactions between particles, (2) the magnitudes of the repulsions due to the overlap of electrical double layers, and (3) the magnitudes of the forces due to the overlap of adsorbed polymer layers.

The sources of attractive and repulsive forces between particles are the attractive and repulsive forces between molecules. Intermolecular forces are all electromagnetic in origin: (a) electrostatic, that is, coulombic and induction forces; (b) electrodynamic, that is, dispersion forces; (c) electron or proton donor–acceptor interactions, for example, hydrogen bonding; and (d) the repulsive overlapping of electron clouds. A summary of the electrostatic and induction contributions to intermolecular forces is given in Appendix D. If either of the interacting molecules has a permanent electric moment, that is, a dipole moment, then the potential energy of interaction depends on their relative orientation. The average energy of interaction is found by integrating the potential as a function of orientation through all orientations multiplied by the Boltzmann probability of that orientation. The orientation-averaged forms are all temperature dependent, as at higher temperatures the rotations of the permanent electric moments reduce net intermolecular forces. The dipole–dipole contributions are often called Keesom energies. The dipole–induced dipole contributions are often called Debye energies.

Dispersion attraction forces are a consequence of the spontaneous fluctuation of the electronic cloud in one material causing a corresponding fluctuation in neighboring materials, leading, on the average, to an attractive force. That this particular attractive force is the predominant one in determining the stability of dispersed systems may be understood from the following brief summary of possible nonbonding intermolecular forces.

London, by using perturbation theory to solve the Schrödinger equation for two (hydrogen) atoms at large distances, showed that the separated atoms attract

---

*An unfortunate coincidence brings two unrelated meanings of the term "dispersion" into juxtaposition in this particular subject. In one usage a dispersion is a stable suspension of a finely divided phase in a phase in which it is immiscible. London's theory of intermolecular forces, based on the interaction of electric fields, makes use of "the dispersion relation" between wavelength and frequency. Because of this feature of his model, London referred to intermolecular forces as "dispersion forces." Another example of two unrelated meanings occurs with the term "phase,"which denotes both a physical state of matter and a property of a wave.

each other with an energy varying as the inverse sixth power of the distance r between them.[1] These forces are strictly quantum mechanical and are called "dispersion forces" because the equations for the attractive energy are expressed in terms of the same oscillator strengths as appear in the equations for the dispersion of light. If the fluctuations in the electric potential of a molecule can be approximated as a simple harmonic oscillator and the distance between the molecules is large compared to molecular size, then the dispersion energy between two such molecules is given by

$$U(r) = -\Lambda_{ab}r^{-6} \qquad\qquad \text{[A.1]}$$

where

$$\Lambda_{ab} = \frac{3}{2}\left\{\frac{h\nu_a h\nu_b}{h\nu_a + h\nu_b}\right\}\alpha_a\alpha_b \qquad\qquad \text{[A.2]}$$

and

ν_a = characteristic frequency of molecule a
ν_b = characteristic frequency of molecule b
α_a = limiting polarizability of molecule a
α_b = limiting polarizability of molecule b

The $h\nu$ terms in Eq. [A.2] are often approximated by setting them equal to the ionization potentials. Since the ionization potentials of most molecules are of the same order of magnitude, the London coefficient for the interaction of two different kinds of molecule, Eq. [A.2], can be approximated by the root mean square:

$$\frac{h\nu_a h\nu_b}{h\nu_a + h\nu_b} \simeq \left[\frac{h\nu_a}{2}\right]^{1/2}\left[\frac{h\nu_b}{2}\right]^{1/2} \qquad\qquad \text{[A.3]}$$

Substituting the approximation [A.3] into Eq. [A.2] gives

$$\Lambda_{ab} \simeq [\Lambda_{aa}\Lambda_{bb}]^{1/2} \qquad\qquad \text{[A.4]}$$

as a good estimate. This same approximation is often used for the estimation of mixed Lennard-Jones parameters and for the calculation of second-virial coefficients of gases (Hirschfelder et al. 1954), and is called Berthelot's principle.[2,3]

The ν's and α's can also be estimated from the frequency dependence of the polarizability near the ionization frequency.[7] Alternatively, the principal frequencies can be estimated from the analysis of the refractive index in terms of a Clausius–Mossotti plot.[8] This latter method is particularly suitable for materials whose ionization potentials are not available.

Several other approximations to calculate the London coefficient Λ_{ab} have been proposed. It can be calculated in principle from the appropriate wave equations.[4-6]

From the analyses of flocculation rates of lyophobic sols, Verwey and Overbeek (1948, p 266) found that when two particles are separated by distances large compared to the wavelength of the ionization potential; dispersion energies of attraction are considerably less than predicted. Verwey and Overbeek postulated that the energy of the dispersion interaction is decreased at large distances because the time of electric field propagation from one body to another and back is such that the fluctuating electric moments become slightly out of phase. Casimir and Polder[9] analyzed the dispersion interactions, including a term for the finite speed of light, and showed that a correction factor is necessary, called "retardation." The retardation correction is a monotonically decreasing function, equal to unity for small distances, and is proportional to the inverse of the separation r for long distances. The form of the retardation expression is complicated but an empirical representation is given by Overbeek.[10]

$$U^r(r) = f(p)\, U(r) \tag{A.5}$$

when
$$0 < p < 3 \quad \text{then } f(p) = 1.10 - 0.14p$$
$$3 < p \quad\quad \text{then } f(p) = 2.45/p - 2.04/p^2$$

where $p = 2\pi r/\lambda_0$
$\lambda_0 = c/v_0$
$v_0 =$ characteristic frequency of the interaction
$c =$ speed of light

Table A.1 lists some characteristic frequencies v_0 from Gregory.[8]

The strength of the dispersion force attraction can be compared with the other types of molecular attractions, using the formulas in Appendix D and the constants of the electrical moments of some simple molecules and ions from Appendix E. Potential energies are compared at the same intermolecular distance, 0.4 nm, and in units of kT (at 25°C) and are given in Table A.2. The

Table A.1   Characteristic Frequencies v_0 of Some Common Materials ($s^{-1} \times 10^{-15}$)

Gases		Liquids		Solids	
He	6.19	H_2O	3.35	Diamond	4.19
Ne	6.26	C_6H_6	2.54	SiO_2 (crystalline)	3.89
Ar	4.12	C_6H_{12}	3.39	SiO_2 (fused)	3.78
Kr	3.57	$n\text{-}C_5H_{12}$	3.39	Polystyrene	2.62
Xe	3.00	$n\text{-}C_6H_{14}$	3.39		
H_2	3.52	$n\text{-}C_7H_{16}$	3.39		
O_2	3.58	$n\text{-}C_8H_{18}$	3.40		
N_2	4.13	$n\text{-}C_9H_{20}$	3.40		
CH_4	3.42	$n\text{-}C_{10}H_{22}$	3.39		
H_2O	3.27				

Table A.2   Representative Intermolecular Interactions

Interaction	Species	Energy
Charge–charge	$Na^+–Cl^-$	$-140\,kT$
Charge–dipole	$Na^+–H_2O$	$-13.4\,kT^a$
Dipole–dipole	Phenol–Phenol	$-0.42\,kT$
Charge–induced dipole	$Na^+–H_2O$	$-1.62kT$
Dipole–induced dipole	Phenol–benzene	$-0.13kT$
Dispersion	Benzene–benzene	$-7.3kT$
Donor–acceptor	Phenol–benzene	$-4.3kT$

F. M. Fowkes, private communication.

[a] The maximum interaction equation is used since the interaction energy is much greater than kT.

donor–acceptor interaction is calculated from Eq. [II.A.17] and has no explicit distance dependence.

The results of these calculations show that the polar interactions, dipole–dipole (Keesom forces) and dipole–induced dipole (Debye forces), are small compared to the others. If the molecules have no net charge, then only the dispersion energy determines the long-range intermolecular attractive potential. Short-range forces are dominated by donor–acceptor interactions.[11]

2.   FORCES OF ATTRACTION BETWEEN PARTICLES

The sources of the attraction between particles must be the same as those between molecules (possibly modified for multibody interactions). Unless the particle itself has a permanent dipole moment, the interactions of the various polar terms cancel. The interactions between charged particles are treated as a separate problem in a later section, since these terms are almost always repulsive. Hence the only significant long-range particle–particle attractive forces arise from the orientation-independent dispersion attractions.

The calculation of the magnitudes of dispersion attractions between particles has been attempted by two different approaches, one based on a molecular model, attributed to Hamaker, and one based on a molar model of condensed media, attributed to Lifshitz. Clerk Maxwell pointed out long ago precisely these two approaches to any scientific problem* (Maxwell, 1873, Volume 2, p 176). He

* Here is Francis Bacon, even earlier (1620), on the same theme: "For that school [Leucippus, Democritus] is so busied with the particles [i.e., molecules] that it hardly attends to the structure; while the others are so lost in admiration [i.e., wonder] of the structure that they do not penetrate to the simplicity of nature. These kinds of contemplation should therefore be alternated and taken by turns; that so the understanding be rendered at once penetrating and comprehensive." *Novum Organum*, Aphorism LVII.

wrote that mathematicians usually begin by considering a single molecule, or unit charge, and then conceive its relation to another molecule or unit charge. But the conception of an elemental unit is an abstraction since our perceptions are related to extended bodies. Is there not, he asked, an alternative and less abstract method in which we begin with the whole rather than building it up from the parts? He went on to illustrate the application of a holistic, or molar, approach to electromagnetism by means of field theory. Similarly, Lifshitz introduced an alternative, molar, approach to the attractive forces between massive particles.

In the molecular model the attraction between particles is calculated by summing the attractive energies between all pairs of molecules in the separate particles, ignoring multibody perturbations. This approximation is equivalent to predicting the spectra of condensed media as the sum of the molecular spectra. Corrections have been introduced to account for such factors as third-body perturbations, the effects of intervening material, and retardation of the dispersion forces (the speed of light is significant on the time scale of electron-cloud fluctuations.)

In the molar model of particle–particle dispersion attractions, the lowering of the zero-point energy of a particle, due to the coordination of its instantaneous electric moments with those of a nearby particle, is calculated by quantum electrodynamics. The expressions derived require as data the dielectric susceptibilities of the particles as a function of frequency and are more complex than those from the Hamaker theory. The complexity of Lifshitz formulas and the difficulty of obtaining the necessary material constants have hampered its use. Ninham and Parsegian, however, have developed a numerical method to approximate the necessary material functions from a few, readily obtained values, and have made the use of this theory practicable for some common materials.

a. Molecular Approach—Hamaker Theory

H. C. Hamaker in 1937 took London's expression for the dispersion attraction between two isolated molecules and integrated it for all the molecules in two separate particles to obtain the dispersion energy of attraction between two macroscopic bodies.[12] The work done, at constant temperature, by the attractive forces in bringing the two particles from infinity to a given separation distance is the Gibbs free energy due to dispersion interactions, ΔG_{12}^{d}:

$$\Delta G_{12}^{d} = -\int_{v_1} dv_1 \int_{v_2} \frac{\Lambda_{12} \rho_1 \rho_2}{r_{12}^{6}} dv_2 \qquad [\text{A.6}]$$

where dv_1 and dv_2 are differential volume elements of v_1 and v_2; and ρ_1 and ρ_2 are the molecular number densities (molecules per unit volume.) To simplify this integration, Hamaker assumed:

(a) The interactions can be considered pair wise, that is, multibody forces are ignored.

(b) The bodies are assumed to have uniform density right to the interfaces.
(c) The interactions of the molecular clouds are instantaneous.
(d) The intervening medium is a vacuum.
(e) All the dispersion force attractions are due to one dominant frequency.
(f) Effects of free charge and permanent dipoles are negligible.
(g) The bodies are not distorted by the attractive forces.

With these assumptions, the term $\Lambda_{12}\rho_1\rho_2$ is a constant and contains all the material dependence of the free energy. The integral separates into a material-dependent term and a geometric integral. The geometric integral has been evaluated for some simple geometries. (Mahanty and Ninham, 1976, pp. 10–22).

(1) The Hamaker Geometries

(a) Planar Parallel Slabs. Two planar parallel slabs of thicknesses b_1 and b_2 separated by a distance H:

$$\Delta G^d_{12} = -\frac{\pi\Lambda_{12}\rho_1\rho_2}{12}[H^{-2} - (H + b_1 + b_2)^{-2} - (H + b_1)^{-2} - (H + b_2)^{-2}]$$

$$[A.7]$$

where ΔG^d_{12} is *the free energy per unit area.* By convention the term $\pi^2\rho_1\rho_2\Lambda_{12}$ is called the Hamaker constant, A_{12}:

$$A_{12} = \pi^2\rho_1\rho_2\Lambda_{12} \qquad [A.8]$$

If b_1 and b_2 are large, then

$$\Delta G^d_{12} = -\frac{A_{12}}{12\pi H^2} \qquad [A.9]$$

The work required to divide a material into two separate free surfaces from an initial intermolecular distance of r_1 to an infinite separation, if the only attraction is the dispersion force, is

$$W^d_{11} = -\Delta G^d_{11} = \frac{A_{11}}{12\pi r_1^2} \qquad [A.10]$$

The work so characterized is the dispersion work of cohesion, equal to twice the surface tension. That is

$$W^d_{11} = 2\sigma^d \qquad [A.11]$$

where σ^d is the theoretical surface tension due to dispersion energies. Hence from Eq. [A.10] and [A.11]:

$$\sigma^d = \frac{A_{11}}{24\pi r_1^2} \qquad [A.12]$$

Equation [A.12] was originally derived by Fowkes[13] and used to calculate the dispersion force contributions to surface tension from Hamaker constants and vice versa. Similarly, if the only force of attraction between two materials is dispersion force, then the work to separate the two phases is given by

$$W^d_{12} = -\Delta G^d_{12} = \frac{A_{12}}{12\pi r^2_{12}} \qquad [A.13]$$

If the London coefficient Λ_{12} is approximated by the root mean square of the individual London coefficients. Λ_1 and Λ_2 (Berthelot's principle), and the intermolecular distance r_{12} is approximated by the root mean square of the individual intermolecular distances.

$$r_{12} = (r_1 r_2)^{1/2} \quad \text{and} \quad A_{12} = (A_{11} A_{22})^{1/2} \qquad [A.14]$$

then,

$$W^d_{12} = \frac{(A_{11} A_{22})^{1/2}}{12\pi r_1 r_2} \qquad [A.15]$$

and

$$W^d_{12} = 2(\sigma^d_1 \sigma^d_2)^{1/2} \qquad [A.16]$$

Equation [A.16] is a key element in Fowkes' treatment of dispersion force contributions to the work of adhesion and shows the rationale of the use of the root-mean-square approximation to calculate interfacial tensions (Chapter II.A.3).

(b) Spheres. Two spheres of radii a_1 and a_2, centers a distance R apart:

$$\Delta G^d_{12} = -\frac{A_{12}}{6}\left(\frac{2a_1 a_2}{R^2-(a_1+a_2)^2} + \frac{2a_1 a_2}{R^2-(a_1-a_2)^2} + \ln\frac{R^2-(a_1+a_2)^2}{R^2-(a_1-a_2)^2}\right) \quad [A.17]$$

where ΔG^d_{12} is *the total free energy (not* per unit area). If $a_1 = a_2$, that is, two equal spheres, then

$$\Delta G^d_{12} = -\frac{A_{12}}{6}\left[\frac{2a^2}{R^2-4a^2} + \frac{2a^2}{R^2} + \ln\left(1-\frac{4a^2}{R^2}\right)\right] \qquad [A.18]$$

Let

$$H = R - (a_1 + a_2) \qquad [A.19]$$

be the least distance between the surfaces of the spheres. When two spheres of any radius approach each other closely,

$$\Delta G^d_{12} = -\frac{A_{12} a_1 a_2}{6H(a_1 + a_2)} \qquad [A.20]$$

And if in addition, $a_1 = a_2$, that is, two equal spheres close together, then,

$$\Delta G_{12}^{d} = -\frac{A_{12}a}{12H} \tag{A.21}$$

If $R \gg a$, that is, two equal spheres far apart, then,

$$\Delta G_{12}^{d} = -\frac{16A_{12}a^6}{9H^6} \tag{A.22}$$

The importance of these equations is that the dispersion attraction between particles is shown to be significant for distances on the order of their own dimensions.

(c) A Sphere of radius a and a semi-infinite slab

$$\Delta G_{12}^{d} = -\frac{A_{12}}{6}\left[\frac{1}{x} + \frac{1}{2+x} + \ln\frac{x}{2+x}\right] \tag{A.23}$$

where ΔG_{12}^{d} is the total free energy change, and

$x = H/a$
$H = $ distance from the surface of the sphere to the plane

If $x \ll 1$, that is, a sphere very close to the plane, then

$$\Delta G_{12}^{d} = -\frac{A_{12}a}{6H} \tag{A.25}$$

which is twice the sphere–sphere interaction as given by Eq. [A.21].

(d) Other Geometries. These integrations can readily be extended to multilayer structures.[14] The interaction energies for most other geometries are orientation dependent and hence less general. Of notable interest, however, is the work of Vold,[15] in which she shows that the preferred orientation of prolate spheroids is end to end, a particularly useful result to explain the flocculation of certain lubricating greases (and possibly the interactions of micelles). She also derived the expression for the Hamaker constant for adsorbed layers on colloidal particles[16] and for the interaction of various anisometric particles.[17] Also useful is the analysis of the interaction of a cone and a flat plate (a model of surface roughness) by Sparnaay,[18] who concludes that the effects of irregularities of a flat surface begin to become significant when their linear dimensions are on the order of 10–20% of the interparticle distance.

(2) The Effects of Intervening Substances. The fundamental Hamaker expression, Eq. [A.6], does not provide for the effects of an intervening medium

since the derivation is for two bodies interacting across a vacuum. The usual approximation is to estimate the effect of an intervening substance, 2, between two bodies of composition 1 and 3, by the principle of Archimedean buoyancy, by which the Hamaker constant is assumed to be

$$A_{123} = A_{13} + A_{22} - A_{12} - A_{23} \qquad [A.26]$$

The constant A_{123} is then used to replace the constant A_{12} in Eqs. [A.7–A.25] when the substances 1 and 3 are separated by substance 2. For instance, the free energy of attraction between two equal spheres of material 1 separated by a thin film of material 2 is

$$\Delta G^d_{121} = -\frac{A_{121}a}{12H} \qquad [A.27]$$

where

$$A_{121} = A_{11} + A_{22} - 2A_{12} \qquad [A.28]$$

Using Eqs [A.2], [A.28], and some algebra, the Hamaker constant A_{121} can be shown to be always positive. If the material-dependent term is always positive, then all the free-energy expressions are negative: hence, particles composed of the same substance always attract each other in a two-component system regardless of the nature of the dispersion medium. This result is called the "deBoer–Hamaker" theorem.

If the root-mean-square approximation for London coefficients, Eq. [A.4], is used in the expression for the Hamaker constant Eq. [A.8], then

$$A_{12} = (A_{11}A_{22})^{1/2} \qquad [A.29]$$

Substituting Eq. [A.29] into Eq. [A.28] gives

$$A_{121} = (A_{11}^{1/2} - A_{22}^{1/2})^2 \qquad [A.30]$$

as a useful approximation. Analogously, from Eqs. [A.26] and [A.29]

$$A_{123} = (A_{11}^{1/2} - A_{22}^{1/2})(A_{33}^{1/2} - A_{22}^{1/2}) \qquad [A.31]$$

If the value of A_{22} is intermediate between the values of A_{11} and A_{33}, then the Hamaker constant is negative; hence, the free energy of interaction is positive and the particles of substances 1 and 3 repel each other through the medium 2. The occurrence of a negative A_{123} is usually accompanied by sensible physical phenomena, for example, a detachment or elution of particles, cells, or macromolecules from a substrate.[19]

(3) The Effect of Retardation. The retardation correction of the dispersion force interaction for two molecules[9] was integrated numerically by Overbeek to obtain

the interaction of two flat plates[10] and presented in the form of a table. In the limit of a fully retarded interaction between two flat plates,

$$U^r(r) = \frac{0.49c}{\pi v_0 H} U(r) \qquad [A.32]$$

where $U(r)$ is the unretarded energy, $U^r(r)$ is the retarded energy, c is the speed of light, v_0 is the natural frequency of interaction, and H is the distance of separation. Clayfield et al.[20] integrated the analytic expression of Casimir and Polder,[9] obtaining expressions that can be evaluated by a computer.

The retardation correction of the free energy of attraction for two equal spheres (Fig. A.1) is taken into account by introducing a correction factor f into Eq. [A.27][20]:

$$\Delta G^d_{121} = -\frac{A_{121} a f}{12H} \qquad [A.33]$$

where a is the radius of the sphere, f is the retardation correction, H is the shortest distance between the spheres, and A_{121} is the Hamaker constant for spheres of material 1 interacting through a film of material 2. When dispersion forces operate over distances approaching a quarter of the characteristic wavelength λ_1 of material 1, the interactions get out of phase and the correction factor f must be used. The value of f is a function of a/λ_1 and H/λ_1 and is readily obtained from a diagram given by Fowkes and Pugh,[21] which was calculated from the exact analytic expression derived by Clayfield et al.[20]

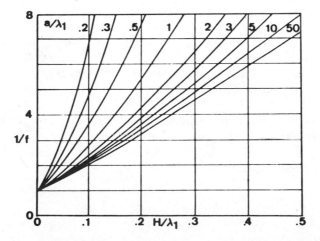

Figure A.1 The retardation correction factor f for dispersion force attractions between equal spherical particles of radius a at separation distance H with dispersion force wavelength λ_1, where the wavelength is the speed of light divided by the natural frequency.[21]

(4) Integration versus Lattice Summation. The Hamaker expressions are derived assuming that the molecular dimensions are much less than the distance between the particles. When the surfaces are close together, this assumption is not valid, and the integrations in the derivations for the Hamaker geometries must be replaced by summations over lattice positions. The usual procedure is to sum over the atoms near the surface and integrate over the more distant atoms. For liquid interfaces this approach has been productive. For solid surfaces, however, the effects of surface roughness and surface heterogeneity are more significant than the finite size of molecules, and calculations of the interactions are less satisfactory.

b. Molar Approach—Lifshitz Theory

The Hamaker approach starts from molecular-pair interactions and works up to the attraction between massive particles, making various simplifying assumptions on the way. An alternative approach is to start with the electrical properties of a massive particle and derive attractive potentials directly. The frequency dependence of the dielectric constant is related to the fluctuation of the electronic clouds of the whole particle. Starting from that premise Lifshitz and co-workers[22,23] calculated the dispersion attractions between particles through a medium, using Feynman diagrams and quantum electrodynamics. The details of that derivation are beyond the scope of this book, but the final formulas for the free energies of interaction for a few important configurations are worth quoting. By means of these equations, the attraction between particles is calculated from the dielectric responses of the materials at frequencies close to the peaks of their absorption spectra. When the time required for light to travel from one particle to another and back is significant on the time scale of these electronic fluctuations, then the attractive forces are diminished and the attractive force is "retarded." In the Hamaker model the free energy of interaction separates into a material-dependent constant (called the Hamaker constant) and a geometry-dependent integral. Lifshitz theory, on the other hand, gives such a separation of terms only for the special case of the interaction between two half-spaces (parallel plates) when retardation is not significant.

(1) Two Half-spaces, Nonretarded Interaction. The free energy of interaction per unit area for two half-spaces with dielectric responses $\varepsilon_1(\omega)$ and $\varepsilon_3(\omega)$ separated by a material of dielectric response $\varepsilon_2(\omega)$ of thickness H (for the nonretarded approximation) is

$$\Delta G_{123}^{nr}(H) = -\frac{A_{123}^{nr}}{12\pi H^2} \qquad [A.34]$$

where

$$A_{123}^{nr} = -\frac{3kT}{2} \sum_{n=0}^{\infty}{}' \int_0^{\infty} x\, dx \ln[1 - \Delta_{12}\Delta_{32}\exp(-x)] \qquad [A.35]$$

$$\Delta_{qr} = \frac{\varepsilon_q(i\xi_n) - \varepsilon_r(i\xi_n)}{\varepsilon_q(i\xi_n) + \varepsilon_r(i\xi_n)} \qquad \text{[A.36]}$$

$$\xi_n = n\frac{4\pi^2 kT}{h} \qquad \text{[A.37]}$$

where k is the Boltzmann constant, T is the absolute temperature, h is Planck's constant, and the prime on the summation indicates that the $n = 0$ term is given half weight. Since the ratios of dielectric constants are used in Eq. [A.36], it does not matter whether absolute values are used or values relative to the permittivity of free space. The integral in Eq. [A.35] can also be expressed as the infinite sum:

$$\int_0^\infty x\,dx \ln[1 - \Delta_{12}\Delta_{32}\exp(-x)] = -\sum_{n=1}^\infty \frac{(\Delta_{12}\Delta_{32})^n}{n^3} \qquad \text{[A.38]}$$

For small dielectric differences, that is, $\Delta_{12}\Delta_{32} \to 0$, the summation in Eq. [A.38] is often approximated by taking only the first term, giving

$$A_{123}^{nr} \simeq \frac{3kT}{2}\sum_{n=0}^\infty{}' \Delta_{12}\Delta_{32} \qquad \text{[A.39]}$$

The triple subscript on the Lifshitz constant A_{123}^{nr} shows the explicit dependence of the free energy of interaction on the intervening medium. This is fundamentally different from the molecular approach in which the effects of the intervening medium were introduced as a crude approximation (Eq. [A.26]).

For two particles of the same material dispersed in a second material, the Lifshitz constant A_{121}^{nr} cannot be negative since the term $\Delta_{12}\Delta_{12}$ is always nonnegative. When the Lifshitz constant is positive, the free energy of interaction decreases as the particles approach each other; and they flocculate spontaneously, unless other forces are important.

When the dielectric constant of the medium, 2, is intermediate between the dielectric constants of substances 1 and 3 at all frequencies, the integrand in Eq. [A.35] is everywhere positive, the Lifshitz constant, A_{123}^{nr}, is negative, and the particles repel each other.

The values needed in Eq. [A.36] are the dielectric responses as a function of imaginary frequencies! These values are not accessible experimentally but they are related to the absorption spectrum, $\varepsilon''(\omega)$, by the Kramers–Kronig relation,

$$\varepsilon(i\xi) = 1 + \frac{2}{\pi}\int_0^\infty \frac{\omega\varepsilon''(\omega)d\omega}{(\omega^2 + \xi^2)} \qquad \text{[A.40]}$$

The function $\varepsilon(i\xi)$ is always real; it is monotone decreasing from the value of the static dielectric constant to a value of one at high frequencies (Landau and

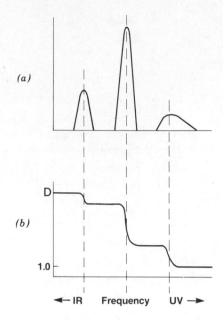

Figure A.2 The relation between (a) the absorption spectrum, $\varepsilon''(\omega)$, as a function of frequency ω and (b) the dielectric spectrum, $\varepsilon(i\xi)$. D is the static dielectric constant.

Lifshitz, 1960). The two functions, $\varepsilon''(\omega)$, the absorption spectrum, and $\varepsilon(i\xi)$ are related as shown in Fig. A.2. The quantity ξ_n is a frequency. At 21°C, ξ_1 is 2.4 $\times 10^{14}$ rad/s, a frequency corresponding to a wavelength of light of about 1.2 μm. As n increases, the value of ξ_n increases and the corresponding wavelength decreases, hence ξ_n takes on more values in the ultraviolet than in the infrared or microwave.

Until recently it was believed that the Lifshitz theory was merely an elegant formalism and the function $\varepsilon(i\xi)$ could not be determined readily. A fundamental achievement of Ninham and Parsegian was to show how to construct the function, $\varepsilon(i\xi)$, from available experimental data.

(2) Method of Ninham and Parsegian. The method of Ninham and Parsegian[14,24] is to approximate the absorption spectrum with a zero-width, infinite-height infrared peak to characterize the low-frequency absorption spectra and a zero-width, infinite-height ultraviolet (or visible) peak to characterize the high-frequency absorption spectra. The frequency dependence of the dielectric susceptibility, $\varepsilon(i\xi)$, can be expressed, therefore, in the terms of Lorentz harmonic oscillators for each absorption peak (infrared, visible or ultraviolet):

$$\varepsilon(i\xi_n) = 1 + \frac{C_{\text{ir}}}{1 + \xi_n^2/\omega_{\text{ir}}^2} + \frac{C_{\text{uv}}}{1 + \xi_n^2/\omega_{\text{uv}}^2} \qquad [\text{A.41}]$$

and

$$D = 1 + C_{ir} + C_{uv} \qquad\qquad \text{[A.42]}$$

where D is the static dielectric constant, ω_{ir} and ω_{uv} are the frequencies of absorption, and C_{ir} and C_{uv} are dielectric constants to be determined. If more than one absorption in the visible or ultraviolet is significant, additional terms can be added. If absorption in the microwave is significant (as for water), it is approximated by Debye rotational relaxation.

Table A.3   Values of the Optical Constants, $C_{ir}\omega_{ir}C_{uv}$, and ω_{uv}

Substance	C_{uv}	ω_{uv} ($\times 10^{16}$ rad/s)	C_{ir}	ω_{ir} ($\times 10^{14}$ rad/s)
Water	0.755	1.899	(See Table A.4)	
Crystalline quartz	1.359	2.032	1.93	2.093
Fused quartz	1.098	2.024	1.70	1.880
Fused silica	1.098	2.033	1.71	1.880
Calcite	1.516	1.897	5.7	2.691
Calcium fluoride	1.036	2.368	5.32	0.6279
Sapphire	2.071	2.017	8.5	1.880
Polymethyl-methacrylate	1.189	1.915	1.2	5.540
Polyvinyl-chloride	1.333	1.815	0.9	5.540
Polystyrene	1.424	1.432	0.2	5.540
	1.447	1.354		
Polyisoprene	1.255	1.565	0.16	5.540
Polytetrafluoro-ethylene	0.846	1.793	0.25	2.270
Normal alkanes				
(number of carbons) 5	0.819	1.877	0.025	5.540
6	0.864	1.873	0.026	5.540
7	0.898	1.870	0.025	5.540
8	0.925	1.863	0.023	5.540
9	0.947	1.864	0.025	5.540
10	0.965	1.873	0.026	5.540
11	0.979	1.853	0.026	5.540
12	0.991	1.877	0.023	5.540
13	1.002	1.852	0.025	5.540
14	1.011	1.846	0.025	5.540
15	1.019	1.845	0.025	5.540
16	1.026	1.848	0.025	5.540

From ref. 24.

The C_j can be determined as follows. The real part of the dielectric response changes only near an absorption peak and is constant between them. Hence for each absorption peak:

$$C_j = \varepsilon_j^b - \varepsilon_j^a \qquad \text{[A.43]}$$

where

ε_j^b = the dielectric constant at frequencies just less than that of the absorption peak

ε_j^a = the dielectric constant at frequencies just greater than that of the absorption peak

The experimental procedure is to measure the absorption spectrum from the ultraviolet through the infrared (or microwave if necessary) and the dielectric constant (square of the refractive index) at convenient frequencies between the absorption peaks. Using the dielectric constants between peaks, the quantities C_{ir} and C_{uv} are calculated; and using the frequencies of the peaks, the quantities $\varepsilon(i\xi_n)$ are calculated for each value of n by means of Eq. [A.41]. From the values of $\varepsilon(i\xi_n)$, the Δ_{qr} are calculated by means of Eq. [A.36], and from those quantities the Lifshitz constant is calculated by means of Eq. [A.35]. Sufficient optical constants to make these calculations are listed for some common materials in Table A.3.

Because of the importance of water in emulsions and suspensions, more care has been taken in characterizing its optical properties so as to have a better

Table A.4   Optical Constants for Water,[a]
$d = 74.8, \; 1/\tau = 6.5 \times 10^{-5} \, \text{eV}$

$\omega_j(\text{eV})$	$f_j(\text{eV})$	$g_j(\text{eV})$
2.07×10^{-2}	6.25×10^{-4}	1.5×10^{-2}
6.9×10^{-2}	3.5×10^{-3}	3.8×10^{-2}
9.2×10^{-2}	1.28×10^{-3}	2.8×10^{-2}
2×10^{-1}	5.69×10^{-3}	2.5×10^{-2}
4.2×10^{-1}	1.35×10^{-2}	5.6×10^{-2}
8.25	2.68	0.51
10	5.67	0.88
11.4	12	1.54
13	26.3	2.05
14.9	33.8	2.96
18.5	92.8	6.26

[a]The first 5 frequencies are in the infrared, the rest are in the ultraviolet. From ref. 25

approximation for $\varepsilon(i\xi_n)$. Parsegian[25] gives the following:

$$\varepsilon(i\xi_n) = 1 + \frac{d}{1 + \xi_n\tau} + \sum \frac{f_j}{\omega_j^2 + \xi_n^2 + g_j\xi_n} \qquad [A.44]$$

where the second term is for the microwave contribution, and the third term is the sum of the infrared and the ultraviolet contributions. The constants d, τ, f_j, ω_j, and g_j are given in Table A.4.

In biological systems the differences in the dielectric permittivities of the various components at ultraviolet frequencies are small, hence the interaction energies are strongly dependent on the infrared and microwave frequencies. In applications of Hamaker theory, only ultraviolet frequencies are used; hence biological systems should be analyzed by Lifshitz theory.

(3) Two Half-Spaces, Including Retardation. Retardation of the dispersion forces is significant when the time required for fluctuating electric fields to propagate from one particle to another is comparable to the period of the fluctuation. When retardation effects are important, the Lifshitz "constant" is given by

$$A_{123}^r = -\frac{3kT}{2} \sum_{n=0}^{\infty}{}' \int_{r_n}^{\infty} x\,dx \ln D \qquad [A.45]$$

where

$$D = [1 - \Delta'_{12}\Delta'_{32}\exp(-x)][1 - \Delta''_{12}\Delta''_{32}\exp(-x)] \qquad [A.46]$$

$$\Delta'_{qr} = \frac{\varepsilon_q x - \varepsilon_r x_q}{\varepsilon_q x + \varepsilon_r x_q} \qquad [A.47]$$

$$\Delta''_{q2} = \frac{x - x_q}{x + x_q} \qquad [A.48]$$

$$x_q = \left[x^2 - r_n^2\left(1 - \frac{\varepsilon_q}{\varepsilon_2}\right) \right]^{1/2} \qquad [A.49]$$

$$r_n = \frac{2H\varepsilon_2^{1/2}\xi_n}{c} \qquad [A.50]$$

H = distance between two half-spaces
$\varepsilon_q = \varepsilon_q(i\xi_n)$

c = speed of light in a vacuum

Note the extra factors, x and x_q, and the change in integration limits from the equation for the nonretarded interaction, Eq. [A.35]. The constant r_n can be thought of as the ratio of the time that light of frequency ξ_n takes to cross the gap and return, $2H\varepsilon_2^{1/2}/c$, to the time of a molecular fluctuation, $1/\xi_n$. The retardation can be neglected when r_n is small. Typically this ratio is small when the distance of separation is less than about 5–10 nm. For the analysis of systems with greater distances of separation, the effect of retardation needs to be included. When retardation is important, these extra terms make the Lifshitz "constant" depend on the interparticle distance. This adds a computational burden but no fundamental difficulty to the calculations. For lipid–water systems of biological interest the interactions are at low frequencies, that is, the microwave or infrared, hence retardation can be ignored.

(4)   Lifshitz Theory for Dispersions in Ionic Solutions, Two Half-Spaces. When the intervening medium between two half-spaces is an ionic solution, the dielectric properties of the solvent are altered. The fluctuating electric fields propagating from each half-space are attenuated by the electrophoretic motion of ions in solution. These ions respond to low frequencies but not to higher frequencies. At room temperature $\xi_n = n \times 2.4 \times 10^{14}$ rad/s. Ions in solution cannot respond to any of these frequencies other than the $n = 0$ term; hence the only term in the calculation of the Lishitz constant that is different for ionic solutions is the first, $n = 0$, term. The change is (Mahanty and Ninham, 1976, p. 202):

$$\Delta'_{qr} = \frac{s\varepsilon_q - x\varepsilon_r}{s\varepsilon_q + x\varepsilon_r} \qquad [A.51]$$

where

$$s = [x^2 - 4\kappa^2 H^2]^{1/2}$$

$$\kappa^2 = \frac{e^2}{D\varepsilon_0 kT} \sum n_{i0} z_i^2 \qquad [B.6]$$

and

$$
\begin{aligned}
e &= \text{electronic charge} \\
n_{i0} &= \text{concentration of ions } i \\
z_i &= \text{charge on ion } i \\
D &= \text{static dielectric constant of the solvent} \\
\varepsilon_0 &= \text{permittivity of free space} \\
k &= \text{Boltzmann constant} \\
T &= \text{absolute temperature} \\
H &= \text{distance between half-spaces}
\end{aligned}
$$

For all the other terms, $n > 0$, the integrand is unchanged. If the ionic strength is high, the $n = 0$ term is completely screened and makes no contribution to the interaction free energy.

The derivation of Eq. [A.51] did not take into account the formation of electric double layers at interfaces. Mitchell and Richmond[26,27] showed that the total free energy of interaction of particles with electrical double layers is given by the sum of the free energies of the electrostatic double layer (derived in the next Chapter) and the free energy, calculated by Eq. [A.51] above, of an electrolyte solution that is homogeneous between the half spaces.

(5) Lifshitz Theory for Other Geometries. Exact, explicit, analytic expressions for other than planar geometries for Lifshitz theory are more difficult to derive than for Hamaker theory. The free energy of interaction does not factor into a material-dependent term and a geometric term. Nevertheless progress has been made in analyzing other significant geometries. Mitchell and Ninham[28] found simple expressions for the limiting cases of two spheres at close separation (no retardation) and two spheres at large distances (no retardation); the interaction of long thin rods, retarded and unretarded, was calculated by Mitchell et al.[29]; Ninham and Parsegian[14] calculated the van der Waals interactions for planar multilayer systems.

A possible approximation for interactions other than planar geometries is a combination of the Hamaker and Lifshitz theories, by using Lifshitz "constants" in the Hamaker geometries. This approximation is best when the dielectric differences are small, that is, when the interactions are between similar materials, or when the distances of separation are small. Smith et al.[30] give a careful analysis of the regimes in which Lifshitz theory deviates significantly from Hamaker theory and where this approximation is inappropriate.

3. OTHER METHODS TO OBTAIN HAMAKER CONSTANTS

a. Direct Measurements of Forces of Attraction

Attempts have been made to measure experimentally the dispersion forces between macroscopic bodies. The most direct approach might seem to be to measure the work necessary to separate two surfaces; or, equivalently, the centrifugal force necessary to separate a particle from a solid surface. This approach is untenable, however, because clean surfaces in close contact form short-range chemical bonds and the energy needed to break these bonds can be significantly greater than dispersion energies. The successful approach, first reported by Derjaguin et al., 1954,[31] is to measure the attractive force between carefully prepared surfaces as they are slowly brought together. Another technique is to measure the disjunctive (disjoining) pressure in a liquid film as the film thins. The early measurements were successful down to separation distances within 100 nm; this distance, however, is still sufficiently large that the dispersion forces are retarded. The results agree with the Lifshitz calculations for retarded interactions. Subsequent measurements, 1968–1972, were made with improved equipment (called Jacob's box) to enable approach to 1.4 nm where the dispersion forces are unretarded. These measurements confirm Lifshitz theory for unretarded interactions.[32,33] Derjaguin et al. in 1982, measuring the interactions of crossed glass fibers in ionic solutions, confirm Lifshitz theory down to molecular distances.[34] All these measurements substantiate the theory.

b. Indirect Measurements of the Forces of Attraction

Heats of vaporization or sublimation, adhesion, capillary phenomena, and so on, all depend on intermolecular forces. When the interactions are predominantly dispersion interactions, these phenomena can be used to determine Hamaker constants. The surface tension of a liquid can be thought of as half of the energy necessary to separate two liquid surfaces of unit interfacial area from intimate contact to infinity. Equation [A.12] can be rewritten in the form

$$A_{11} = 24\pi r_1^2 \sigma^d \qquad \text{[A.52]}$$

If this approximation is extended to the work of adhesion between two different materials at an interface, Eq. [A.13], then

$$A_{12} = 12\pi r_{12}^2 W_{12}^d \qquad \text{[A.53]}$$

where r_{12} is the equilibrium intermolecular distance between the two liquids.

Any theory that predicts the dispersion force contribution to the surface tension from other physical properties can be used in conjunction with Eq. [A.52] to calculate the Hamaker constant for that material. A particularly interesting example is due to Croucher[35] who took the Davis and Scriven equation for surface free energy, the corresponding-states relations of Prigogine, and Eq. [A.52] to derive a relation between the Hamaker constant and the bulk thermodynamic properties of a material:

$$A_{11} = \frac{3kT}{4(1 - bV^{-1/3})} \qquad \text{[A.54]}$$

where

$$b = (m/n)^{1/(n-m)} \qquad \text{[A.55]}$$

m, n are the exponents of the interaction potential (e.g., $(m, n) = (6, 12)$ for the Lennard-Jones potential); and the reduced volume V is given implicitly by the relation

$$(\alpha T)^{-1} = -\frac{m}{3} + \frac{n - m}{3(V^{(n-m)/3} - 1)} + \frac{b}{3(V^{1/3} - b)} \qquad \text{[A.56]}$$

where α is the coefficient of thermal expansion. For the 6–∞ potential, $b = 1$ and

$$V = \left(\frac{3 + 7\alpha T}{3 + 6\alpha T}\right)^3 \qquad \text{[A.57]}$$

Table A.5 Hamaker Constants A_{11} for Some Common Materials[a]

Substance	Hamaker Constant $A_{11}(J)(\times 10^{20})$	Method	Reference
Gold	45.3	a	36
Silver	39.8	a	36
Germanium	29.9	a	36
Silicon	25.5	a	36
Selenium	16.2	a	36
Alumina	15.4	a	36
Magnesia	10.5	a	36
Water	4.35	a	36
Ionic crystals	6.3–15.3	a	37
Oxides	10.6–15.5	a	37
Metals	16.2–45.5	a	37
Polymers	6.15–6.6	a	37
Acetone	4.20	b	38
Carbon tetrachloride	4.78	b	38
Chlorobenzene	5.89	b	38
Cyclohexane	4.82	b	38
Decane	5.45	b	38
Diethyl ether	4.30	b	38
1,4-Dioxane	5.26	b	38
Dodecane	5.84	b	38
Ethyl acetate	4.17	b	38
Hexadecane	6.31	b	38
Hexane	4.32	b	38
Methyl ethyl ketone	4.53	b	38
Octane	5.02	b	38
Pentane	3.94	b	38
Toluene	5.40	b	38
Natural rubber	8.58	b	38
Polybutadiene	8.20	b	38
Polybutene-1	8.03	b	38
Polydimethylsiloxane	6.27	b	38
Polyethylene oxide	7.51	b	38
Polyisobutylene	10.10	b	38
Polystyrene	9.80	b	38
Polystyrene	6.57	a	36
Polyvinyl acetate	8.91	b	38
Polyvinyl chloride	10.82	b	38
Polypropylene oxide	3.95	b	38

[a]Method of calculation: a = molar, b = thermodynamic.

Table A.6 Hamaker Constants A_{121} for the Nonretarded Interaction across a film of Water (2)[a]

Substance	Hamaker Constant, A_{121} (J) ($\times 10^{20}$)	Method	Reference
Gold	33.5	a	36
Silver	26.6	a	36
Germanium	16.0	a	36
Silicon	11.9	a	36
Selenium	4.77	a	36
Alumina	4.12	a	36
Magnesia	1.60	a	36
Crystalline quartz	1.70	a	24
Fused quartz	0.833	a	24
Fused silica	0.849	a	24
Calcite	2.23	a	24
Calcium fluoride	1.04	a	24
Sapphire	5.32	a	24
Ionic crystals	0.31–4.85	a	37
Oxides	1.76–4.17	a	37
Metals	3–33.4	a	37
Polymers	(0.35)	a	37
Polyisoprene	0.743	a	24
Polymethyl methacrylate	1.05	a	24
Polystyrene, average of 2	0.931	a	24
Polystyrene	0.27	a	36
Polystyrene	0.35	a	37
Polytetrafluoroethylene	0.333	a	24
"Teflon FEP"	0.381	a	24
Polyvinyl chloride	1.30	a	24
Pentane	0.336	a	24
Hexane	0.360	a	24
Heptane	0.386	a	24
Octane	0.410	a	24
Nonane	0.435	a	24
Decane	0.462	a	24
Undecane	0.471	a	24
Dodecane	0.502	a	24
Tridecane	0.504	a	24
Tetradecane	0.514	a	24
Pentadecane	0.526	a	24
Hexadecane	0.540	a	24

[a]Method of calculation: a = molar

Therefore, from the coefficient of thermal expansion and an estimate of the form of the interaction pair potential (often taken as 6–∞), the Hamaker constant can be calculated.

Any process that is strongly influenced by the long-range dispersion attractions can be a resource for the calculation of Hamaker constants. The coagulation rate of dispersions is such a process, where the rate of flocculation (which is measurable) depends, in part, on dispersion force attractions. The Hamaker constant can be estimated from an analysis of the variation in the stability of a dispersion as electrolyte is added (Verwey and Overbeek, 1948; also Chapter III.D).

Another process influenced by long-range dispersion attractions is the spreading of a liquid film on a solid when the spreading coefficient is positive. The Hamaker constant can be estimated from ellipsometric measurements of the equilibrium film thickness (Section II.A.12).

4. VALUES OF HAMAKER CONSTANTS FOR SOME COMMON MATERIALS

Table A.5 gives the value of the Hamaker constant (unretarded) for some common materials.

The Hamaker constant for the interaction of two different materials can be calculated or measured in the same manner as for two identical materials or it can be estimated by Berthelot's principle from the data in Table A.5 by the relation

$$A_{12} = (A_{11}A_{22})^{1/2} \qquad [A.29]$$

The Hamaker constants for the interaction of two materials across a third can be

Table A.7   Hamaker Constants A_{123}, for the Nonretarded Interaction across a Film of Water (2)^a

Substance	Hamaker Constant, A_{123} (J) ($\times 10^{20}$)	Method	Reference
Polystyrene(1):Au(3)	2.98	a	36
Polystyrene(1):Ag(3)	2.67	a	36
Polystyrene(1):Se(3)	1.15	a	36
Selenium(1):Magnesia(3)	2.83	a	36
Selenium(1):Silicon(3)	7.72	a	36
Gold(1):Selenium(3)	12.2	a	36

^{a}Method of calculation: a = molar.

calculated or measured or estimated by the principle of Archimedean buoyancy:

$$A_{123} = A_{13} + A_{22} - A_{12} - A_{23} \qquad\qquad\qquad \text{[A.26]}$$

$$\text{or } A_{123} = (A_{11}^{1/2} - A_{22}^{1/2})(A_{33}^{1/2} - A_{22}^{1/2}) \qquad\qquad \text{[A.31]}$$

Hamaker constants, A_{121}, for the nonretarded interaction across a film of water are listed in Table A.6; and Hamaker constants, A_{123}, for the nonretarded interaction across a film of water are listed in Table A.7.

CHAPTER IIIB
Forces of Repulsion

1. ELECTROSTATIC REPULSION

a. Origin of Charge in Aqueous Systems

Quincke[39-41] and later investigators showed that electrokinetic phenomena and, accordingly, the presence of surface charges at the solid–liquid interface exist in nearly all systems, especially when the liquid is water. Generally an interface in distilled water is negatively charged. On contact with other liquids, however (e.g., oil of turpentine), the surface is frequently positive; still other liquids (e.g., ether, petroleum ether, or oil of hartshorn) produce no surface charge. The surface charge may be acquired by surface ionization or by preferential adsorption of anions or cations. In water, mineral oxides and sulfides (e.g., silica, iron oxide, arsenic sulfide) become positively charged at low pH and negatively charged at high pH, with a zero charge at an intermediate pH, which is known as the point-of-zero charge (PZC) or isoelectric point. The PZC of acidic oxides occur at a low pH and for a basic oxide, such as Al_2O_3, at a high pH.

All dispersed particles in water spontaneously acquire electric charges by two principal mechanisms: dissociation of ionogenic groups or preferential adsorption of ions, usually anions, from solution. Examples of the first mechanism are the pH dependence of the charge on metal oxides and the dependence of the charge of the silver halides on the concentration of silver ions in solution. These are represented by the following equilibria:

lattice–SiOH = lattice–SiO$^-$ + H$^+$
lattice–AlOH = lattice–Al$^+$ + OH$^-$
lattice–AgCl = lattice–Cl$^-$ + Ag$^+$ or   lattice–AgCl = lattice–Ag$^+$ + Cl$^-$

The point-of-zero charge is defined as the negative of the Briggsian logarithm of the activity of the potential-determining ion at which the net charge on the dispersed particle is zero. At the PZC the electrostatic repulsion between the particles is lost. Typical values of the PZC for various materials are given in Tables B.1 and B.2. The values quoted, especially for the oxides, should be used with caution, as the value for a particular material is influenced by its source, method of preparation, pretreatment, and the presence of trace impurities. At values of pH less than PZC, the particles are charged positively; and at values of pAg less than PZC, the silver halide particles are charged positively.

An example of preferential adsorption of surface-active anions by positively charged particles of silver iodide is shown in Fig. B.1, which reports the zeta potential as a function of the equilibrium concentration of various solutes in water. The sign and magnitude of the zeta potential reflect the sign and magnitude of the charge on the particle. The adsorption of the anions causes reversal of the sign of the charge on the particles at concentrations far below the CMC values of the solutes.

A more complex example of preferential adsorption of a potential-determining

Table B.1   Point-of-Zero Charge of Some Oxides

Substance	PZC = pH
Quartz, SiO_2	2–3.7
Cassiterite, SnO_2	4.5
Rutile, TiO_2	6.0
Hematite (natural), Fe_2O_3	4.8
Hematite (synthetic)	8.6
Corundum, Al_2O_3	9.0
Magnesia, MgO	12

From ref. 42.

Table B.2   Point-of-Zero Charge of Some Ionic Solids

Substance	PZC
Fluorapatite, $Ca_5(PO_4)_3(F, OH)$	pH = 6
Hydroxyapatite, $Ca_5(PO_4)_3(OH)$	pH = 7
Calcite, $CaCO_3$	pH = 9.5
Fluorite, CaF_2	pCa = 3
Barite (synthetic), $BaSO_4$	pBa = 6.7
Silver iodide, AgI	pAg = 5.6
Silver chloride, AgCl	pAg = 4
Silver sulphide, Ag_2S	pAg = 10.2

From ref. 42.

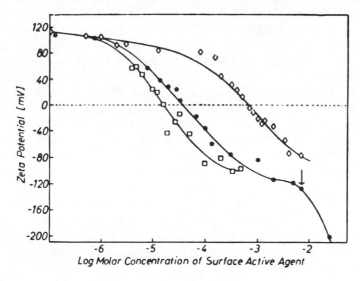

Figure B.1 Zeta potential as a function of concentration of various surface-active solutes in water (AgI–water interface.) ◇ Sodium decyl sulfate; ● sodium dodecyl sulfate; □ sodium tetradecyl sulfate.[43]

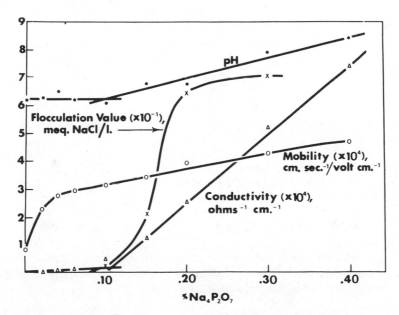

Figure B.2 A graphical correlation of properties with weight percent TSPP added.[44]

ion is the deflocculation of kaolin by tetrasodium pyrophosphate (TSPP), $Na_4P_2O_7$. Kaolin particles have the form of a stack of platelets. In the presence of water, the flat faces carry negative charges, while the ragged edges of the platelets carry positive charges. Before being dispersed, the particles are flocculated by a face-to-edge electrostatic bonding. The change of charge on a particle on adding TSPP is reflected by a change in various properties of the suspension. A graphical correlation of the measured properties of kaolin suspensions as a function of percent TSPP by weight of kaolin is shown in Fig. B.2. The electrophoretic mobility of the particles increases rapidly with the first small additions of TSPP. The rapid increase in particle mobility at TSPP additions up to about 0.05% is interpreted as the result of specific adsorption of anion on the positively charged edges of the kaolin. The electrical conductivity, the flocculation value and the pH, all indicate that TSPP continues to be adsorbed up to about 0.10%, but the electrophoretic mobility does not increase as rapidly as at first. This is interpreted as evidence of continued adsorption of phosphate on the negatively charged faces of the kaolin particles, but with less influence on the net charge and on the mobility.

b.   Gouy–Chapman Theory

Adsorbed ions leave their counterions in the solution, and these excess charges cluster near the surface forming an electrical double layer. The double layer consists of two regions: an inner region composed of adsorbed counterions (the Stern layer) and a diffuse region containing the remainder of the excess counterions. In the diffuse region (Gouy–Chapman layer) the counterions are distributed according to a balance between their thermal motion and the forces of electrical attraction.

The model used to describe the ion distribution functions and the potential of the region near the charged interface was developed independently by Gouy[45] and Chapman[46] by combining the Poisson equation for the second derivative of the electric potential as a function of charge density with the Boltzmann equation for the charge density as a function of potential. The problem is formulated as: Given a surface charge, find the spatial distribution of its counter ions. The assumptions are:

(a) The surface charge is continuous and uniform.
(b) The ions in the solution are point charges.

Despite the dubiety of these assumptions, the Poisson–Boltzmann equation has been found to agree with observation over a wide range of experimental conditions.[47-49]

Let the electrical potential be Φ_0 at the surface and $\Phi(x)$ at a distance x from the surface in the electrolyte solution. Taking the surface to be positively charged for argument's sake, the Boltzmann equation gives

$$n_i = n_{i0} \exp\left(-\frac{z_i e\Phi(x)}{kT} \right)$$

[B.1]

where n_i is the concentration of ions of kind i at a point where the potential is $\Phi(x)$; n_{i0} is the concentration in the bulk of the solution; z_i is the valency including the sign of the charge: positive for cations, negative for anions.

The charge density ρ in the solution is the algebraic sum of the ionic charges per unit volume:

$$\rho = \sum z_i e n_i = \sum z_i e n_{i0} \exp\left(-\frac{z_i e\Phi}{kT}\right) \qquad [B.2]$$

The Poisson equation relates the charge density to the Laplacian of the electric potential (Appendix H):

$$\nabla^2\Phi = -\rho/D\varepsilon_0 \qquad [B.3]$$

where D is the dielectric constant and ε_0 is the permittivity of free space, namely, 8.854×10^{-12} C/V m. The negative sign in Eq. [B.3] results from the charge density in the solution, ρ being negative for a positively charged surface.

Combining Eq. [B.2] and [B.3] gives the differential equation (the Poisson–Boltzmann equation) for the potential Φ as a function of the coordinates:

$$\nabla^2\Phi = -\frac{1}{D\varepsilon_0}\sum z_i e n_{i0} \exp\left(-\frac{z_i e\Phi}{kT}\right) \qquad [B.4]$$

No exact analytical solution has been found for Eq. [B.4] that is valid for all values of the parameters. The common approximations used to simplify Eq. [B.4] are the special cases of (i) a small surface potential or (ii) a symmetrical electrolyte:

(i)   Exact Solutions of the Poisson–Boltzmann Equation for Small Potentials and Any Electrolyte. Equation [B.4] can be solved approximately when the surface potential is small, that is $|z_i e\Phi/kT| \ll 1$, as, for example, when $z_i\Phi$ is less than 25 mV at room temperature. The procedure is to expand the exponential: the first term, $\sum z_i e n_{i0}$, is zero because the solution has a negligible net charge; and only the second term of the series is retained:

$$\nabla^2\Phi = \kappa^2\Phi \qquad [B.5]$$

where

$$\kappa^2 = \frac{e^2}{D\varepsilon_0 kT}\sum n_{i0}z_i^2 \qquad [B.6]$$

$$\kappa = 3.288(\sum n_{i0}z_i^2)^{1/2}\ \text{nm}^{-1}$$

for water at 25°C when the n_{i0} are in moles per liter and z_i is the charge per ion. Note that κ has units of reciprocal length; and $1/\kappa$ is often called the Debye length or the double-layer thickness. Equation [B.5] can be solved for different geometries. For the planar interface

$$\Phi = \Phi_0 \exp(-\kappa x) \tag{B.7}$$

For the sphere of radius a

$$\Phi = \frac{a\Phi_0}{r} \exp[-\kappa(r-a)] \tag{B.8}$$

where r is the distance from the center of the sphere. In these approximations the potential drops exponentially with distance. At a distance from the surface equal to $1/\kappa$, the potential has fallen by a factor of $1/e$, that is, 43.4%, and the number of counterions in the diffuse double layer has decreased by an even larger fraction; which perhaps justifies the designation of $1/\kappa$ as the "thickness' of the double layer.

The surface charge density has the same sign as the surface potential; the net charge of the diffuse layer, which is the charge contained by the volume subtended by unit area extending from the surface ($x = 0$ or $r = a$) to infinity, has the same numerical value but the opposite sign. The surface charge density σ is given in general by

$$\sigma = -\int_a^\infty \rho\,dr = \int_a^\infty D\varepsilon_0 \nabla^2\Phi\,dr \tag{B.9}$$

The surface charge density on a plane with low surface potential is found by using Eq. [B.7], the Poisson equation [B.5] in planar coordinates (Appendix C), and Eq. [B.9]:

$$\sigma = D\varepsilon_0 \kappa \Phi_0 \tag{B.10}$$

The surface charge density on a sphere is found by using Eq. [B.7], Poisson's Equation [B.3] in spherical coordinates (Appendix C), and Eq. [B.9]:

$$\sigma = \int_a^\infty D\varepsilon_0\left(\frac{d^2\Phi}{dr^2} + \frac{(2/r)\,d\Phi}{dr}\right)dr = -D\varepsilon_0\left(\frac{d\Phi}{dr}\right)_{r=a} \tag{B.11}$$

When the surface potential on the sphere is small, the surface charge density is found by differentiating Eq. [B.8] with respect to r and evaluating at $r = a$, to give

$$\sigma = \frac{D\varepsilon_0\Phi_0(1 + \kappa a)}{a} \tag{B.12}$$

The charge on a sphere, Q, is $4\pi a^2$ times the surface charge density σ; therefore, the charge on a sphere with low surface potential is

$$Q = 4\pi a D\varepsilon_0\Phi_0(1 + \kappa a) \tag{B.13}$$

Equation [B.13] was used by Debye and Hückel to calculate the potential around an ion. Equations [B.10] and [B.12] show that the surface potential Φ_0, the surface charge density σ, and the ionic composition of the solution through its influence on κ are interrelated. The surface charge density σ is affected only by the adsorption of potential-determining ions. If inert electrolyte is added, κ increases; hence the surface potential decreases without any change of surface charge density.

(ii)   Solutions of the Poisson–Boltzmann Equation for Symmetrical z–z Electrolytes (e.g., 1–1, 2–2, 3–3,...). For the special case of a symmetrical electrolyte, Eq. [B.4] reduces to

$$\nabla^2\Phi = -\frac{2zen_0}{D\varepsilon_0}\sinh\left(-\frac{ze\Phi}{kT}\right) \qquad [B.14]$$

with no restriction on the magnitude of the surface potential. Equation [B.14] has only been solved analytically for the planar interface. For a spherical particle the Poisson–Boltzmann equation has no tractable analytical solution, so it must be solved by numerical integration. The most detailed calculations, which include asymmetrical electrolytes covering a wide range of values of κ and Φ_0, were published by Loeb, Wiersema and Overbeek (1961) in the form of tables (LWO tables) for the potential, charge density, and free energy of the electrical double layer around a single spherical colloidal particle. Stigter[50] has given accurate interpolation functions for the LWO tables. A detailed description of various approximate solutions of the Poisson–Boltzmann equation for spherical symmetry is reported by Dukhin and Derjaguin.[51]

For the planar interface and symmetrical electrolyte, the solution of the Poisson–Boltzmann equation [B.14], is (Verwey and Overbeek, 1948)

$$\Phi = \frac{2kT}{ze}\ln\frac{1 + \gamma\exp(-\kappa x)}{1 - \gamma\exp(-\kappa x)} \qquad [B.15]$$

for which

$$\sigma = (8n_0D\varepsilon_0kT)^{1/2}\sinh\frac{ze\Phi_0}{2kT} \qquad [B.16]$$

where

$$\kappa = \left(\frac{2n_0ze}{D\varepsilon_0kT}\right)^{1/2} \text{ for a symmetrical electrolyte}$$

$$= 3.288\, zc^{1/2}\, \text{nm}^{-1} \text{ at } 25°C \qquad [B.17]$$

where

c = the bulk electrolyte concentration (moles/liter) in water

$$\gamma = \tanh\frac{ze\Phi_0}{4kT}, \qquad [B.18]$$

A useful approximation for γ for small surface potentials is

$$\gamma \simeq \frac{ze\Phi_0}{4kT} \qquad \text{[B.19]}$$

These same equations are used to approximate the solution of the Poisson–Boltzmann equation for asymmetrical electrolytes. This simplification causes little error in the description of colloidal phenomena because the valency of the ion of the same charge as the particle is unimportant and its value of z can be set equal to the valency of the counterions. For example, if $CaCl_2$ is used as a flocculant for a negatively charged sol, z is taken as 2.

Added inert electrolyte increases the value of κ (Eq. [B.17]), thus decreasing the value of $1/\kappa$, the double layer thickness, and causing the double layer to contract. At the same time the added electrolyte decreases the surface potential Φ_0 (Eq. [B.16]), as it does not affect the surface charge density. Hence, the effect of added electrolyte on the potential curve is twofold.

A useful approximation for Eq. [B.15] is the solution for the potential at a large distance from a particle carrying any charge, large or small. Equation [B.15] in the limit of large x reduces to

$$\Phi = \frac{4\gamma kT}{ze}\exp(-\kappa x) \qquad \text{[B.20]}$$

Adding the requirement that the surface potential be small, then introducing Eq. [B.19], gives $\Phi = \Phi_0 \exp(-\kappa x)$.

c. Stern Theory

The inadequacies of the Gouy–Chapman theory become evident when observations are made of the properties of electrodes where surface charge density, surface potential, and capacitance can be measured. The electrical double layer at the mercury–solution interface is the one most thoroughly studied. This system has a well-defined interfacial geometry, a wide range of potentials, and the possibility of working with expanding surfaces to minimize contamination. Other materials can be dispersed as stable sols with an electrode reversible to the dispersed phase. An example is an aqueous suspension of silver iodide, for which the potential-determining mechanism is known and stable AgI electrodes exist.[52]

The capacitance C (per unit area) of the double layer at a planar interface is (Hunter, 1981, p. 33)

$$C = \frac{d\sigma}{d\Phi_0} \qquad \text{[B.21]}$$

By means of the Gouy–Chapman theory, the capacitance of a planar interface can be evaluated, using Eq. [B.16], by

$$C = D\varepsilon_0 \cosh\frac{ze\Phi_0}{2kT} \qquad\qquad [B.22]$$

The observed capacitances are always much less than the theoretical capacitance obtained from Eq. [B.22]. The source of this difference is ascribed to the neglect of the finite size of ions in the Gouy–Chapman model, especially for the counterions near the charged surface.

Stern[53] introduced ionic dimensions into the model of the double layer for its inner localized portion (the Stern layer). The constitution of a portion of the particle, and the Stern layer, along with the diffuse double layer associated with the surface of the particle, is depicted diagrammatically in Fig. B.3, for (a) a negatively charged sol of silver iodide and (b) a positively charged sol of silver iodide (Weiser, 1949, p. 215). The negatively charged sol is stabilized with a slight

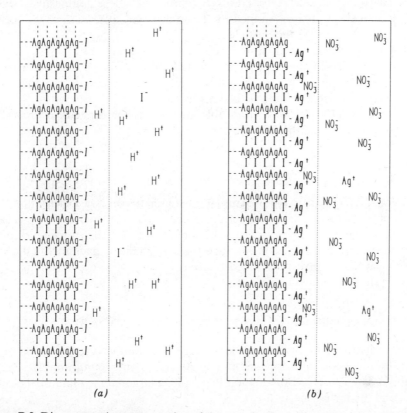

(a)   (b)

Figure B.3 Diagrammatic representation of the constitution of a portion of a particle of silver iodide. (a) Negatively charged by adsorbed iodide ions; (b) positively charged by adsorbed silver ions (Weiser, 1949, p. 215).

excess of hydrogen iodide in solution, and the positively charged sol is stabilized with a slight excess of silver nitrate in solution. Consider first the negatively charged particle: roman letters Ag and I represent the two ions in the lattice, while I^- represents the potential-determining ion adsorbed out of the solution. The Stern layer contains a few counterions (H^+) and the remainder of the counterions are in the diffuse double layer. The diagram depicts electrical neutrality: The surface of the negatively charged particle has a net represented charge of $15I^-$ adsorbed on it, with $3H^+$ in the Stern layer, and $14H^+$ plus $2I^-$ in the diffuse double layer, to give no total excess of charge. The positively charged particle is represented as charged by adsorbed silver ions, Ag^+. Again the positively charged particle has a net represented charge of $15Ag^+$ adsorbed on it, with $3NO_3^-$ in the Stern layer, and $14NO_3^-$ plus $2Ag^+$ in the diffuse double layer, to no total excess of charge. The depicted Stern layer is separated from the depicted diffuse double layer by a dotted line. The Stern layer moves with the particle, but the region of shear, which is operative only when the particle moves with respect to its medium, is not exactly at the boundary of the Stern layer but a short distance within the diffuse double layer. The excess ions in the diffuse double layer beyond the plane of shear are not trapped by the electrostatic attraction of the surface, and do not move with the particle.

The total surface potential Φ_0 is divided into a potential drop Φ_s over the diffuse part of the double layer and a potential drop, $\Phi_0 - \Phi_s$ over the Stern layer. The Gouy–Chapman theory applies without change to the diffuse part of the double layer, which originates at a short distance from the charged surface. The theoretical capacity of the total double layer, C_t, is the capacity of the diffuse layer, C_d, and that of the Stern layer, C_s, in series:

$$C_t = \frac{C_d C_s}{C_d + C_s}$$ [B.23]

Equation [B.23] explains the discrepancy between the observed capacitance and the impossibly high values of the Gouy–Chapman model; because if C_d is large, the total capacitance given by Eq. [B.23] reduces to C_s, the capacitance of the Stern layer. The measured capacitance is then the smaller capacitance of the Stern layer. Sparnaay[54] gives a detailed analysis of the ion size corrections to the Poisson–Boltzmann equation.

2. ANALYSIS OF INTERACTING PLANAR DOUBLE LAYERS

We have discussed the analysis of isolated electrical double layers. To analyze the forces of repulsion in a dispersion, a treatment of interacting double layers is required. The Poisson–Boltzmann differential equation for symmetrical electrolytes and planar geometry is

$$\frac{d^2\Phi}{dx^2} = -\frac{\rho}{D\varepsilon_0} = -\frac{2zen_0}{D\varepsilon_0}\sinh\frac{ze\Phi}{kT}$$ [B.24]

For two parallel plates separated by a distance $2d$, the following boundary conditions must be met by the solution to the differential equation:

$$\text{at } x = d, \qquad \Phi = \Phi_d \quad \text{and} \quad \frac{d\Phi}{dx} = 0$$

$$\text{at } x = 0, 2d \quad \Phi = \Phi_0$$

The solution can be represented by an integral (Verwey and Overbeek, 1948, p. 66)

$$-\kappa d = \int_{\Phi_0}^{\Phi_d} \frac{(ze/kT)d\Phi}{\{2[\cosh(ze\Phi/kT) - \cosh(ze\Phi_d/kT)]\}^{1/2}} \qquad \text{[B.25]}$$

This integral leads to an elliptical integral of the first kind for which tables are available; it can therefore be solved numerically for Φ_d if Φ_0 is known. Equation [B.25] is not very illuminating. What is sought is some explicit analytic expression giving potential as a function of separation so that the force of repulsion can be calculated. To reach this goal either of the two following approximations can be used.

a. The Langmuir Approximation

The repulsive potential between plane parallel double layers has been calculated numerically by Overbeek[10] from the change in the free energy as two surfaces approach each other from infinite distance; an equivalent analytic derivation was obtained by Langmuir[55] by an essentially physicochemical approach, based on the balance between osmotic and electrical forces.

The electrical potential of two overlapping electrical double layers at a

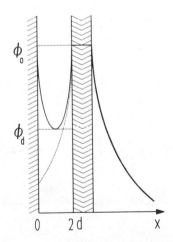

Figure B.4 Schematic representation of the electric potential between two plates in comparison with that for a single double layer.

separation $2d$ is shown in Fig. B.4. To maintain an equilibrium, the electrical forces of attraction tending to condense the whole dispersion are balanced by the resulting increase of osmotic pressure. At any point the additional osmotic pressure Π due to the excess ions balances the pressure due to the electric field, that is,

$$d\Pi = -\rho \, d\Phi \qquad \text{[B.26]}$$

Equation [B.26] can be integrated by first introducing for Poisson equation for ρ, Eq. [B.3], in planar coordinates:

$$d\Pi - D\varepsilon_0 \frac{d^2\Phi}{dx^2} \cdot d\Phi = 0 \qquad \text{[B.27]}$$

which on integration gives

$$\Pi - \frac{D\varepsilon_0}{2} \left(\frac{d\Phi}{dx} \right)^2 = \text{constant} \qquad \text{[B.28]}$$

Equation [B.28] says that at every point between the plates shown in Fig. B.4, the algebraic sum of the osmotic pressure and the pressure due to the electric field (Maxwell stresses) is constant. This constant can be evaluated at the midpoint between the plates where $d\Phi/dx$ is equal to zero; hence the integration constant is Π_d, the excess osmotic pressure at the midpoint. Figure B.5 shows these relations. The pressure that drives the plates apart is the osmotic pressure at this midpoint, which is the same as the total repulsive pressure (also called the disjoining or disjunctive pressure) at every point between the plates.

The problem now becomes to evaluate Π_d, the excess osmotic pressure at the midpoint, which can be done by means of the van't Hoff equation:

$$\Pi = nkT \qquad \text{[B.29]}$$

where n is the number of excess ions per unit volume. The excess concentration of ions at the central plane is determined by the electric potential at that point, Φ_d, from the Boltzmann equation:

$$n = n_0 \left[\exp\left(-\frac{z_- e\Phi_d}{kT} \right) - 1 + \exp\left(-\frac{z_+ e\Phi_d}{kT} \right) - 1 \right] \qquad \text{[B.30]}$$

Equations [B.29] and [B.30] embody ideal-solution approximations. For symmetrical electrolytes

$$n = 2n_0 \left(\cosh \frac{ze\Phi_d}{kT} - 1 \right) \qquad \text{[B.31]}$$

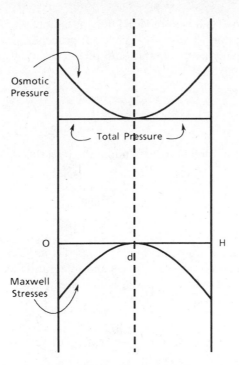

Figure B.5 The osmotic pressure (repulsive), the Maxwell stresses (attractive but zero at midpoint), and their sum (a constant) in overlapping electrical double layers between two flat plates ($H = 2d$).

Combining Eq. [B.29] and [B.31] gives

$$\Pi_d = 2n_0 kT \left(\cosh \frac{ze\Phi_d}{kT} - 1 \right) \qquad \text{[B.32]}$$

Equation [B.32] was also obtained from Lifshitz theory by Mitchell and Richmond.[26] Equation [B.32] is an expression for the *force* per unit area between the plates as a function of their separation distance. The potential energy is then found by integration with respect to distance. This potential energy is the work associated with any isothermal and reversible process of building up the double layer between two plates, and therefore equals the variation of the free energy with particle separation.

To obtain the repulsive potential energy between the two plates from Eq. [B.32] requires the integration of the pressure Π_d as a function of the plate separation. The free energy of repulsion, ΔG_R, is given by

$$\Delta G_R = \int_d^\infty \Pi_x \, dx \qquad \text{[B.33]}$$

The integration requires an expression for Φ_d as a function of plate separation d; but the precise relation is complex. The usual procedure is to approximate the overlapping potentials as the simple sum of the potentials of the undisturbed double layers (Eq. [B.20])

$$\Phi_d = \frac{8\gamma kT}{ze}\exp(-\kappa d) \qquad\qquad [B.34]$$

Equation [B.32] is simplified by expanding the hyperbolic cosine and retaining only the first term:

$$\Pi_d = 2n_0 kT\left(\frac{ze\Phi_d}{kT}\right)^2 \qquad\qquad [B.35]$$

Substituting Eq. [B.34] into Eq. [B.35] gives Π_d as a function of distance. Substituting this relation into Eq. [B.33] and integrating leads to the following approximation for the repulsive potential for highly charged parallel plates at relatively large separations, $H = 2d$:

$$\Delta G_R = \frac{64 n_0 kT}{\kappa}\gamma^2 \exp(-2\kappa d) \qquad\qquad [B.36]$$

Equation [B.36] is the key relation for evaluating the energy of repulsion between two dispersed particles. Combining this with a suitable expression for the attractive energy between dispersed particles gives a relation for the stability of the dispersion in terms of measurable or calculable values, and is the essence of the DLVO theory.

b. The Debye–Hückel Approximation

A second approximation to evaluate the repulsive energy ΔG_R for two parallel plates is to use the Debye–Hückel approximation for slightly charged plates. This is appropriate when $ze\Phi_0/kT \ll 1$.

Expanding the hyperbolic cosine terms in Eq. [B.25] and keeping only the leading terms gives

$$-\kappa d = \int_{\Phi_0}^{\Phi_d} \frac{(ze/kT)d\Phi}{[(ze\Phi/kT)^2 - (ze\Phi_d/kT)^2]^{1/2}} \qquad\qquad [B.37]$$

Integrating gives

$$\kappa d = \ln\left(\frac{ze\Phi_0/kT + [(ze\Phi_0/kT)^2 - (ze\Phi_d/kT)^2]^{1/2}}{ze\Phi_d/kT}\right) \qquad\qquad [B.38]$$

Rearranging [B.38] gives

$$\frac{ze\Phi_d}{kT} = \frac{ze\Phi_0}{kT \cosh \kappa d} \qquad \text{[B.39]}$$

Squaring Eq. [B.39], using the approximation that

$$\left(\frac{ze\Phi_d}{kT}\right)^2 \simeq \cosh\left(\frac{ze\Phi_d}{kT}\right) - 1 \qquad \text{[B.40]}$$

and substituting gives

$$\cosh\left(\frac{ze\Phi_d}{kT}\right) - 1 = \left(\frac{ze\Phi_0}{kT \cosh \kappa d}\right)^2 \qquad \text{[B.41]}$$

Substituting this equation into Eq. [B.32] gives

$$\Pi_d = \frac{2n_0 z^2 e^2 \Phi_0^2 \text{cosech}^2 \kappa d}{kT} \qquad \text{[B.42]}$$

Substituting this expression for the repulsive pressure between two charged plates into the formula for the total repulsive potential, Eq. [B.33], and integrating gives

$$\Delta G_R = \kappa D \varepsilon_0 \Phi_0^2 [1 - \tanh(\kappa d)] \qquad \text{[B.43]}$$

This expression is an approximation for Eq. [B.36] with the qualification that the surfaces are only slightly charged. Equation [B.36] is not more complicated and is more general, hence it is more frequently used.

3.   ANALYSIS OF INTERACTING SPHERICAL DOUBLE LAYERS

In many colloidal systems such as emulsions, inorganic sols, soap solutions, and certain polymer solutions, the particles are better represented by a spherical model than by plane parallel plates. In principle, the repulsive energy for spherical particles can be calculated in the same way as for plane parallel plates, starting with the expressions for the isolated spherical double layer. The Poisson–Boltzmann equation, however, for a spherical particle has only been solved numerically, so that the repulsive energies for the overlap of spherical double layers also must be found numerically.

Various approximations are available for special cases: (a) where the particles are large compared to the double layer thickness[56]; or (b) where the potential between the particles is approximated as the sum of the potentials of the undisturbed spherical double layers[57]; or (c) where it might be permissible to use the low-potential (the Debye–Hückel approximation) form of the Poisson–Boltzmann equation.[58]

a. The Derjaguin Approximation

Each sphere is considered to have a stepped surface, and the electrostatic repulsion between two charged spheres is approximated as the sum of a series of separate repulsions between parallel rings of graded diameters, facing each other in matched pairs. The interaction between the spheres is thus reduced to the sum of interactions between parallel walls at different distances of separation. By making the rings infinitesimal the summation becomes an integration. For each pair of rings the repulsive interaction is the product of the facing areas, $2\pi h\, dh$, and the repulsion interaction, $\Delta G_R^{plane}(H)$ per unit area of two flat parallel plates at a distance H. The total repulsion, $\Delta G_R^{sphere}(H_0)$, for the two spheres of radius a, a distance H_0 apart, is obtained by integrating from $h = 0$ to $h = \infty$.

$$\Delta G_R^{sphere}(H_0) = 2\pi \int_0^\infty \Delta G_R^{plane}(H)h\, dh \qquad [\text{B.44}]$$

From Fig. B.6

$$\tfrac{1}{2}(H - H_0) = a - [a^2 - h^2]^{1/2} \qquad [\text{B.45}]$$

Differentiating Eq. [B.45] and rearranging gives

$$2h\, dh = dH[a^2 - h^2]^{1/2} \qquad [\text{B.46}]$$

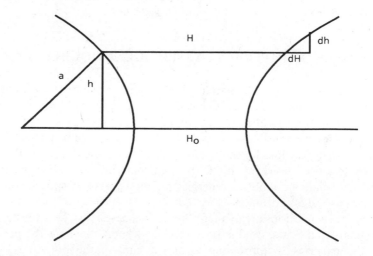

Figure B.6 The Derjaguin approximation for the interaction of two spheres of radius a at a distance H_0 apart ($H_0 = 2d$).

Since $h^2 \ll a^2$

$$2h \, dh \simeq a \, dH \qquad \text{[B.47]}$$

Combining Eq. [B.44] and [B.47] gives

$$\Delta G_R^{\text{sphere}}(H_0) = \pi a \int_{H_0}^{\infty} \Delta G_R^{\text{plane}}(H) \, dH \qquad \text{[B.48]}$$

Equation [B.48] gives the Derjaguin method of calculating the interaction of two equal spheres from the interaction per unit area of two parallel planes, with the added assumptions of thin double layers and large particles.

The three solutions already described for the interaction of plane surfaces, namely, the tabulated solutions to Eq. [B.25], the Langmuir approximation Eq. [B.36], and the Debye–Hückel approximation Eq. [B.43], can each be used with the Derjaguin model of interacting spheres, to obtain solutions that are applicable, respectively, under the same conditions that hold for each of its planar precursors. (Verwey and Overbeek, 1948, pp. 137–142).

Substituting Eq. [B.36], the Langmuir approximation, for the repulsion between two parallel plates into Eq. [B.48] and integrating gives an approximation for the repulsion between two large and equal spheres ($H_0 = 2d$):

$$\Delta G_R = \frac{64 n_0 \pi a k T \gamma^2}{\kappa^2} \exp(-\kappa H_0) = \frac{16 \pi a D \varepsilon_0 k^2 T^2 \gamma^2}{n_0 z^2 e^2} \exp(-\kappa H_0) \qquad \text{[B.49]}$$

Substituting Eq. [B.43] into Eq. [B.48] and integrating, gives the Debye–Hückel approximation for the repulsion between two parallel plates, gives a good approximation for the repulsion between two large and equal spheres of relatively low potential, at separations that are large compared to $1/\kappa$ but small compared to the size of the spheres:

$$\Delta G_R = 2 \pi a D \varepsilon_0 \Phi_0^2 \ln[1 + \exp(-\kappa H_0)] \qquad \text{[B.50]}$$

This equation applies to emulsion droplets, for instance, but would not be as suitable for particles of colloidal size and high potential. For the latter situation Verwey and Overbeek used Derjaguin's method with the exact numerical approximation for the repulsion between parallel planes, solutions to Eq. [B.25], to obtain more exact numerical solutions for the repulsion potential between equal spheres.[10]

b. The Debye–Hückel Approximation

When κa is large, $\kappa a > 10$, the radius of curvature is larger than the thickness of the double layer, and Derjaguin's approximation is reasonable. When κa is small, $\kappa a < 1$, a different approach is required. Here the conditions are favorable for the Debye–Hückel approximation, and the solution of Eq. [B.25] becomes practicable. Verwey and Overbeek (1948, p. 152) found a relation between the charge on spherical particles and the surface potential. From this relation they were able to calculate the energy of interaction for constant surface potential and constant surface charge density. The results were presented in Tables. A simple approximate expression is

$$\Delta G_R = \frac{4\pi a^2 D \varepsilon_0 \Phi^2 \exp(-\kappa H_0)}{2a + H_0} \qquad [B.51]$$

The conditions for the use of this solution are rarely met in an aqueous medium, but the equation is well suited for suspensions of small particles in nonaqueous media where double-layer thicknesses are large.

c. Constant Surface Charge Density

When sorption equilibrium between the substrate and its potential-determining ions is rapid, the surface potential, which is determined by the activity of the ions in solution, remains constant as two charged surfaces interact. This implies that the surface charge changes during the interaction. For the surface charge to change, material has to diffuse to and from the surface. On the other hand, slow equilibrium, such as might occur when soap, pyrophosphate, or proteins are adsorbed, causes the interaction to take place at constant surface charge. Fortunately, however, the resulting repulsion at constant surface charge density does not vary much from the repulsion at constant surface potential, so the difference may be ignored.[10] Neither extreme condition is likely to occur in a real system, but rather some intermediate, with both surface charge density and surface potential changing slightly.

CHAPTER IIIC

Stability of Systems

1. ELECTROCRATIC AND LYOCRATIC DISPERSIONS

Colloidal dispersions can be divided into two major classes according to their mode of stabilization: (a) lyophilic colloids that acquire stability by solvation of the interface, where the term "solvation" includes all degrees of interaction from mere physical wetting to the formation of adherent thick layers of oriented solvent molecules, which may display highly viscous or gelated properties; (b) lyophobic colloids that are stabilized by an electrostatic repulsion between particles, arising from ions that are either adsorbed on to or dissolved out of the surface of the solid.

Hauser (1939, p. 28) named the two classes of colloid systems lyocratic (governed by solvation) and electrocratic (governed by electrostatics), respectively. The stability of biocolloids and synthetic macromolecules can be attributed to interaction with the solvent (hydration, solvation). Stable dispersions of lyophilic colloids require more than physical wetting: Without a strong anchoring bond the adhesion of the stabilizer is not large enough to withstand relatively small shearing forces; consequently an acid–base interaction combined with the steric effect of a large lyophilic group on the stabilizer is required for stability (Section IV.C.5). Hauser's lyocratic dispersions would nowadays be said to be sterically stabilized, but the old name is still suggestive. To form a stable dispersion of an electrocratic colloid, a surface charge and a surrounding electric double layer are required. In water the two classes of dispersion differ in their sensitivity to added electrolyte. A small quantity of soluble salts added to an electrocratic dispersion causes the particles to coagulate. Lyocratic colloids are

247

less sensitive to added electrolyte, although they too can be coagulated by a high concentration of salts, which changes the hydration forces.

In reality, the two mechanisms are not mutually exclusive: They frequently occur together, although often one evidently predominates and the other is relatively minor. The adsorbed ions that are responsible for the electric charge also provide a hydrated layer on the surface of the particle. The electrostatic repulsion may extend over tens of nanometers, depending on the ionic strength of the solution; the repulsion between hydrated layers extends far less. Steric stabilization, provided by adsorbed polymers of high molecular weight, can extend much farther.

2. ELECTROSTATIC REPULSION VERSUS DISPERSION FORCE ATTRACTION

The significant properties of the electric double layer that are experimentally available are the zeta potential ζ and the thickness of the double layer, $1/\kappa$. Neither of these properties can be measured directly: The zeta potential is computed from measurements of electrokinetic phenomena at infinite dilution of particles, and thickness of the double layer is computed from the ionic strength of the medium. The stability of an electrocratic dispersion depends on both properties. The zeta potential by itself is useful to predict the resistance of a dispersion to flocculation by electrolytes, by determining the "critical zeta potential," that is, the value of ζ below which the suspension is coagulated; but zeta potential alone does not predict the effects of other stresses such as those that affect temporal stability (shelf life), thermal stability, or stability under mechanical shock or shear. The actual determinant of stability is the balance between the electrostatic repulsion as calculated by double-layer theory and the attraction as calculated by dispersion force theory. This concept was developed independently by Derjaguin and Landau in the U.S.S.R. and Verwey and Overbeek in the Netherlands and is consequently known as the DLVO theory.

a.   DLVO Theory: Two Flat Plates

Taking the simplest case of two flat plates, the variation with distance H_0 of the free energy of repulsion is given by

$$\Delta G_R = \frac{64 n_0 k T}{\kappa} \gamma^2 \exp(-\kappa H_0)$$

[B.36]

The energy of dispersion force attraction between two flat plates through medium 2 is

$$\Delta G_A = -\frac{A_{121}}{12 \pi H_0^2}$$

[A.9]

Combining these two equations gives the total free energy per unit area of the interaction between two flat plates as a function of distance:

$$\Delta G_\mathrm{T} = \frac{64 n_0 kT}{\kappa} \gamma^2 \exp(-\kappa H_0) - \frac{A_{121}}{12\pi H_0^2} \qquad [\mathrm{C.1}]$$

b.   DLVO Theory: Two Equal Spheres of Radius a

The total energy of interaction between two equal spheres is also obtained by adding the electrostatic repulsion and the dispersion force attraction. The energy of repulsion between two spheres with thin double layers at shortest distance H_0 between their surfaces, by the Derjaguin approximation, Eq. [B.49], is

$$\Delta G_\mathrm{R} = \frac{64 n_0 \pi a k T \gamma^2}{\kappa^2} \exp(-\kappa H_0) \qquad [\mathrm{B.49}]$$

and the free energy of attraction through medium 2 per sphere is given by Eq. [A.27]:

$$\Delta G_\mathrm{A} = -\frac{A_{121} a}{12 H_0} \qquad [\mathrm{A.27}]$$

Therefore the total energy of interaction is

$$\Delta G_\mathrm{T} = \frac{64 n_0 \pi a k T \gamma^2}{\kappa^2} \exp(-\kappa H_0) - \frac{A_{121} a}{12 H_0} \qquad [\mathrm{C.2}]$$

The variation of total energy of interaction with distance has, on the whole, the same characteristics as that for flat plates, with the exception of the effect of particle size, a. The interaction between two flat plates has no size dependence, whereas the interaction between spheres is directly proportional to the particle size. Thus the height of the barrier to coagulation of spherical particles is 10 times greater for particles of $1 \, \mu m$ compared to particles of $0.1 \, \mu m$. *In general, electrostatic stabilization is more important the larger the particle size, and in practice it is inoperative for microemulsions or colloidal sols.*[59]

3.   APPLICATIONS OF DLVO THEORY TO DISPERSIONS

Figure C.1, computed from measurements of zeta potential and ionic strength for dispersions of platelets of kaolin clay, shows the variation of the potential energy of interaction with the distance of separation between the platelets for different values of the ionic strength in the aqueous medium.[60] The increase in the height of the maximum with decreasing ionic strength demonstrates that the barrier is produced by electrical repulsion.

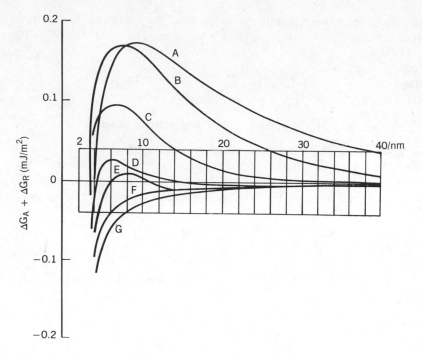

Figure C.1 Total potential energy of interaction between flat plates of kaolin clay as a function of separation at pH = 10.4 for several values of ionic strength of an indifferent electrolyte in the aqueous medium; assuming $\psi_0 = \zeta$ and $A_{121} = 2 \times 10^{-19}$ J. From ref. 60.

A: $\kappa = 0.05\,\text{nm}^{-1}$ $\zeta = -96\,\text{mV}$
B: $\kappa = 0.1\,\text{nm}^{-1}$ $\zeta = -80\,\text{mV}$
C: $\kappa = 0.2\,\text{nm}^{-1}$ $\zeta = -65\,\text{mV}$
D: $\kappa = 0.3\,\text{nm}^{-1}$ $\zeta = -57.3\,\text{mV}$
E: $\kappa = 0.35\,\text{nm}^{-1}$ $\zeta = -54.5\,\text{mV}$
F: $\kappa = 0.4\,\text{nm}^{-1}$ $\zeta = -53\,\text{mV}$
G: $\kappa = 0.5\,\text{nm}^{-1}$ $\zeta = -52\,\text{mV}$

The top curve represents a low concentration of electrolyte for which a repulsive energy barrier amounting to about 50 times kT prevents two platelets coming together. As the concentration of electrolyte is increased in the medium, the energy barrier declines steadily to zero and a coagulation concentration can be identified as the value of n_0 at which the potential energy barrier opposing coagulation just disappears; this concentration is called the *critical coagulation concentration* (CCC). An approximate estimate of the CCC can be obtained by using the expressions above for the free energies of repulsion and attraction. When the potential energy barrier just disappears: $d(\Delta G_A + \Delta G_R)/dH_0 = 0$ and $\Delta G_A + \Delta G_R = 0$. Using Eq. [C.1] and solving these simultaneous equations for H_0 gives $\kappa H_0 = 1$. Substituting $\kappa H_0 = 1$ into Eq. [C.1] and using Eq. [B.17] for κ

gives an expression for the CCC:

$$\text{CCC (molecules/cm}^3) = \frac{(4\pi\varepsilon_0)^3(2^{11}3^2 D^3 k^5 T^5 \gamma^4)}{\pi \exp(4)e^6 A_{121}^2 z^6} \qquad \text{[C.3]}$$

Taking $T = 298\,\text{K}$, $D = 80$, leads to the critical coagulation concentration as a function of Hamaker constant, A_{121}, valency, and surface potential:

$$\text{CCC (moles/liter)} = \frac{8.74 \times 10^{-39}\gamma^4}{z^6 A_{121}^2} \qquad \text{[C.4]}$$

where the Hamaker constant is in joules. When the surface potential is sufficiently high, γ is near unity and is independent of the valency and the surface potential. At lower surface potentials, γ can be approximated by Eq. [B.19] and the CCC then varies with the fourth power of the ξ potential and the inverse square of the valency.

At high surface potentials $\gamma \to 1$ and the coagulation values of uni-, di-, and trivalent ions, according to Eq. [C.4], occur in the ratio $1:(\frac{1}{2})^6:(\frac{1}{3})^6 = 100:1.6:0.13$. This strong dependence of the critical coagulation concentration on the valency (z) is known as the Schulze–Hardy rule, found empirically by Schulze[61] and Hardy.[62] They observed that the coagulating power of the dominating ion increases with its valency roughly in the proportion of $1:100:1000$ for uni-, di-, and trivalent ions. Thus if the univalent ion requires 100 units of concentration for coagulation, the divalent ion would require 1 unit and the trivalent would require 0.1 unit. The DLVO theory provides the theoretical rationale for these observations and their agreement is a major success of the theory.

The effect of electrolyte concentration in reducing the repulsive energy barrier, which is shown clearly in Fig. C.1, has practical applications in understanding the use of alum to coagulate dispersed solids in waste water and in other industrial processes such as papermaking. The alum provides polyvalent ions in solution and effectively compresses the double layer. An example of the precipitation of dispersed clay solids by electrolyte is the sediment plume of the Mississippi River as it mixes with the sea water of the Gulf of Mexico: A sharp demarcation exists between the muddy water of the river and the clear water of the Gulf, as the high electrolyte concentration in the sea coagulates the suspended sediment carried in by the river. (See the cover of *Science* for October 25, 1985, for a photograph of the sediment plume).

Another feature of the potential curves of interaction is shown in curves D and E, Fig. C.1, in the presence of a shallow minimum, known as the secondary minimum, at a relatively large distance of separation ($d = 15$–$25\,\text{nm}$). The presence of a secondary minimum is due to the longer range of the dispersion force attraction compared to that of electrostatic repulsion. The depth of the minimum depends mainly on the attractive force which, for two spheres, is proportional to the size of the particles; therefore the depth of the secondary minimum is also proportional to the size of the particles. Hence effects of the secondary minimum are more likely to occur with larger particles. If the depth of the minimum is several times kT, flocculation results on collision. The nature of flocculation in the secondary minimum differs from the coagulation in the primary minimum in that the equilibrium distance is several times the Debye length, and the particles adhere without uniting. The floc, therefore, is easily redispersed. Practical use is made of this property in paints, where a loosely flocculated structure entails a yield point. This confers two advantages: It prevents running and dripping during application and also prevents a hard sediment from forming in time at the bottom of the can.

4.   COLLOID STABILITY AND COMPLEX ION CHEMISTRY

The work of Matijević and his school[63] shows that the effect of electrolytes on the stability of dispersions in aqueous media is not fully described unless the possibility of complex ion formation in the solution is included. For example, in using aluminium salts as coagulating agents, the nature of the counterion depends on the pH of the solution: at low pH the counterion is Al^{+3}; as the pH is increased the counterions are higher order complexes formed by hydrolysis.

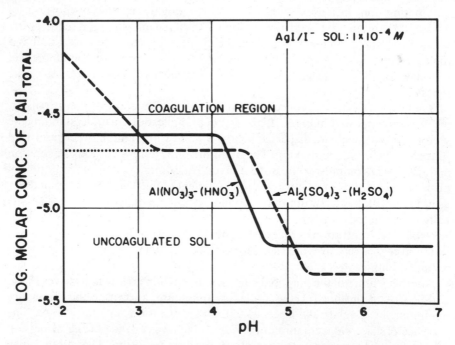

Figure C.2 Critical coagulation concentrations of aluminum nitrate, acidified with HNO_3 when necessary (———), and of aluminium sulfate, acidified with H_2SO_4 when necessary (————), as a function of pH for a negatively charged silver iodide sol.[63]

Complex ions in the medium have two main effects: They change the ionic strength of the medium, and they can change the charge on the particle by adsorption. The high valency of most complex ions does more to increase the ionic strength than the loss of concentration on complexing does to decrease it. The high molecular weight of complex ions also promotes their adsorption. Coordination sometimes occurs on the surface of the particle (chemisorption) instead of in the medium.

Figure C.2 shows the CCC of two aluminum salts on a negatively charged silver-iodide sol as a function of pH. For each salt the CCC decreases at pH values above about 4. At low pH the counterion is Al^{+3}: higher order complexes, such as $Al_8(OH)_{20}^{+4}$ or $Al_7(OH)_{17}^{+4}$, form at higher pH. Because of their higher valency, these complexes are more effective coagulants (lower CCC). The previously held idea that Al^{+3} forms complexes of lower charge with hydroxyl ion as the pH increases leads to the erroneous conclusion that aluminum salts are poor coagulants at high pH.

At a sufficiently high concentration of counterions the suspension can become stable again, as adsorption of the charged counterion reverses the charge. This concentration is called the critical stabilization concentration (CSC). Because

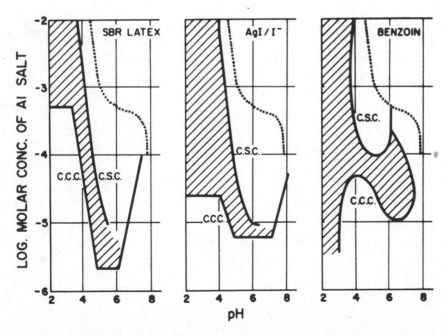

Figure C.3 Phase diagrams of three lyophobic sols, showing stability domains as a function of $Al(NO_3)_3$ (or $AlCl_3$) concentration and pH; styrene–butadiene rubber (SBR) latex (left); silver iodide sol (middle); and benzoin sols prepared from powdered Sumatra gum (right). The formation of aluminium hydroxide precipitate in the absence of sol particles is shown (······).[63]

stability depends on both pH and concentration of counterion, a plot of these variables is a kind of phase diagram revealing stability domains. Figure C.3 represents the behavior of three different sols in the presence of aluminum salts in solution. The lower regions of these diagrams represent stable sols; the shaded regions represent coagulated sols beyond critical coagulation concentrations but at less than critical stabilization concentrations; and the upper regions represent sols stabilized by charge reversal.

Matijević warns that many current textbooks and articles quote data in which the chemistry of complex ions is ignored in explaining colloidal stability.

CHAPTER IIID

Kinetics of Coagulation

No lyophobic sol is absolutely stable against coagulation. The dispersed particles or droplets will, with sufficient collisions, ultimately coagulate and separate by sedimentation or creaming. The relative stability of a dispersed system depends on the rate of particle collisions and the probability of a collision leading to sticking. An approximate kinetic theory for the rate of particle collisions in a dispersion was first proposed by Smoluchowski for particles without interparticle forces (rapid coagulation). Later, Fuchs included the effects of particle–particle attraction and repulsion (slow coagulation) to estimate the initial rate of particle collisions based on the concentration of particles, the solvent viscosity, temperature, particle size, and the potential energy of two particles as a function of separation. The approach used by both was to describe the particle–particle collision rate by the same kinetic equation as that for bimolecular reactions. For a monodispersed system of particles, the rate of change in the number of particles per unit volume is given by the second-order rate equation:

$$\frac{-dn}{dt} = k_2 n^2 \qquad \text{[D.1]}$$

where n is the number of particles per unit volume, t is time, and k_2 is the rate constant. The term dn/dt is the rate of formation of two-particle clusters. Equation [D.1] has the solution

$$1/n = 1/n_0 + k_2 t \qquad \text{[D.2]}$$

where n_0 is the initial concentration of particles. The time $t_{1/2}$ for the concentration to reach to one-half of its initial value is given by

$$t_{1/2} = \frac{1}{k_2 n_0}$$ [D.3]

The form of Eq. [D.1] implies that only the production of doublets is considered and that these doublets form no larger clusters.

We now attempt, following Overbeek,[64] to evaluate k_2 from properties of the dispersed system, that is, properties of the particles and of the medium. Suppose that the particles are in Brownian motion and approach each other by diffusion plus any forces of attraction (e.g., van der Waals) or repulsion (e.g., electrostatic), then the flux of particles diffusing across a spherical surface of area $4\pi r^2$ may be described by Fick's first law of diffusion in spherical coordinates:

$$J = 4\pi r^2 D \left[\frac{dn}{dr} + \frac{n}{Df} \frac{dU_{121}}{dr} \right]$$ [D.4]

where

$$J = \text{flux of particles to a sphere of area } 4\pi r^2$$
$$dn/dr = \text{concentration gradient of particles}$$
$$D = \text{relative diffusion constant}$$
$$U_{121} = \text{potential energy of interaction between two particles in medium 2}$$
$$f = \text{friction constant due to viscosity}$$

The relative diffusion constant is taken to be

$$D = D_1 + D_2$$ [D.5]

where D_1 and D_2 are the single-particle diffusion constants given by the Einstein relation for the diffusion constants of monodisperse hard spheres:

$$Df = kT \text{ or } D = \frac{kT}{6\pi\eta a}$$ [D.6]

Assuming the particle concentration is described by the Boltzmann distribution,

$$n = n_r \exp\left(-\frac{U_{121}}{kT} \right)$$ [D.7]

where n_r is the concentration of particles in the absence of a pair-interaction potential, that is, when U_{121} equals zero. Equation [D.4] is simplified to give

$$dn_r = \frac{(3\eta a J/kT) \exp(+U_{121}/kT) \, dr}{r^2}$$ [D.8]

The boundary conditions for Eq. [D.8] are

$$
\begin{aligned}
&\text{at } r = 2a \quad n_r = 0 \\
&\text{at } r = \infty \quad n_r = n_0 \quad \text{and} \quad U_{121} = 0
\end{aligned}
$$

Hence,

$$
n_0 = \frac{3\eta a J}{kT} \int_{2a}^{\infty} \frac{\exp(U_{121}/kT)dr}{r^2} \tag{D.9}
$$

Rearranging for J gives

$$
J = \frac{kTn_0}{3\eta a \displaystyle\int_{2a}^{\infty} \exp(U_{121}/kT)dr/r^2} \tag{D.10}
$$

J is now the flux of particles to the surface of a sphere of radius a. The rate of two-particle collisions in the dispersion is greater than the flux given by Eq. [D.10] by a factor of n_0. For each collision two singles are lost, and a two-particle cluster is formed. Hence the rate of change in the number of particles is

$$
-\frac{dn}{dt} = 2Jn_0 \tag{D.11}
$$

Combining Eqs. [D.1] and [D.11] gives

$$
k_2 = 2J/n_0 \tag{D.12}
$$

and combining Eqs. [D.10] and [D.12] gives

$$
k_2 = \frac{4kT}{3\eta W} = \frac{6 \times 10^{-12}}{W} \text{ cm}^3/\text{s for water at } 25°C \tag{D.13}
$$

where the factor W is given by

$$
W = 2a \int_{2a}^{\infty} \frac{\exp(U_{121}/kT)dr}{r^2} \tag{D.14}
$$

The factor W is called the stability ratio.[65] When U_{121} is zero everywhere except where particles touch, $W = 1$ and Eq. [D.13] gives the Smoluchowski rate constant for rapid coagulation. If the particles repel each other over a certain distance, W becomes larger, the rate of coagulation becomes smaller, hence the dispersion is more stable.

The above model due to Fuchs embodies several simplifying assumptions:

(a) Only binary collisions are taken into account.
(b) The particles are uniform in size.

(c) The relative diffusion during coagulation is the sum of the single-particle diffusion coefficients.
(d) The effect of viscous flow of the medium on close approach of particles is ignored.

These assumptions have been discussed by several writers[66,67] (Vold and Vold, 1983, p. 261). In general, however, they introduce minor errors compared to the uncertainties that still beset the expressions for attractive and repulsive forces.

Substituting Eq. [D.13] into Eq. [D.3] gives the half-life:

$$t_{1/2} = \frac{3\eta W}{4kTn_0} = \frac{2 \times 10^{11} W}{n_0} \quad \text{s for water at } 25°C \qquad [D.15]$$

Equation [D.15] shows that the half-life of a dispersion depends inversely on the initial concentration of particles. When that concentration is very small, the dispersion may appear to be stable simply because of the few collisions.

An important application of this result is to calculate the interparticle forces from the stability ratio W as determined from the measured half-life of coagulation and the viscosity of the medium. A useful approximation is to replace the energy barrier by a rectangular step of height U_{max} and width $1/\kappa$, so that

$$\begin{aligned} U(r) &= -\infty, & r &< 2a \\ U(r) &= U_{max}, & 2a &< r < 2a + 1/\kappa \\ U(r) &= 0, & r &> 2a + 1/\kappa \end{aligned} \qquad [D.16]$$

Substituting this potential into the expression for the stability ratio, Eq. [D.14], gives

$$W \sim \frac{1}{2\kappa a} \exp \frac{U_{max}}{kT} \qquad [D.17]$$

The calculation of the stability ratio from the more realistic energy curves obtained from DLVO theory, Eq. [C.1] or [C.2], can be approximated in other ways: namely, graphically or by a series expansion of the exponential term or by approximating the maximum by a Gaussian function.

Equation [D.17] shows the approximately exponential dependence of the stability ratio, and hence the stability of the dispersion, on the magnitude of the repulsion between particles.

The Smoluchowski model is a limiting case of no electrostatic repulsion: that is, the stability ratio is $W = 1$. A relation that may be deduced from Eq. [D.17] is the effect of concentration of electrolyte on the value of W when W is large: since κ is proportional to the square root of the concentration of electrolyte, Eq. [B.6], then

$$\log W = -k_1 \log c + k_2 \qquad [D.18]$$

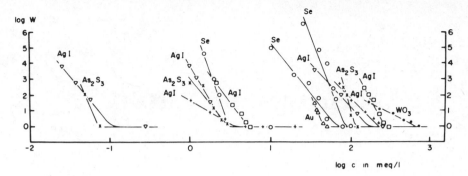

Figure D.1 Effect of electrolyte concentration on the stability ratio of various lyophobic sols. The three groups of curves refer to univalent, divalent, and trivalent electrolytes (from the right) (Miller and Neogi, 1985, p. 131).

This linear equation is verified by experiments as seen in Fig. D.1. The stability ratio is determined by measuring the half-life of the dispersion by its change of turbidity with time. The separation of the effects according to the valency of the counterions is in reasonable agreement with the Schulze–Hardy rule.

References for Part III

1. London, F. The general theory of molecular forces, *Trans. Faraday Soc.* **1937**, *33*, 8–26.
2. Berthelot, D. Sur le mélange des gaz, *C. R. Hebd. Seances Acad. Sci.* **1898**, *126*, 1703–1706.
3. Berthelot, D. Sur le mélange des gaz, *C. R. Hebd. Seances Acad. Sci.* **1898**, *126*, 1857–1858.
4. Pitzer, K. S. Inter- and intramolecular forces and molecular polarizability, *Adv. Chem. Phys.* **1959**, *2*, 59–83.
5. Slater, J. C.; Kirkwood, J. G. The van der Waals forces in gases, *Phys. Rev.* **1931** (2), *37*, 682–697.
6. Lennard-Jones, J. E. Interatomic Forces, *Spec. Publ., Indian Assoc. Adv. Sci.* **1939**, *8*, 1–44.
7. Richmond, P. The theory and calculation of van der Waals forces, *Colloid Sci.* **1975**, *2*, 130–172.
8. Gregory, J. The calculation of Hamaker constants, *Adv. Colloid Interface Sci.* **1969**, *2*, 396–417.
9. Casimir, H. B. G.; Polder, D. The influence of retardation on the London–van der Waals forces, *Phys. Rev.* **1948** (2), *73*, 360–372.
10. Overbeek, J. Th. G., "The interaction between colloidal particles," in *Colloid Science*, Kruyt, H. R., Ed.; Elsevier: New York, 1952; Vol. 1, pp. 245–277.
11. Marmo, M. J.; Mostafa, M. A.; Jinnai, H.; Fowkes, F. M.; Manson, J. A. Acid-base interaction in filler-matrix systems, *Ind. Eng. Chem., Prod. Res. Dev.* **1976**, *15*, 206–211.
12. Hamaker, H. C. The London–Van der Waals attraction between spherical particles, *Physica (Utrecht)* **1937**, *4*, 1058–1072.
13. Fowkes, F. M. Attractive forces at interfaces, *Ind. Eng. Chem.* **1964**, 56(12), 40–52; also *Chemistry and Physics of Interfaces*; Ross, S., Ed.; American Chemical Society: Washington, D.C., 1965, pp. 1–12.

14. Ninham, B. W.; Parsegian, V. A. Van der Waals interactions in multilayer systems, *J. Chem. Phys.* **1970**, *53*, 3398–3402.

15. Vold, M. J. The van der Waals interaction of anisometric colloidal particles, *Proc. Indian Acad. Sci., Sect.A* **1957**,*46*, 152–166.

16. Vold, M. J. The effect of adsorption on the van der Waals interaction of spherical colloidal particles, *J. Colloid Sci.* **1961**, *16*, 1–12.

17. Vold, M. J. Van der Waals' attraction between anisometric particles, *J. Colloid Sci.* **1954**, *9*, 451–459.

18. Sparnaay, M. J. Four notes on van der Waals forces, *J. Colloid Interface Sci.* **1983**, *91*, 307–319.

19. van Oss, C. J.; Visser, J.; Absolom, D. R.; Omenyi, S. N.; Neumann, A. W. The concept of negative Hamaker coefficients II. Thermodynamics, experimental evidence and applications, *Adv. Colloid Interface Sci.* **1983**, *81*, 133–148.

20. Clayfield, E. J.; Lumb, E. C.; Mackey, P. H. Retarded dispersion forces in colloidal particles: Exact integration of the Casimir and Polder equation, *J. Colloid Interface Sci.* **1971**, *37*, 382–389.

21. Fowkes, F. M.; Pugh, R. J. Steric and electrostatic contributions to the colloidal properties of nonaqueous dispersions, *A.C.S. Symp. Ser.* **1984**, *240*, 331–354.

22. Lifshitz, E. M. The theory of molecular attractive forces between solids, *Sov. Phys.– JETP (Engl. Transl.)* **1956**, *2*, 73–83.

23. Dzyaloshinskii, I. E.; Lifshitz, E. M.; Pitaevskii, L. P. The general theory of van der Waals forces, *Adv. Phys.* **1961**, *10*, 165–209.

24. Hough, D. B.; White, L. R. The calculation of Hamaker constants from Lifshitz theory with applications to wetting phenomena, *Adv. Colloid Interface Sci.* **1980**, *14*, 3–41.

25. Parsegian, V. A. Long-range van der Waals forces, *Phys. Chem.: Enriching Top. Colloid Surf. Sci.* **1975**, 27–72.

26. Mitchell, D. J.; Richmond, P. A general formalism for the calculation of free energies of inhomogeneous dielectric and electrolytic systems, *J. Colloid Interface Sci.* **1974**, *46*, 118–127.

27. Mitchell, D. J.; Richmond, P. The force between two charged dielectric half spaces immersed in an electrolyte, *J. Colloid Interface Sci.* **1974**, *46*, 128–131.

28. Mitchell, D. J.; Ninham, B. W. Van der Waals forces between two spheres, *J. Chem. Phys.* **1972**, *56*, 1117–1126.

29. Mitchell, D. J.; Ninham, B. W.; Richmond, P. Van der Waals forces between cylinders. I. Nonretarded forces between thin isotropic rods and finite size corrections, *Biophys. J.* **1973**, *13*, 359–369.

30. Smith, E. R.; Mitchell, D. J.; Ninham, B. W. Deviations of the van der Waals energy for two interacting spheres from the predictions of Hamaker theory, *J. Colloid Interface Sci.* **1973**, *45*, 55–68.

31. Derjaguin, B. V.; Titijevskaia, A. S.; Abricossova, I.I.; Malkina, A. D. Investigations of the forces of interaction of surfaces in different media and their application to the problem of colloid stability, *Discuss. Faraday Soc.* **1954**, *18*, 24–41.

32. Tabor, D.; Winterton, R. H. S. Surface forces: Direct measurement of normal and retarded van der Waals forces, *Nature (London)* **1968**, *219*, 1120–1121.

33. Israelachvili, J. N.; Tabor, D. Measurement of van der Waals dispersion forces in the range 1.4 to 130 nm, *Nature (London), Physical Sci.* **1972**, *236*, 106.

34. Rabinovich, Ya. I.; Derjaguin, B. V.; Churaev, N. V. Direct measurements of long-range surface forces in gas and liquid media, *Adv. Colloid Interface Sci.* **1982**, *16*, 63–78.

35. Croucher, M. D. A simple expression for the Hamaker constant of liquid-like materials, *Colloid Polym. Sci.* **1981**, *259*, 462–466.

36. Bargeman, D.; Van Voorst Vader, F. Van der Waals forces between immersed particles, *J. Electroanal. Chem. Interfacial Electrochem.* **1972**, *37*, 45–52.

37. Visser, J. On Hamaker constants: A comparison between Hamaker constants and Lifshitz-Van der Waals constants, *Adv. Colloid Interface Sci.* **1972**, *3*, 331–363.

38. Croucher, M. D.; Hair, M. L. Hamaker constants and the principle of corresponding states, *J. Phys. Chem.* **1977**, *81*, 1631–1636.

39. Quincke, G. Über eine neue art elektrischer ströme, *Ann. Phys. Chem. (Poggendorf's)* **1859**, *107*, 1–47.

40. Quincke, G. Über die fortführung materieller theilchen durch strömende elecktricität, *Ann. Phys. Chem. (Poggendorf's)* **1861**, *113*, 513–598.

41. Quincke, G. Über capillaritäts-erscheinungen an der gemeinschaftlichen oberfläche zweier flüssigkeiten, *Ann. Phys. Chem. (Poggendorf's)* **1870**, *139*, 1–89

42. Fuerstenau, D. W. The adsorption of surfactants at solid-water interfaces, *Chem. Biosurfaces* **1971**, *1*, 143–176.

43. Ottewill, R. H.; Watanabe, A. Studies on the mechanism of coagulation. Part 2. The electrophoretic behavior of positive silver-iodide sols in the presence of anionic surface-active agents, *Kolloid Z.* **1960**, *171*, 132–139.

44. Olivier, J. P.; Sennett, P. Electrokinetic effects in kaolin-water systems. I. The measurement of electrophoretic mobility, *Clays Clay Miner.* **1967**, *15*, 345–356.

45. Gouy, G. Sur la constitution de la charge électrique a la surface d'un électrolyte, *J. Phys. Theor. Appl.* **1910** (4), *9*, 457–467.

46. Chapman, D. L. A contribution to the theory of electrocapillarity, *Phil. Mag.* **1913**, (6), *25*, 475–481.

47. Grahame, D. C. The electrical double layer and the theory of electrocapillarity, *Chem. Rev.* **1947**, *11*, 441–501.

48. Lau, A.; McLaughlin, A; McLaughlin, S. The adsorption of divalent cations to phosphatidylglycerol bilayer membranes, *Biochim. Biophys. Acta* **1981**, *645*, 279–292.

49. Pashley, R. M. Hydration forces between mica surfaces in electrolyte solutions, *Adv. Colloid Interface Sci.* **1982**, *16*, 57–62.

50. Stigter, D. Functional representation of properties of the electrical double layer around a spherical colloid particle, *J. Electroanal. Chem. Interfacial Electrochem.* **1972**, *37*, 61–64.

51. Dukhin, S. S.; Derjaguin, B. V. Equilibrium double layer and electrokinetic phenomena, *Surf. Colloid Sci.* **1974**, *7*, 49–272.

52. Lyklema, J.; van Leeuwen, H. P. Dynamic properties of the AgI-solution interface: implications for colloid stability, *Adv. Colloid Interface Sci.* **1982**, *16*, 127–137.

53. Stern, O. Zur Theorie der elektrolytischen doppelschicht, *Z. Elektrochem. Angew. Phys. Chem.* **1924**, *30*, 508–516.

54. Sparnaay, M. J. Ion-size corrections of the Poisson-Boltzmann equation, *Electroanal. Chem. Interfacial Electrochem.* **1972**, *37*, 65–70.

55. Langmuir, I. The role of attractive and repulsive forces in the formation of tactoids, thixotropic gels, protein crystals and coacervates, *J. Chem. Phys.* **1938**, *6*, 873–896.

56. Derjaguin, B. V.; Landau, L. Theory of the stability of strongly charged lyophobic sols and of the adhesion of strongly charged particles in solutions of electrolytes, *Acta Physicochim. URSS* **1941**, *14*, 733–762.

57. Reerink, H.; Overbeek, J. Th. G. The rate of coagulation as a measure of the stability of silver iodide sols, *Discuss. Faraday Soc.* **1954**, *18*, 74–84.

58. Wiese, G. R.; James, R. O.; Yates, D. E.; Healey, T. W. Electrochemistry of the colloid-water interface, in *Physical Chemistry*; Butterworths: London, 1976; Ser. 2, Vol. 6, pp. 53–102.

59. Koelmans, H.; Overbeek, J. Th. G. Stability and electrophoretic deposition of suspensions in nonaqueous media, *Discuss. Faraday Soc.* **1954**, *18*, 52–63.

60. Hunter, R. J.; Alexander, A. E. Surface properties and flow behavior of kaolinite., Part I: Electrophoretic mobility and stability of kaolinite sols, *J. Colloid Sci.* **1963**, *18*, 820–832.

61. Schulze, H. Schwefelarsen in wässeriger lösung, *J. Prakt. Chem.* **1882**, (2), *25*, 431–454.

62. Hardy, W. B. A preliminary investigation of the conditions that determine the stability of irreversible hydrosols, *Proc. Roy. Soc. London* **1900**, *66*, 110–125.

63. Matijević E. Colloid stability and complex chemistry, *J. Colloid Interface Sci.* **1973**, *43*, 217–245.

64. Overbeek, J. Th. G. Kinetics of flocculation, in *Colloid Science*; Kruyt, H. R., Ed.; Elsevier: New York, 1952: Vol. 1, pp. 278–301.

65. von Smoluchowski, M. Versuch einer mathematischen theorie der koagulationskinetik kolloider lösungen (Mathematical theory of the kinetics of the coagulation of colloidal solutions). *Z. Physik. Chem., Stoechiom. Verwandschaftsl.* **1917**, *92*, 129–168.

66. Spielman, L. A. Viscous interactions in Brownian coagulation, *J. Colloid Interface Sci.* **1970**, *33*, 562–571.

67. Honig, E. P.; Roebersen, G. J.; Wiersema, P. H. Effect of hydrodynamic interaction on the coagulation rate of hydrophobic colloids, *J. Colloid Interface Sci.* **1971**, *36*, 97–109.

PART IV
DISPERSED-PHASE SYSTEMS

CHAPTER IVA

Emulsions

1. DEFINITIONS AND GLOSSARY OF TERMS

An emulsion is a dispersion of one liquid in another in which it is immiscible. The particle sizes of the dispersed phase lies between a few hundred nanometers and a few micrometers. Stable emulsions require the presence of a third component (the emulsifying agent); but practical emulsions seldom consist of only three components. Polycomponent systems are not readily accessible to mathematical descriptions, so the understanding (technology) of emulsions is largely a matter of empirical rules. Like other specialities, emulsion technology had developed its own language and definitions. The two immiscible liquids that constitute an emulsion are referred to as "oil" and "water," as these are proverbial representatives of two such liquids. Within an emulsion, one liquid phase is in the form of droplets and is therefore distinguished from the other phase. A number of different terms, listed in Table A.1, are used to express this distinction.

Table A.1 Terminology of Phases

Phase 1	Phase 2
Droplet	Serum
Dispersed	Medium
Discontinuous	Continuous
Internal	External

Emulsions appear as two types: water droplets dispersed in oil, designated W/O; and oil droplets dispersed in water, designated O/W. Read W/O as water in oil, and O/W as oil in water. A common example of an O/W emulsion is milk, and a common example of a W/O emulsion is butter. When one type of emulsion is altered to the other, the process is called "inversion." When an emulsion separates into its two constituent phases, it is said to be "broken." Because a density difference may exist between the two phases, the dispersed phase may rise or sink within the medium; these processes are called "creaming" or "sedimenting" (or "settling"), respectively, and they are not the same as breaking of the emulsion.

2. DETERMINATION OF EMULSION TYPE

The simplest way to determine the type of an emulsion is to see whether a small volume of the emulsion mixes readily with water; if it does so, the continuous phase of the emulsion is aqueous. This test may be performed under a microscope using a glass rod to mix the water and the emulsion. Conversely a W/O emulsion mixes readily with oil and not with water. Concentrated emulsions are highly viscous and, even when water is the continuous phase, may not readily mix with water. Such emulsions are likely to defy any test for type that can be suggested.

Another test for emulsion type is the electrical conductivity. O/W emulsions almost always have high conductivity, whereas W/O emulsions have low conductivity. Very concentrated emulsions may invert on addition of more of their internal phase or on other change of conditions. Such inversion is accompanied with a large reduction of viscosity, or "thinning."

3. COALESCENCE OF EMULSION DROPLETS

We have already seen (Section II.A.5) that the Helmholtz free-energy change for the process of coalescence of two droplets at constant volume, temperature, and composition is negative and that therefore coalescence is spontaneous. The situation is different for emulsion droplets, as the reduction of area consequent on coalescence leaves less room for adsorbed solute, which therefore has to be returned to the solution. But the adsorbate had arrived at the interface spontaneously, and its removal or desorption requires energy. Therefore, for the net change of free energy on coalescence of emulsion droplets to be positive, and the emulsion to be sufficiently stable so as not to coalesce spontaneously, the third component should have a negative free energy of adsorption large enough to overcome any effect of area reduction. This means that the necessary third component of a stable emulsion is a surface-active solute. Another consequence of this condition, a concomitant of adsorption, is the reduction of interfacial tension occasioned by the third component. If the net free energy of coalescence of an emulsion is positive, spontaneous emulsification is thermodynamically possible. Nowadays, with many powerful emulsifying agents available, spontaneous emulsification is commonplace and widely accepted; but so improbable did it appear to the savants of 60 years ago that when J. W. McBain announced this discovery (or rather rediscovery, for it was known since 1878) to the Royal Society, it was dismissed by the chairman, Sir W. B. Hardy, with the comment, "Nonsense, McBain!"

4.   BANCROFT'S RULE

W. D. Bancroft pointed out that the type of emulsion is derived, at least in part, from the nature of the emulsifying agent. He formulated the rule that, in the making of an emulsion, the liquid in which the agent is the more soluble becomes the continuous phase. Thus, for example, a water-soluble substance like sodium oleate emulsifies oil in water, and an oil-soluble substance like calcium oleate emulsifies water in oil. The rule tends to break down when high concentrations of internal phase are emulsified, but it may well be that the larger relative quantity of internal phase leaches emulsifier from the continuous phase. Bancroft provided no explanation of this rule, which yet has been found of wide general validity. The following considerations help to make Bancroft's rule more intelligible.

In an aqueous medium a preponderantly hydrophilic solute will have either an ionic charge or a polyethylene-oxide moiety, which because of its ability to attach water by hydrogen bonding creates a thick gelated layer on the aqueous side of

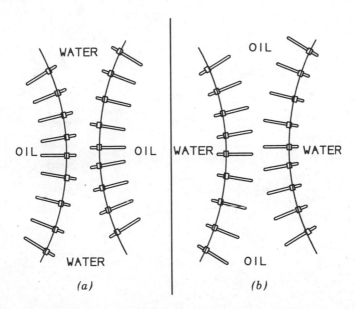

Figure A.1 A lipophilic solute adsorbed at the oil–water interface is more effective in keeping apart water drops in an oil medium (b) than in keeping apart oil drops in a water medium (a).[1]

the interface. The electric charge entails an electrostatic field, by which a force of repulsion for like charges extends outwards from the interface into the aqueous medium. Therefore, these hydrophilic characteristics confer on the solute molecule either an electrostatic or a steric field of repulsion that extends farther into the aqueous than into the nonaqueous medium. The converse holds for molecules of preponderantly lipophilic solute. Here again, both steric and electrostatic fields of mutual repulsion are possible. The mechanism of charge separation in lipophilic solvents is now better understood (Section IV.C.4). It depends on the solubilization of counterions by oil-soluble or inverse micelles, which are created by the preponderantly lipophilic solute. In nonaqueous media, because of the low dielectric constant, the repulsive field of force can extend for a considerable distance from the interface. Steric repulsion in a nonaqueous medium is created by the same mechanism as in an aqueous medium, namely, by adsorption of solute molecules in an orientation of minimum potential energy at the interface. If solute molecules are represented schematically in terms of their respective fields of force rather than by a literal interpretation of the actual volumes of their atomic constituents, we should expect to see for the preponderantly hydrophilic molecule a larger territory in the aqueous medium, and for the preponderantly lipophilic molecule a larger territory in the nonaqueous medium. The former molecules are therefore more effective in separating oil drops in an aqueous medium than in separating water drops in an oil medium; the latter molecules are more effective in separating water drops in an oil medium than in separating oil

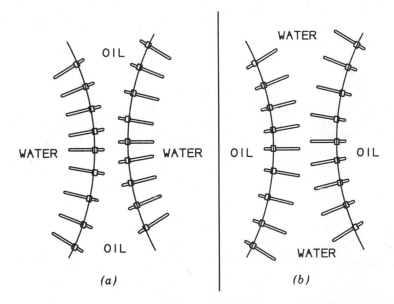

(a)   (b)

Figure A.2 A hydrophilic solute adsorbed at the oil–water interface is more effective in keeping apart oil drops in a water medium (b) than in keeping apart water drops in an oil medium (a).[1]

drops in an aqueous medium. These comparisons are shown schematically in Figs. A.1 and A.2. The destabilization of an emulsion or a dispersion can be achieved by destroying or diminishing the repulsive field of force, or territory, created by the interfacially active solute.

5.　AMPHIPATHIC PARTICLES AS EMULSION STABILIZERS

Effective resistance to coalescence of emulsion drops is conferred by certain finely divided solids. The basic sulfates of iron, copper, nickel, zinc, and aluminium in moist condition emulsify petroleum oil in water; carbon black, rosin, and lanolin stabilize water in kerosene and in benzene. Any insoluble substance that is wetted more readily by water than by oil, if sufficiently finely divided, will serve to emulsify oil in water. In some cases it is possible to see under a microscope the coating of solid particles that surrounds the oil drop. These emulsions resist spontaneous demulsification since the interface is mechanically stronger than an interface built from soluble components. Excellent emulsions may be formed by adding lime, or lime-water, to the normal sulfates of iron or copper; on adding kerosene emulsification is produced by slight agitation.[2] Besides their ease of manufacture and the absence of spontaneous demulsification, these emulsions are not decomposed by adding caustic soda. When the copper salt is used, the emulsion has the insecticidal properties of Bordeaux mixture,* without the disadvantage of settling out as a hard-packed sediment.

A general resemblance to Bancroft's rule is demonstrated by the properties described above, as the moist surfaces of the inorganic salts are predominantly hydrophilic and so tend to promote O/W emulsions, while the surfaces of carbon black and rosins are predominantly lipophilic and so tend to promote W/O emulsions. This resemblance can be demonstrated by a simple model of a spherical particle of radius a sited at the interface between liquids 1 and 2. When the particle is at the interface, the area immersed in liquid 1 is A_{1s} the interfacial energy as F_{1s} and the area immersed in liquid 2 is A_{2s} with an interfacial energy of F_{2s}; and it has displaced an area A_{12} of the interface with an interfacial energy of σ_{12}. The equilibrium position of the particle is determined by the minimum in free energy:

$$dF = F_{1s}\,dA_{1s} + F_{2s}\,dA_{2s} + \sigma_{12}\,dA_{12} = 0 \qquad [\text{A.1}]$$

The change in area can be calculated by simple geometry and the equation simplified to

$$F_{2s} - F_{1s} = \sigma_{12}\cos\theta \qquad [\text{A.2}]$$

which is the Young–Dupré equation for contact-angle equilibrium. The particle

*Bordeaux mixture is an aqueous suspension of copper sulfate and lime.

will seek a position such that θ becomes the equilibrium contact angle. Equation [A.1] tacitly disregards buoyancy forces.

Equation [A.1] can be satisfied by two positions of the sphere, giving two complementary values of θ, of which, however, only one is consistent with the Young–Dupré equation, [A.2]. The two solutions are given by

$$\cos \theta = 1 - h/r \qquad \text{[A.3]}$$

Where the true angle of contact is acute, $h/r < 1$; and where the true angle of contact is obtuse, $h/r > 1$. Figure A.3 shows that the liquid making the smaller angle of contact with the particle contains the bulk of the particle. The emulsion droplets are stabilized by virtue of the steric repulsion of the bulk of the particles in the continuous phase, that is, by the same mechanism described in Section IV.C.5 for molecular stabilizers of suspensions. The liquid with the smaller angle of contact has the larger work of adhesion to the particle surface. We may, therefore, rephrase Bancroft's rule for applications to emulsions stabilized by solid particles: the liquid with the larger work of adhesion to the particle is the continuous phase.

The angle of contact at the three-phase boundary between two immiscible liquids and a solid substrate can be calculated if the surface tensions of the two liquids in the presence of both of their vapors, the interfacial tension between them, and the contact angles that each of them makes separately with the substrate are all known. If these contact angles are measured with the purpose of using this calculation, the substrate should be in equilibrium with the combined vapors of *both* liquids. The works of adhesion of each liquid on the substrate are given by (Eq. [IIA.51] and [IIA.52])

$$W_1^{\text{adh}} = F_{\text{sv}} + \sigma_1 - F_{1\text{s}} = \sigma_1 \left(1 + \cos \theta_1\right) \qquad \text{[A.4]}$$

$$W_2^{\text{adh}} = F_{\text{sv}} + \sigma_2 - F_{2\text{s}} = \sigma_2 \left(1 + \cos \theta_2\right) \qquad \text{[A.5]}$$

The Young–Dupré equation for the three-phase boundary of the two liquids

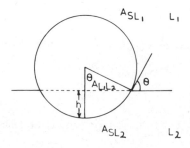

Figure A.3 Solid sphere at the liquid–liquid interface (Becher, 1983, p. 274).

in equilibrium with the substrate is

$$F_{2s} - F_{1s} = \sigma_{12} \cos \theta_3 \qquad \text{[A.6]}$$

where θ is contact angle measured through liquid 1. Substituting [A.4] and [A.5] into [A.6] gives

$$\sigma_{12} \cos \theta = (\sigma_2 - \sigma_1) - (W_2^{adh} - W_1^{adh}) \qquad \text{[A.7]}$$

from which

$$\sigma_{12} \cos \theta = \sigma_1 \cos \theta_1 - \sigma_2 \cos \theta_2 \qquad \text{[A.8]}$$

As an example, consider a sphere of PTFE at the interface between white oil and water. The surface tension of water and its contact angle against PTFE are 72.9 mN/m and 115.7°; the surface tension of white oil and its contact angle against PTFE are 28.9 mN/m and 50.0°; the interfacial tension between white oil and water is 51.3 mN/m. An application of Eq. [A.8] gives the angle of contact of white oil on the PTFE sphere at the three-phase boundary, that is, in the presence of water, as 12°. The sphere would, therefore, be situated at the oil–water interface, immersed in the water to a depth of about 2% of its radius.

6.   THE HLB SCALE

Bancroft's rule points out the importance of the emulsifier in determining the type of emulsion. An attempt to give the rule quantitative expression was initiated in 1949 by Griffin[3-4] of the Atlas Powder Company (now ICI Americas). This company markets a number of nonionic surface-active solutes used as emulsifiers in food industries. Some of these are water soluble and, in accordance with Bancroft's rule, used as stabilizers of O/W emulsions; others are oil soluble and are used as stabilizers of W/O emulsions. Griffin created a continuous series of emulsifying agents, ranging from 100% oleic acid, predominantly lipophilic, to 100% sodium oleate, predominantly hydrophilic, by making mixtures of known composition from these two substances. The relative proportions of the two ingredients determine the balance between the hydrophilic and lipophilic properties, or HLB, of the mixture. A value of 1 was assigned arbitrarily to oleic acid and a value of 20 to sodium oleate; intermediate values were based on the relative amounts of each constituent in the composition, as follows:

$$HLB = 1 W_1 + 20 W_2 \qquad \text{[A.9]}$$

where W_1 is the weight fraction of oleic acid and W_2 is the weight fraction of sodium oleate. Each composition was tested as an emulsifying agent by adding 1 g to a mixture of 50 ml of a refined white oil and 50 ml of water and shaking on a standard shaker for a fixed time. The emulsions were poured into viewing tubes,

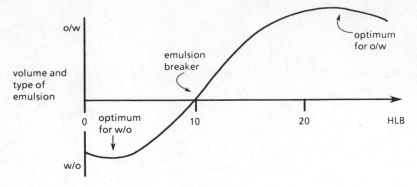

Figure A.4 Variation of type and amount of residual emulsion with HLB number of emulsifier.

held for a period of time, and compared as to the volume and type of stable emulsion remaining. Where two types of emulsion were present in the same tube, only the one of larger volume was noted. A presentation of typical data is shown in Fig. A.4.

The compositions made with oleic acid and sodium oleate serve as reference standards for commercial emulsifiers, which are tested in exactly the same way. The commercial emulsifier is assigned the HLB number corresponding to the reference sample that it most closely matches. Those that form W/O emulsions have a low HLB number, and those that form O/W emulsions have a high HLB number. After much experience, the HLB number was found to be associated with other applications: A particular number was best for a particular application of a surface-active solute. A summary of the HLB range required for different purposes is given in Table A.2.

The original method of determining HLB numbers is so profligate of time that shorter routes to the same end have been suggested. One approach is to calculate the HLB number of a molecule by adding contributions from its constituent

Table A.2 Summary of Applications at Different Ranges of HLB

HLB Range	Application
3.5–6	W/O emulsifier
7–9	Wetting agent
8–18	O/W emulsifier
13–15	Detergent
15–18	Solubilizer

From ref. 5.

Table A.3   Davies' HLB Group Numbers

	Group Number
Hydrophilic Groups	
—O-SO$_3^-$ Na$^+$	38.7
—COO$^-$K$^+$	21.1
—COO$^-$Na$^+$	19.1
N (tertiary amine)	9.4
Ester (sorbitan ring)	6.8
Ester (free)	2.4
—COOH	2.1
—OH (free)	1.9
—O—	1.3
—OH (sorbitan ring)	0.5
Lipophilic Groups	
—CH—	
—CH$_2$—	0.475
CH$_3$—	
=CH—	
Derived Groups	
—CH$_2$CH$_2$O—	0.33
—CHCH$_3$CH$_2$O—	−0.15

From ref. 5.

groups, using the empirical relation obtained by Davies[5]:

$$HLB = 7 + \sum (\text{hydrophilic group numbers}) - \sum (\text{lipophilic group numbers})$$

The HLB group numbers are given in Table A.3. Extensive bibliographies on HLB are given by Becher and Griffin in *McCutcheon's Detergents and Emulsifiers* (1985) and Becher (1985) in *Encyclopedia of Emulsion Technology, Volume 2, Applications.*

The agreement between HLB numbers determined experimentally and those calculated from group numbers is satisfactory. The concept underlying the HLB number, as it is based on Bancroft's rule, is informative about the emulsion type but not about emulsion stability. Emulsion stability is the result of a complex interaction of properties, which include droplet size, interfacial viscosity, the magnitude of electrostatic and steric repulsion, internal volume, and so on. For this reason the application of the HLB concept cannot be expected to provide the whole answer to practical emulsion problems.

7. THE PHASE-INVERSION TEMPERATURE

The HLB number of an emulsifier corresponds to a composition on a graded scale from those that stabilize W/O emulsions to those that stabilize O/W emulsions. The HLB number of an emulsifier varies with temperature because the relative solubilities of the lipophile and the hydrophile vary with temperature. The variation of solubility with temperature is most profound for nonionic emulsifiers as their solubility in water depends on hydrogen bonding. At higher temperatures hydrogen bonding is weakened by thermal forces and the emulsifier is less soluble in water. Common nonionic emulsifiers are water soluble at low temperatures, where they stabilize O/W emulsions; and are oil soluble at higher temperatures, where they stabilize W/O emulsions. The temperature at which the emulsifier changes from an O/W stabilizer to a W/O stabilizer is called the *phase-inversion temperature* or PIT. The PIT of an emulsifier is the temperature at which its lipophilic nature and its hydrophilic nature just balance. The PIT, therefore, is a measure by which the hydrophile–lipophile balance of emulsifiers, particularly nonionics, can be classified. The advantage of the PIT over the HLB number is that it is easier to determine experimentally. Both the HLB number and the PIT are functions of the composition of the oil and the ionic strength of the water. The study of the effects of these variables on emulsion stability is rendered simpler by use of the PIT.

The HLB number and the PIT are correlated. Figure A.5 shows the relation of the HLB number to the PIT of a variety of nonionic emulsifiers for cyclohexane and water emulsions. At temperatures above the PIT, the emulsifier stabilizes the W/O emulsion; at temperatures below the PIT, the emulsifier stabilizes the O/W emulsion. The relation between the HLB number and PIT is determined for a variety of oils and ionic strengths (Shinoda and Friberg, 1986).

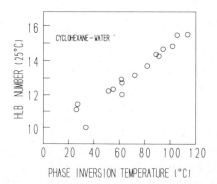

Figure A.5 The correlation between HLB numbers of nonionic surface-active solutes and their PIT in cyclohexane and water emulsions, at 3% by weight of emulsifier (Shinoda and Friberg, 1986, p. 63).

8.   MECHANICAL PROPERTIES OF THE INTERFACE

Emulsifying agents are adsorbed at the oil–water interface, which may become so crowded with solute molecules that it develops rheological properties different from those of either bulk phase. The interface may be non-Newtonian, plastic for example, while the oil and the water phases remain Newtonian. Great stability is conferred on an emulsion by a plastic interface, which is solid at low shearing stresses and so prevents coalescence of droplets on collision. Many instruments have been designed to measure the rheology of an interface. The simplest of these will hardly do more than register the presence of a plastic interface; a more sophisticated design is capable of providing quantitative information that can be interpreted in terms of absolute film properties.

The rheological behavior of an interface can be measured either by shearing by a ring or disc in the plane of the surface while the area remains constant, or by expanding or contracting the interface. An example of the first method is the use of a torsional pendulum, which is damped by immersing its bob, in the form of a double cone so as to have one cone in the upper liquid, the other cone in the lower liquid, and the widest part of the bob, that is, where the cones are joined, at the interface. The torsion wire is supported at the end of a shaft that can be given a partial turn by a lever. The damping of the oscillations is measured by a light beam reflected from a mirror attached to the wire. The rate at which the amplitude of the oscillation decreases is logarithmic when the pendulum is placed in a Newtonian fluid.[6] Normally the two bulk liquid phases in which the cones are immersed are Newtonian fluids, but the interface may not be so. The bob is so designed that the damping influence of the interface is maximized. A plastic interface would affect the damping behavior sufficiently to throw it out of its regular semilogarithmic pattern; the amplitudes of the oscillations decrease more rapidly. Plasticity is characterized by a "yield point," that is, the critical shearing stress above which flow begins. At certain portions of its oscillatory cycle, the motion of the pendulum produces a shearing stress that is less than the yield point; the result is as much as if the free oscillation were halted from time to time by an applied force (Fig. A.7).

An instrument capable of measuring the rheology of an interface at nearly constant shearing stress is a two-dimensional analogue of the Couette type of viscosimeter. The sample is contained in a cylindrical vessel on top of a turntable, geared down to 0.5 rpm or less. The bob is placed at the interface, and carefully positioned (for reproducibility) by means of a cathetometer. The torque communicated to its supporting wire by the rotation of the turntable is measured by a strain gauge. A instrument of this type is available from Zed Instruments Ltd. (Appendix J). Figure A.6 shows details of the apparatus. A plot of the angle of deflection versus the rate of rotation is a straight line through the origin for a Newtonian interface; shear thinning and yield point character is revealed on the appropriate rheogram.

Another type of instrument is reported by Biswas and Haydon[7]. It uses an electromagnetic device to drive a ring placed at the interface at constant applied

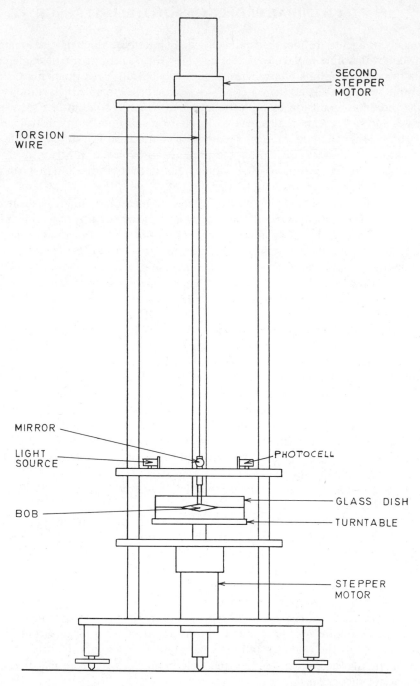

Figure A.6 Schematic of a surface viscosimeter with an oscillating circular knife edge. (Courtesy of Zed Instruments Ltd.)

278

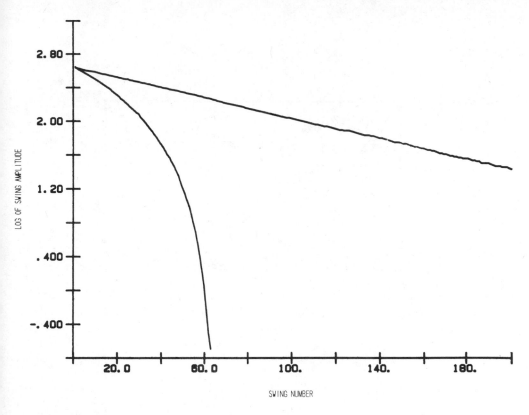

Figure A.7 The semilogarithmic damping of the torsion pendulum for Newtonian and shear-thinning interfacial flow.

stress or to hold a constant position at varying stress. The current in the electromagnet is a direct measure of the force applied to the ring at any time. The motion imparted is a two-dimensional analogue of Couette flow between concentric cylinders and is analyzed in a similar way. As well as measuring the coefficient of viscosity, the instrument can be used in two types of quasi-static experiments (i.e., those in which the shear rate is so small that inertial effects can be ignored): creep recovery and stress relaxation.

Instruments of similar design are used to measure the rheology of liquid–air surfaces (Section IV.B.6b).

9. VARIATION WITH CONCENTRATION OF INTERNAL PHASE

The closest packing of uniform spheres fills about 74% of space; each sphere has 12 nearest neighbors, and about 26% of the space is void. Applied to emulsions, it seemed that some hindrance would be met on trying to emulsify more than 74% by volume of internal phase. So insistent on this point was

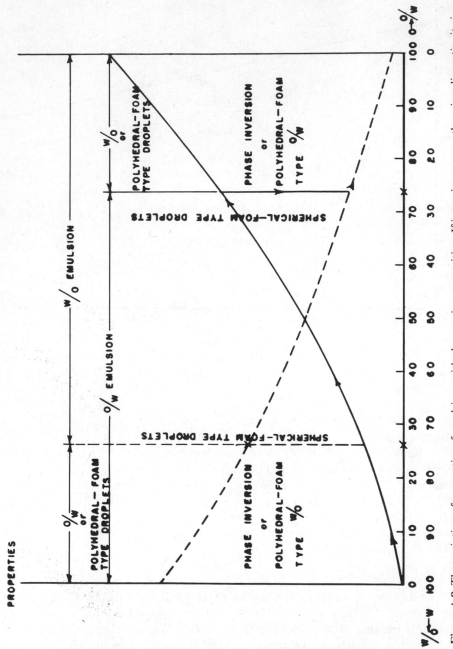

Figure A.8 The variation of properties of emulsions with changes in properties, as they change from one curve to the other. Above 74% there is a discontinuity in properties, as they change from one curve to the other. Above 74% there is either a phase inversion or the droplets are deformed to polyhedra[9] (Manegold, 1952, p. 23).

Wolfgang Ostwald that he stimulated an Englishman, S. U. Pickering to disprove it. Pickering succeeded in making emulsions containing 99 + % internal phase. In such systems the continuous phase must be in the form of thin lamellae, and the stability of the emulsion depends on these lamellae being heavily gelated and incapable of flow. This property is secured by the high concentration of soap in the water which brings the system into a mesophase region of the phase diagram. The lamellae are liquid crystals, immobilizing water within their structure, and forming a gel. Pickering proved his point, but the concentration of 74% by volume internal phase nonetheless influences the physical properties of an emulsion. Even though emulsions are seldom composed of droplets of uniform size, droplets begin to make close contact at between 60 and 70% volume concentration; and close contact begins to affect some of their sensible properties.

Manegold (1952) has published a useful diagram (Fig. A.8) to represent the dependence of some properties of emulsions on volume concentration. The diagram is a summary of general behavior, which cannot be relied on to predict properties of a specific emulsion. Above a volume concentration of about 70 the properties of an emulsion become either discontinuous (inversion) or remain continuous, with the production of deformed droplets in contact. The deformation of the spheres is created by concentrations of internal phase greater than about 70% by volume. This type of emulsion is called "polyhedral foam type," using an analogy with the structure of drained foam, and, like such a foam, has a large structural viscosity. Ostwald called such emulsions "liquid–liquid foams," Bütschli (1892) pointed out their analogy to protoplasm, and Sebba[8] has suggested that the secret of life lies in the behavior of immiscible liquids under the influence of their interfacial forces.

a. Rheology of Emulsions

A readily observed property of emulsions is viscosity. At concentrations below 74% the spheres are not in contact, and the flow of the emulsion is not unduly impeded by interference of the droplets with one another. In more concentrated emulsions, interference occurs and the resistance to flow becomes more marked. Finally, the droplets may be packed so close to each other that flow is seriously impeded, which is made manifest by a high viscosity, requiring a large shearing stress to overcome the structure built up by many spheres in contact. Should the emulsion invert on addition of more of the internal phase, as frequently happens, the inversion is marked by a sudden reduction of viscosity. The transition is a remarkable transformation from a consistency like ointment to that of a thin cream.

Measurements of the viscosities of emulsions illustrate the changes that occur with concentration of internal phase. Figure A.9 shows the data of Richardson[10] for both O/W and W/O types. The significant point is the enormous increase of viscosity of the O/W emulsion at concentration above about 70% by volume. The

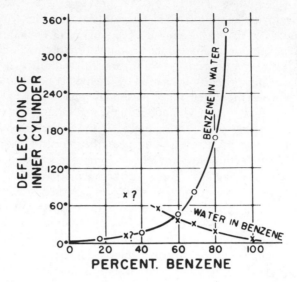

Figure A.9 The viscosity of two types of emulsion as a function of composition.[9,10]

other type of emulsion shows inversion at higher concentrations, with a discontinuity in the observed property.

In emulsions with less than 70% internal phase, the size of the droplets has little effect on the viscosity. As the concentration approaches that of polyhedral foam type where the viscosity would be expected to increase sharply, a wider distribution of droplet sizes makes it possible to include larger volume concentrations before structural viscosity becomes marked. An ordered arrangement of closed-packed spheres forms readily and spontaneously when the spheres are uniform in size; a relatively small number of outsized spheres prevents such an arrangement from forming, thus reducing the average number of nearest neighbors, which increases the free motion of the spheres past each other. Homogenizing such an emulsion narrows the size distribution and results in increased viscosity.

Concentrated emulsions and foams have many practical applications, which take advantage of their peculiar rheological properties, such as a high viscosity compared to that of their pure constituents, a yield stress, and shear-thinning behavior. Concentrated emulsions also provide a means to transport highly viscous fluids as emulsified droplets. Asphalt emulsions, used as protective coatings in road paving and as a means to mix asphalt with fibers and other forms of particulate matter, are more convenient to handle than bulk asphalt. Concentrated emulsions are used in the cosmetic industry as skin conditioners, hand creams, hair-grooming gels, suntan lotions, and so on; in the food industry, as gravies and salad dressings; and foams are used as whipped cream and meringue.

Princen[11] modeled monodisperse, concentrated emulsions, and foams as infinitely long, uniform cylindrical drops (or bubbles.) He took into account the volume fraction of the dispersed phase, the droplet radius, the interfacial tension, the thickness of the interstitial films, and the contact angle associated with the films. He found the yield stress and the shear mŏdulus to be proportional to the interfacial tension and inversely proportional to the radius of the droplets. The yield stress increases sharply with increasing volume fraction; the shear modulus increases as the square root of the volume fraction. The effect of a finite contact angle is to decrease the shear modulus and, in most cases, to increase the yield stress. Finally, the effect of a finite film thickness is to increase both the yield stress and the shear modulus. These results are in general agreement with rheological measurements of concentrated emulsions and foams, and are valuable in suggesting scaling laws.

In addition to the foregoing, the following trends in the rheology of emulsions have been identified[11]:

(a) The yield stress and apparent viscosity increase with the volume fraction of the dispersed phase and with decreasing drop (or bubble) size.

(b) The yield stress of typical emulsions with volume fractions of dispersed phase above 0.90 is of the order of a few hundreds to a few thousand millinewtons per square meter. It is much smaller for foams, presumably because of the larger size of the dispersed unit.

(c) The observed rheology varies with the nature of the wall surface in contact with the sample.

(d) Both emulsions and foams can be destroyed at sufficiently high rates of shear.

b.   Electrical Conductivity of Emulsions

The electrical conductivity of an emulsion can be used to follow changes in emulsion composition and to determine emulsion type. If interfacial conductivity is significant, the electrical conductivity of the emulsion will vary with droplet size and distribution, and may then be used to follow any change of these properties with time. Many emulsions, however, show electrical conductivity that is independent of particle size, and thus may be treated by simple electrical theory. Maxwell (1873, Volume 1, p. 440) developed the following expression (ignoring interfacial effects) for the specific electrical conductivity L of a two-phase system, of which the specific electrical conductivity of the outer phase is L_1 and of the inner phase is L_2:

$$\frac{L_1 - L}{L + 2L_1} = \frac{V(L_1 - L_2)}{L_2 + 2L_1}$$

[A.10]

or in its equivalent form:

$$L = \frac{L_1[2L_1 + L_2 - 2(L_1 - L_2)V]}{2L_1 + L_2 + (L_1 - L_2)V}$$

[A.11]

where V is the volume fraction. Since this relation specifies the inner and the outer phase, it yields two equations for any pair of liquids, one for the O/W type and another for the W/O type of emulsion that they can form between them. For example, if the liquids have values of L that are 2×10^{-3} and $2 \times 10^{-4} \, \Omega^{-1} \, cm^{-1}$, respectively, then the conductivities of the W/O and O/W types of emulsion are

$$L_{W/O} = 2 \times 10^{-4} \left(\frac{4 + 6V}{4 - 3V} \right) \tag{A.12}$$

$$L_{O/W} = 2 \times 10^{-3} \left(\frac{7 - 6V}{7 + 3V} \right) \tag{A.13}$$

The values chosen here as illustrations are not far from those of two actual liquids

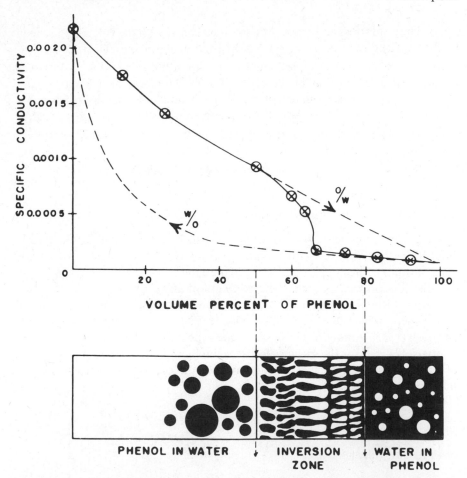

Figure A.10 The specific conductivity of aqueous potassium iodide and phenol emulsions as a function of composition (Manegold, 1952, p. 30).

investigated by Eucken and Becker (Manegold, 1952, pp. 27–31): an aqueous solution of potassium iodide and water-saturated phenol, at 19.6°. In Fig. A.10 the two theoretical equations, indicated by dashed lines, are compared with experimentally observed values. The observations move from the theoretical values for an O/W emulsion to those for a W/O emulsion, indicating inversion. The inversion is not abrupt but exists through a zone of concentration. Theory and experiment agree well with each other. The formation of multiple emulsions, that is, O/W/O or W/O/W, would invalidate the application of this theory. Detailed analyses of dielectric properties of emulsions and related systems are given elsewhere.[12]

10.   MEASUREMENT OF EMULSION STABILITY

An indication of the thermodynamic work done in creating an emulsion is provided by the area of interface produced. As the emulsion ages, the area of interface decreases. The most fundamental way to express emulsion stability is by means of the variation of interfacial area with time. The interfacial area may be expressed as square centimeters per cubic centimeter of emulsified liquid. For an average size droplet of diameter $2\,\mu m$, the interfacial area would be $30,000\,cm^2/cm^3$, and would decrease to $20,000\,cm^2/cm^3$ when the average droplet size had increased to $3\,\mu m$. Rather than measure area directly, it may be calculated from the particle size.

A direct way to get the average particle size is by means of the microscope. Size distribution curves may be obtained by measuring about 400 droplets, but better accuracy is obtained if the number measured is even larger. Levius and Drommond[13] found a Bausch and Lomb camera lucida more convenient to use than photomicrography. They report that they traced the outlines of from 400 to 800 drops per determination: "after some practice this operation could be performed in less than 45 minutes." Digital imaging techniques are now available: See Appendix J for manufacturers. Results reported by this optical method vary from 28,000 to $12,000\,cm^2/cm^3$. Emulsions were prepared by premixing in a Waring blender for 10 min and then passing through a hand-operated lever-type homogenizer. Some degree of uniformity is produced by this method, but the emulsions are relatively coarse. Emulsions prepared by the method of phase inversion, for example, produce much smaller particles, but they are then not visible under the microscope and some other method such as quasi-elastic light scattering (Section I.A.3) must be used.[14]

The advantages of using the microscope are that it is direct, inexpensive, and yields a size frequency distribution as well as a value for the interfacial area. Some methods give only an average particle size, and an interfacial area, if wanted, is based on that value. Where it is applicable, the use of the microscope is to be preferred to any other method.

Particle size counters, such as the electrozone sensors and the photozone sensors, work quite well for O/W droplets larger than a micron in diameter (Appendix J).

We must turn to other methods for smaller emulsion drops. Van der Waarden[15,16] reported the preparation of emulsions of medicinal oil in water, stabilized with from 5 to 35% of alkylbenzene sulfonates in the oil, prepared by the method of phase inversion. As the content of emulsifying agent was increased, average particle sizes decreased to the range of 20–90 nm. The interface contained all the emulsifying agent, and simply increased in area as more agent was present in the system. These emulsions were centrifuged for half an hour at an acceleration of $25,000g$ without any visible creaming of emulsified oil. The particle sizes were below the resolving power of a microscope, but could be measured by light-scattering methods. The methods of light-scattering, however, require extreme dilution of the emulsion. The dilution may have a destabilizing effect, unless the emulsion serum is used as a diluent.

11.   MAKING EMULSIONS

The use of high-shear mixers, colloid mills, homogenizers, and sonic and ultrasonic dispersers to make emulsions are described in Chapter I.E. Other methods are given below.

a.   The Method of Phase Inversion

The method of phase inversion takes advantage of interfacial tension as a mode of creating small particles out of extended liquid films. When the internal phase is at high concentration, the external phase is attenuated into a continuous thin sheet. Depending on the emulsifier used, this sheet may be unstable and breaks up into small droplets as the internal phase coalesces. The droplets of the final emulsion are much smaller than those of the parent emulsion. If this technique is to be used, the emulsifier must be able to stabilize, at least temporarily, the type of emulsion opposite to the one that is finally desired. The amphipathic nature of the emulsifier makes this possible. The advantage of the technique is that an emulsion of fine droplets can be made with the least amount of mechanical action and its attendant heat.

To make an O/W emulsion, the emulsifier is dissolved in the oil, and water is added slowly as the emulsion circulates through a colloid mill. As more water is added, the W/O emulsion acquires ointment like consistency at concentrations of water about 60–70%. The thick emulsion is transferred to a mixing tank and the final amount of water is added with gentle stirring. The viscosity drops suddenly, marking the transition to an O/W emulsion.

The inversion of the W/O emulsion is found occasionally to take place at concentrations much less than 60–70% internal phase, even as low as 25%. When this occurs, the inverted emulsion has a high concentration and consequently a high viscosity. The usual loss of viscosity on inversion is here a gain of viscosity; but the inverted emulsion has water as the external phase and so can imbibe additional water more readily than before its inversion. The driving force for inversion in such cases is not the close contact of water drops but the massive migration of a predominantly hydrophilic emulsifier out of the oil and into the water.[17]

b.   Phase-Inversion Temperature (PIT) Method

Small oil droplets can be formed by emulsifying just below the PIT of the emulsifier. As the temperature is raised toward the PIT, the interfacial tension becomes continuously lower: an example of Lundelius' rule. When oil is emulsified in water 2–4 degrees below the PIT, the low interfacial tension enables small droplets to form. Once the fine emulsion is made, it is cooled quickly to room temperature (Shinoda and Friberg, 1986, p. 129).

c.   Condensation Methods

Colloidal suspensions of insoluble solids, such as arsenic sulfide, barium sulfate, and silver iodide, are made by precipitation reactions in which particle size is limited by promoting a large number of nucleating centers. Fine suspensions are better made by precipitation from solution than by comminution of macroscopic particles. An analogous method to make a fine emulsion is to solubilize an internal phase in micelles. The internal phase may be introduced as a vapor, which nucleates heterogeneously on dust or in micelles, or as a liquid. To obtain small droplets, a large concentration of micelles is required. Since this method depends on a degree of solubility of the internal phase in the medium, in order that molecules may reach the micelles, the same solubility promotes Ostwald ripening, leading to instability. We have already seen how this instability is overcome (Section II.E.3e) by adding an insoluble co-emulsifier to set up a counteracting osmotic pressure within the droplets.

d.   Intermittent Milling

An emulsion can be made by shaking two phases together in a test tube. Briggs[18,19] found that emulsification of some systems is much more efficient if the shaking is interrupted by rest periods. For instance, 60% by volume of benzene in 1% aqueous sodium oleate is completely emulsified with only 5 shakes by hand in about 2 min, if after each shake an interval of 20–25 s is allowed. If the shaking is not interrupted, about 3000 shakes in a machine, lasting about 7 min, is required. The rest intervals provide the time required for the stabilizer to diffuse to newly created interface; without rest intervals, shaking promotes coalescence by collisions of unstabilized droplets. In industrial processing, intermittent milling is frequently used. Continuously recirculating systems, which have large holding volumes and small milling volumes, produce intermittent milling automatically, with efficiency and economy.

e.   Electric Emulsification

Mechanical methods of emulsification extend the internal phase as threads or films, which are then pinched off as droplets by surface tension (Rayleigh instability). Charging the interface electrically produces electrohydrodynamic instability, which promotes the formation of more threads and films as electric charges repel each other. The technique is to eject the internal phase through a fine capillary into the medium. The capillary is held at a high potential with respect to ground. A spray of fine droplets emerges from the tip and is dispersed vigorously throughout the medium.

f.   Special Methods

A stable emulsion can be produced by means of a chemical reaction to produce a stabilizer at the interface. If a fatty acid is dissolved in the oil phase and a base is dissolved in the aqueous phase, these reactants combine to form a soap at the interface where the two phases meet. This method ensures that the stabilizer is concentrated at the interface.

Another method is to dissolve the stabilizer in the *internal* phase before bringing the two phases together. Bancroft's rule states that the stabilizer is more soluble in the external phase, and therefore it migrates through the interface during the mixing process. This operation also ensures that the stabilizer is concentrated at the interface, especially during the processing.

12.   BREAKING EMULSIONS

Many situations arise where the separation of the two phases of an emulsion is desired. Coarse or macroemulsions (droplet diameters greater than 1000 nm) separate on standing, especially if the density difference between the two liquid phases is large. If the droplets come together without coalescence, the emulsion becomes more concentrated and the process is called *creaming*, by analogy with the separation of cream from milk. Complete phase separation requires that the droplets coalesce, which may indeed occur spontaneously in the cream if the droplets are not well stabilized. Further steps have to be taken if coalescence is to be promoted. These steps may be mechanical, thermal, or chemical. The subject is reviewed by Menon and Wasan[20] and by Lissant.[21]

a.   Mechanical Demulsification

The rate of demulsification can be promoted by agitation, as in a blender. Many emulsions are sensitive to high shear, which throws the droplets into one another with consequent coalescence. Centrifugation is another mechanical method to accelerate creaming or breaking.

Passing the emulsion through a filter bed whose surfaces are wetted by the internal phase often leads to a separation. This process is more effective with W/O emulsions and hydrophilic filter beds than with O/W and lipophilic beds. The same effect can sometimes be obtained more readily by mixing particles of a hydrophilic solid with a W/O emulsion or particles of low surface energy with an

O/W emulsion. The latter procedure, again, is less often successful. Ultrafilters employ membranes with pores less than 5 nm in diameter. An O/W emulsion is passed along a tubular membrane under pressure; water diffuses through the membrane and flows out.[22]

b.   Thermal Demulsification

Most emulsions are less stable at higher temperatures, as the adsorption of the stabilizer decreases with temperature. In some cases the emulsifying agent is thermally decomposed. Emulsions stabilized with agents containing polyethylene oxide or polypropylene oxide are sensitive to an increase in temperature, as these polymeric moieties become insoluble at higher temperatures.

O/W emulsions stabilized electrostatically, but without protective colloid, are subject to breaking on freeze–thaw cycling. The separation of ice allows the electrolyte in the aqueous medium to become concentrated, thus reducing the electrostatic repulsion between droplets. The freeze–thaw process usually needs to be done slowly and repeatedly for best results.

c.   Chemical Demulsification

An emulsion will often break if the emulsifying agent is chemically altered. Emulsions stabilized with alkali-metal soaps are broken on adding acid or metal ions, such as iron or calcium, which convert soaps to water-insoluble fatty acids or metallic salts. Emulsions stabilized with anionic agents can be broken by adding a cationic detergent, such as a quaternary ammonium salt, which converts the agent to a water-insoluble complex. A more subtle chemical effect is to alter the HLB of the emulsifying agent by adding a surface-active solute with a very different HLB number. This procedure depends on the two solutes being able to comicellize, which allows them to blend so intimately that the mixture behaves as a unit of intermediate HLB. Thus an O/W emulsion stabilized with a agent of high HLB may be vulnerable to the addition of an agent of low HLB.

Another way to attack a stable emulsion is to replace the emulsifying agent with a surface-active solute of greater adsorptive potential but less stabilizing effect. For example, petroleum emulsions owe their great stability to the mechanical strength of the interface, provided by asphaltenes of high molecular weight. These can be displaced by the more surface-active petroleum sulfonaphthenic acids, which weaken the interface sufficiently to break the emulsion.[23] Canevari and Fiocco applied nonionic surface-active solutes such as the polyethoxyalkenes to break Athabasca-tar-sand froths.[24]

The emulsifying agent can be made to desorb from the droplets by the addition of a water-miscible organic solvent, such as methanol, ethanol, or acetone, to an O/W emulsion; or the addition of an oil-soluble Lewis base or Lewis acid, whichever is appropriate to interact with the emulsifying agent and solubilize it. In an analogous way, emulsions stabilized with adsorbed solid particles can be broken by the addition of a solvent that wets the particles and removes them from the interface.

Among chemical methods to destabilize emulsions, is the addition of electrolyte to increase the ionic strength of the medium, thus reducing the repulsion between droplets of an electrostatically stabilized emulsion. The most effective electrolyte for this purpose is a salt with multivalent ions.

13.   MICROEMULSIONS AND MINIEMULSIONS

Microemulsions result from a large free energy of adsorption of the surface-active components combined with a low interfacial tension. The Helmholtz free-energy change for any process that increases the interfacial area at constant volume, temperature, and composition, is given by

$$\frac{dF}{dA} = \sigma_{12} - W^{\text{des}} \qquad \text{[A.14]}$$

where σ_{12} is the interfacial tension and W^{des} is the work of desorption per unit area. The work of desorption is the resultant of various components, such as changes of entropy, changes of surface charge density, and molecular interactions between constituents of the interfacial film. When the work of desorption is sufficiently large and σ_{12} is sufficiently small, $dF/dA < 0$ and the interfacial area increases spontaneously. A spontaneous increase in total interfacial area leads to a spontaneous decrease in average droplet size. For this reason microemulsions are thermodynamically stable. The small particle size accounts for their being translucent.

A large work of desorption and a small interfacial tension is achieved practically by combining an ionic surface-active solute with an insoluble fatty derivative. This insoluble component contains electron donor or acceptor groups, such as hydroxyl or chloride, that allow it to partition itself between the core of the droplet and the interface. The component is now appropriately named a co-emulsifier although its insolubility is still a sine qua non. The essential feature of this microsystem is the spontaneous adsorption of co-emulsifier at the interface. Such materials are strongly adsorbed because of an acid–base interaction with water at the interface, and because they reduce the electrostatic repulsion between the ionic heads of the primary emulsifier.

Adsorption of the co-emulsifier may move the composition of the interface into a liquid-crystal regime that confers additional stability to the microemulsion. For styrene emulsions the stability conferred by a fatty-alcohol co-emulsifier decreases in the following order with respect to the chain length of the alcohol: $C_{16} > C_{18} > C_{14} > C_{12} > C_{10}$ (similar to the order of the Ferguson effect and perhaps for the same reason, see Section II.E.2). This order reflects the way in which the liquid-crystal regime in a polycomponent system changes with the chain length of the co-emulsifier.

In general a microemulsion consists of four components: water, oil, surface-active solute, and co-emulsifier such as a fatty alcohol, although other suitable nonionic surface-active solutes may be effective. The droplet size of a microemulsion is in the range of 10–150 nm and more often 10–60 nm. In this range the microemulsion is monodisperse. The droplets are too small to be effectively stabilized by electrostatic repulsion, but they are small enough that a surface layer 2–3 nm thick, provided by conventional nonpolymer emulsifiers, confers steric stabilization. The volume fraction of the dispersed phase varies over a fairly wide range (20–80%).

Swollen micelles, microemulsions, miniemulsions, and (macro)emulsions form a continuum of properties determined by the size of the droplets, which in turn is determined by the nature of the adsorbed solute. Microemulsions combine a property of micelles, inasmuch as they form spontaneously and a property of emulsions, inasmuch as the interior of the droplet is bulk phase. Symposia on micellization, solubilization, and microemulsions are held frequently (Mittal, 1977).

Miniemulsions droplets occur in the range 100–1000 nm. They may be produced by extreme comminution. Because of their large interfacial area, high concentrations of emulsifier are required. If the internal phase has low solubility in the medium, miniemulsions may be stabilized by the usual electrostatic or steric mechanisms. If the internal phase is slightly soluble in the medium, however, miniemulsions even though stabilized with respect to coalescence, are subject to degradation by diffusion. When, for example, a component such as styrene, which is slightly water soluble, is homogenized in water containing a large concentration of emulsifier, microdroplets are formed at first; the styrene then diffuses through the water from smaller to larger droplets because of the differences in Laplace pressures (Ostwald ripening.) Higuchi and Misra[25] discovered that a small amount of a water-insoluble component, such as hexadecane, incorporated into the styrene droplet arrests the Ostwald ripening. The mechanism is as follows. Styrene dissolves in the micelle containing the water-insoluble component, which cannot diffuse appreciably; meanwhile Laplace pressure (also called capillary pressure because it arises from differences in interfacial curvature) causes the styrene to diffuse from smaller to larger droplets. The concomitant reduction of styrene concentration in the smaller droplets reduces their osmotic pressure with respect to the larger ones. The difference in osmotic pressure thus created, counteracts the capillary pressure. Miniemulsions of styrene, and in fact of all monomers that are slightly soluble in water (as most of them are, so that the application of this principle to emulsion polymerization is evident), may be stabilized in this way. Without the added insoluble component (costabilizer) the smaller drops could not persist.

These effects may be treated by simple thermodynamics.[26] Consider the case of an miniemulsion of a completely water-insoluble compound 2 to which is added a slightly water-soluble compound 1. Starting with an miniemulsion of, say, hexadecane (2) in water, how much styrene (1) would be absorbed by the preformed droplets of hexadecane? For an ideal solution in the droplet

$$\frac{RT}{V_m} \ln \Phi_i + \frac{2\sigma}{r_i} = 0 \qquad \qquad [A.15]$$

where V_m is the partial molar volume of styrene, Φ_i is the mole fraction of styrene in the ith droplet of equilibrium radius r_i, and σ is the interfacial tension. Equation [A.15] is the thermodynamic condition for the stability of droplets of various size as a function of the amount of hexadecane in each droplet. Introducing corrections for nonideality gives the Morton equation (Flory, 1953, p. 512, Eq. 30):

$$\frac{RT}{V_m} \ln \Phi_1 + \frac{RT}{V_m}\left(1 - \frac{1}{j_2}\right)\Phi_2 + \frac{RT}{V_m}\Phi_2^2\chi_1 + \frac{2\sigma}{r} = 0 \qquad [A.16]$$

where j_2 is the ratio of the molar volumes of 2 to 1, χ_1 is the Flory interaction parameter related to the enthalpy of mixing, and Φ_1 and Φ_2 are the volume fractions.

Equation [A.16] demonstrates the possible effect on the equilibrium droplet size of varying the molar volume of the insoluble component. For simplicity, take a uniform dispersion with $T = 300$ K, $\chi_1 = 0.5$, $V_m = 100$ cm^3, $\sigma = 5$ mN/m, and $r_0 = 100$ nm. The ratio of the volume of styrene absorbed V_1 to the volume of hexadecane V_2 as calculated by Eq. [A.16] for various values of j_2 is given in Table A.4.

Table A.4 shows that the swelling capacity r/r_0 is strongly affected by the value of j_2. The smaller the molecular weight of the insoluble component, the greater the swelling. An explanation of the swelling effect is that the large energy of mixing, mainly entropy of mixing, due to the interaction of component 1 with the insoluble component 2, balances the increase in interfacial energy of the enlarged droplet. Ugelstad et al.[27] demonstrated experimentally the strong effect of hexadecane on the stability and particle size of miniemulsions of styrene. Styrene was added with mild stirring to miniemulsions of small amounts of hexadecane in water, stabilized with excess of an anionic or cationic emulsifier, and with initial particles in the submicron size. The styrene diffused rapidly through the water

Table A.4   Effect of the ratio of Molar Volumes of Emulsified Components on the Swelling Capacity of Droplets

j_2	V_1/V_2	r/r_0
1	4000	15.9
2	1350	11.1
5	355	7.1
10	125	5.0
∞	4.5	1.7

and was absorbed in the hexadecane droplets to form a stable miniemulsion with droplets in the range of 200–1000 nm in radius. The amount of styrene absorbed per unit volume of hexadecane was 100–200 times.[26]

14.   EMULSION POLYMERIZATION

Emulsion polymerization originated in the wartime need for a substitute for rubber, which comes in the form of a dispersion of natural rubber in an aqueous serum and bears a strong resemblance to the appearance of milk, whence arises the name, latex (Latin for milk). A fairly evident approach to make a synthetic latex is to emulsify a monomer or a mixture of monomers is water using conventional emulsifying agents. For gaseous monomers, such as butadiene, the emulsification is done under sufficient pressure to liquefy them. Polymerization is initiated by a water-soluble initiator, such as sodium or potassium persulfate. The quality of the product depends critically on the concentration of emulsifier (soap.) A typical formulation contains about 5–8% soap, almost all of which, during the course of the polymerization, leaves the aqueous phase for the polymer–water interface.

At the start of the reaction, the components are present in the following forms:

(a) Soap micelles swollen with solubilized monomer to an average diameter of about 5–10 nm.
(b) Emulsion droplets of monomer stabilized by soap, about 1–3 μm.
(c) The aqueous medium containing the initiator, dissolved ions to control final stability, and a small amount of dissolved monomer.

Free radicals are generated in the aqueous phase and migrate to the interfaces. Polymerization starts in the aqueous phase to form macroradicals that are sorbed by micelles and continue to polymerize on or in them. About 10^{18} micelles/cm^3 and about 10^{11} emulsion droplets/cm^3 are present so that almost all the free radicals are captured by micelles rather than by emulsion drops. As polymerization continues, the micelles get larger, by diffusion of monomer from the more concentrated emulsion to the ertswhile micelle. The reaction is stopped by "stripping" the latex of excess monomer under vacuum. In the final product the latex is often found to be "soap starved," that is, the final area per molecule of the soap at the polymer–water interface is far from saturated. More emulsifier is then added as a post-stabilizer. In general, the amount of surface-active solute determines the number of micelles and hence the number of particles; the amount of monomer determines the size of the particles; and the amount of initiator controls the molecular weight of the polymer.

CHAPTER IVB

Foams

1. PROPERTIES OF FOAMS

Foams are coarse dispersions of gas in a relatively small amount of liquid. Foam cells vary in size from about 50 μm to several millimeters. Foam densities range from nearly zero to about 700 g/L, beyond which gas emulsions rather that foams are found. Foams are related to *concentrated* emulsions, sometimes called biliquid foams, in which the dispersed phase is another liquid rather than a gas. In solid foams the continuous phase is a solid; examples are foamed plastics, breads, and cakes. Certain characteristics of foam are desired for particular applications; for example, shampoos, shaving creams, and bubble bath compositions form slow-draining and long-lived foams. For fire extinguishing, foams should resist destruction during contact with the fuel and on exposure to high temperatures. The low density of foam ensures that it floats on burning oil or gasoline. In other processes, for example, machine laundering, distillation, fractionation, and solvent stripping, too much foam has to be avoided. The control, inhibition, or destruction of foam is often important in industrial processes. Foam properties depend primarily on the chemical composition and properties of the adsorbed films and are affected by numerous factors such as the extent of adsorption from solution at the liquid–gas surface, the surface rheology, diffusion of gas out of and into foam cells, size distribution of the cells, surface tension of the liquid, and external pressure and temperature.

Much of the scientific work on foams deals with the behavior of liquid films. In examining the interference colors in soap films, in 1672, Hooke observed depressions in the films, and Newton described black films of different shades. In

294

1869 Plateau's studies of flow behavior in films led to the concept of a surface shear viscosity.

2. GEOMETRY OF BUBBLES AND FOAMS

The two laws of bubble geometry (Platcau's laws) that hold for all assemblies of bubbles and for the morphology of foams are based on the minimizing of surface area of liquid films, which is the direct result of the tension of liquid surfaces. These laws are (a) along an edge, three and only three liquid lamellae meet; the three lamellae are equally inclined to one another all along the edge; hence, their mutual or dihedral angles of inclination equal 120°; (b) at a point, four and only four of those edges meet; the four edges are equally inclined to one another in spacc; hence the angle at which they meet is the tetrahedral angle (109° 28″ 16′.) Plateau's laws describe all foam structures where nonspherical bubbles are in contact. If the structure is disturbed by the rupture of a liquid lamella, the bubbles rearrange in such a way as to reconform to Plateau's laws. Clusters of a few bubbles, known as composite bubbles, demonstrate the laws clearly. They also show striking structural similarities to assemblies of biological cells produced by mitosis (Thompson, 1942).

When large numbers of small bubbles are produced together, a foam is formed. Foams that retain a relatively large amount of liquid between the bubbles are said to be "wet"; the bubbles are then spherical. As the liquid drains away, the bubbles form plane or nearly plane surfaces of contact and, except for rounding at the angles, become irregular polyhedra called foam cells. Because of the variety of cell sizes, these polyhedra are not uniform, but nevertheless the rules of bubble geometry still apply throughout. If the foam is made by blowing bubbles one at a time with gas at uniform pressure so that every bubble is the same size, the spheres thus produced are mobile and will tend to arrange themselves in a close-packed array, in which each bubble has 12 nearest neighbors. As the liquid drains away, the bubbles become almost regular polyhedra. The condition of 12 nearest neighbors means that the polyhedral form they take is that of a dodecahedron. A foam morphology composed of uniform dodecahedra cannot meet the laws of bubble geometry.[28] Consequently a foam constructed in this way readjusts its structure to entrain some additional bubbles that destroy its regularity. A structure of uniform polyhedra all of the same shape and size can be built, however; but not if the foam is produced so as initially to entail 12 nearest neighbors. We can imagine such an idealized foam. Kelvin[29] pointed out that the appropriate polyhedron for this function, one that divides space into equal cells with minimum surface and without voids, is the "minimal tetrakaidecahedron," which has curved edges meeting at the vertices at the tetrahedral angle, six plane faces that are curvilinear squares, and eight hexagonal faces that are nonplanar but have zero net curvature, as shown in Fig. B.1.

A foam made of such elements would have the lowest potential energy with respect to the form of the cells; but observations of the structure of foams reveal

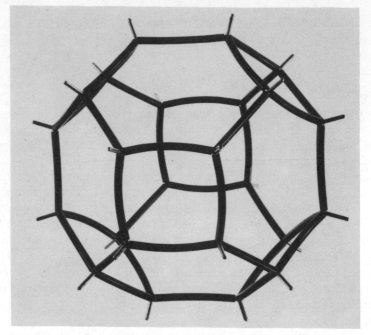

Figure B.1 Kelvin's unit foam cell that can tesselate with minimum wall area.[28]

that the majority of the lamellae are pentagonal, whereas the Kelvin cell has only quadrilateral and hexagonal faces. The probable explanation is that in the usual course of foam production, going from spherical to polyhedral bubbles, the structure is influenced toward 12 nearest neighbors rather than the 14 nearest neighbors of a foam composed of Kelvin cells.

Foam is a system that stores potential energy in the compressed gas inside the bubbles and in the extended surface of the liquid that encloses the gas. Foams can change with time, losing their potential energy both by expansion of the contained gas and by contraction of the liquid films. This loss is brought about spontaneously by the rupture of the lamellae between foam cells, and also by the transfer of gas from smaller to larger cells by diffusion. Some liquid foams are resistant to lamellae rupture but none is immune to gas diffusion.

3.   EQUATION OF STATE OF FOAM

The state variables of a gas are four: moles and temperature, volume and pressure. When the gas is incorporated in a foam, two further state variables are operative: surface area and surface tension. The equation of state of a gas relates the four state variables so that only three are independent; were an equation of state of a foam to exist, it would relate the six state variables so that only five are independent. The question has more than theoretical interest because one of the state variables, the area of the extended liquid surface, is not readily measured and is an important mechanical property of foam. Also, the variation with time of the area of liquid surface in a foam is a fundamental index of the stability of the foam. A universal equation of state would prove, therefore, of some practical interest to facilitate the determination of the surface area of liquid in a foam and its rate of decay.

The simplest derivation of an equation of state of foam is the application to a single spherical bubble or radius r, in which p denotes internal pressure, P denotes external pressure, σ denotes surface tension, n denotes number of moles of gas, and T denotes temperature in K. Laplace's equation gives

$$p - P = 4\sigma/r \qquad \text{[B.1]}$$

For a sphere of volume V and area A,

$$V/A = r/6 \qquad \text{[B.2]}$$

For an ideal gas,

$$pV = nRT \qquad \text{[B.3]}$$

Combining the above three equations gives

$$PV + \tfrac{2}{3}\sigma A = nRT \qquad \text{[B.4]}$$

Equation [B.4] is an equation of state, but proved only for a special case. Derjaguin[30] and Ross[31,32] found the same equation holds for several other special cases. While it is likely that this equation applies to all foam in thermal equilibrium, a general proof is still lacking.[33]

Equation [B.4] can be applied in several different ways. For any spontaneous isothermal process of foam degradation, as long as no gas is lost to the atmosphere and the external pressure remains constant,

$$3P\,\Delta V + 2\sigma\,\Delta A = 0 \qquad \text{[B.5]}$$

For foam degradation at constant volume, the equation gives

$$3V\,\Delta P + 2\sigma\,\Delta A = 0 \qquad \text{[B.6]}$$

The changes of pressure in a closed vessel containing foam can be monitored readily, and by means of Eq. [B.6] the rate of change of area in the foam can be found:

$$\frac{dA}{dt} = -\frac{3V}{2\sigma}\frac{dP}{dt} \qquad \text{[B.7]}$$

The measurement of the pressure change and the use of Eq. [B.7] to give the change of surface area is more convenient and precise than direct observation of the area.[34]

4. THEORIES OF FOAM STABILITY

Pure liquids do not form stable foams: They allow entrained air to escape with no delay other than what is required for the Stokesian rate of rise, which is controlled by the diameter of the bubble of dispersed air and the viscosity of the bulk liquid. Certain solutes are able to stabilize thin lamellae of liquid: If these solutes are present, the escape of entrained bubbles is more or less retarded, and a head of foam is produced. Theories of foam postulate plausible mechanisms to account for this behavior, with the ultimate objective of understanding the phenomenon so thoroughly that predictions can be made about the expected influence of a given solute prior to actual observation. One may say at the outset that this final goal is not yet attained.

a. Thermodynamic Stability

Foams are coarse dispersions of gas in a relatively small amount of liquid. Pure liquids cannot sustain a stable foam. A simple thermodynamic argument makes this statement understandable. For a two-component system (pure liquid plus a completely insoluble gas) that has sufficient surface area to make the surface energy a significant contribution to the total energy, the Helmholtz function is given by

$$dF = -S\,dT - p\,dV + \sigma\,dA + \mu_1\,dn_1 + \mu_2\,dn_2 \qquad \text{[B.8]}$$

where σ is the surface tension of the liquid and A is the surface area. Integrating Eq. [B.8] at constant T, V, and n_i gives

$$\Delta F = \sigma\,\Delta A \qquad \text{[B.9]}$$

where ΔF is the change of the Helmholtz free energy for bubble coalescence.

If we apply Eq. [B.9] to the coalescence of foam cells, we see that a decrease of the Helmholtz function results from a decrease of area; hence, a foam composed of an insoluble gas in a pure liquid is thermodynamically unstable. Experience shows that such foams have only a momentary existence. The foams we are accustomed to see, therefore, which can remain stable for minutes, hours, or even indefinitely if carefully protected, cannot be created from pure liquids. So universal is this principle that one can estimate the purity of distilled water by watching bubbles that form in it: If the bubbles are stable, the water is contaminated. To ensure the thermodynamic stability of foam, the component that acts as the foam stabilizer must have some specific properties: Equation [B.9] requires additional terms besides the $\sigma \Delta A$ term; also, these terms must be positive so that they ultimately change the sign of the whole expression for ΔF. These requirements can be met by a solute that is spontaneously surface active. The surface activity of a solute is defined as its ability to lower the surface or interfacial tension of the solvent. According to the Gibbs adsorption theorem, this ability requires that the solute be positively adsorbed at the surface or interface. Positive adsorption implies that work must be done to transfer solute molecules from the surface to the bulk phase; this work provides the additional free-energy terms that are required. When bubbles or droplets coalesce, surface area is reduced, and at the same time the solute molecules segregated at the surface are transferred back into the bulk liquid. If the loss in free energy resulting from the $\sigma \Delta A$ term were more than made up by the gain in free energy arising from the transfer of solute from surface to bulk phase, then spontaneous coalescence would not occur, and the foam would be thermodynamically stable. It remains mechanically fragile, nevertheless, unless the walls between the cells are solid, such as those that exist in polyurethane and similar polymer foams.

Nakagaki[35] has advanced a thermodynamic theory of foam formation and stability in terms of foam volume (foaminess) and foam life. For dilute solutions of low viscosity, foam stability and foaminess are directly proportional to the work required to transfer excess solute from the film surface into the bulk liquid. This argument, although it shows one of the conditions necessary for the production of stable foam, is still incomplete; it lacks a term to express the effects of gravity, and it does not take into account the diffusion of gas from smaller to larger foam cells, due to differences of pressure that necessarily exist because of curvature of the film surfaces. A complete thermodynamic description of a foam has, in fact, never been written. Each factor that contributes to foam stability may, however, be identified and discussed in a qualitative way.

The presence of a surface-active solute is, as we have seen, required for thermodynamic stability of a liquid film and hence of a foam. But it does not by itself ensure the persistence of the film against the pervasive action of gravity, as well as other stresses that might tend to destroy the film. Rayleigh[36] pointed out a second consequence of having a surface-active agent present in the solution, namely, local gradients of surface tension induced by gravity:

> Imagine a vertical soap film. Could the film continue to exist if the [surface] tension were equal at all its parts? It is evident that the film could not exist for more than a moment; for the interior part, like the others, is acted on by gravity, and if no other forces are acting, it will fall 16 feet in a second. If the [surface] tension above be the same as below, nothing can prevent the fall. But observation proves that the central parts do not fall, and thus that the [surface] tension is not uniform, but greater in the upper parts than in the lower. A film composed of pure liquid can have but a very brief life. But if it is contaminated, there is then a possibility of a different [surface] tension at the top and at the bottom, because the [surface] tension depends on the degree of contamination.

b. Marangoni–Gibbs–Rayleigh Theory

A thermodynamic "explanation" does not provide any physical mechanism. Having stated that adsorption of the solute is a requirement for stability, it leaves unsaid a description of *how* the adsorbed layer acts to stabilize foam. The earliest attempt to offer such an explanation, based on ideas developed successively by Marangoni,[37-39] Gibbs,[40] and Rayleigh,[36] has best withstood criticism through the years. This theory refers the stability of foam to an elasticity or restoration of liquid lamellae which depends on the existence of an adsorbed layer of solute at the liquid surface and the effect of this adsorbed layer in lowering the surface tension of the solution below that of the solvent. The two effects, surface segregation or adsorption, and the lowering of the surface tension are concomitant: An observed reduction of surface tension due to the addition of a solute is evidence, admittedly indirect but no less certain than were it given by direct observation, that the solute is segregated at the surface. The degree of segregation in a two-component system is measured as excess moles of solute per unit area of surface, designate Γ_2^G, and is equal to the variation of the surface tension with the chemical potential of solute; that is,

$$\Gamma_2^G = -\frac{d\sigma}{d\mu_2} \qquad [B.10]$$

where $d\sigma$ is the change of the surface tension caused by changes in the chemical potential of solute, $d\mu_2$.

c. The Marangoni Effect

When local areas of a foam lamella are expanded, as would happen for example when a bubble of air pushes through a liquid surface, new areas of surface are created where the instantaneous surface tension is large, because the adsorbed layer has not had sufficient time to form. The greater surface tension in these new areas of surface exerts a pull on the adjoining areas of lower tension, causing the surface to flow toward the region of greater tension. The viscous drag of the moving surface carries an appreciable volume of underlying liquid along with it, thus offsetting the effects of both gravitational and capillary drainage (explained below) and restoring the thickness of the lamella. The same mechanism explains how liquid lamellae withstand mechanical shocks, such as the passage through the foam of lead granules, mercury drops, iron filings, or steel spheres, all of which have been used by one or another investigator to test the resiliency of lamellae.[41] The lamella survives because the local increase of surface tension, where the impinging solid deforms the surface, causes flow toward the weakened region. Aged lamellae are less fluid than younger and thicker ones and so are more readily broken, as are lamellae made from solutions in which the surface tension gradient is small, for example, oil solutions or very dilute aqueous solutions of surface-active solutes.

What force checks the downward flow of the central liquid in Lord Rayleigh's example? The force originates in the higher surface tension in the upper part of the film. Where a higher tension appears, movement of the surface at that point immediately ensues: Surrounding areas of lower tension surface are pulled in toward the high-tension spot until the difference has been effaced. The movement of the surface layer from areas of low to areas of high surface tension is accompanied by the motion of relatively thick layers of underlying fluid, amounting to several microns in depth, so that the downward flow of the central liquid is offset to some extent by a counterflow of liquid at the surface. Flow in response to a surface tension gradient is called the Marangoni effect (see also Section II.A.16); it was recognized also by Gibbs. A simple experiment readily demonstrates this effect. On a metal sheet, such as a silver tray, pour out a thin layer of water. When a piece of ice is held below the tray, local cooling of the water above causes its surface tension to increase. Immediately, the motion of the surrounding areas of lower tension becomes evident by the heaping up of the water just above the spot where the ice is held. When the ice is moved slowly across the bottom of the tray, the little mound of water follows it faithfully.

d.   Stresses on Foams

Stresses that create areas of higher tension on a liquid film are always at hand. The first of these is gravity-induced drainage. The drainage of liquid through the foam structure takes place between two coherent layers provided by the surface-active solute at the liquid–gas surface. The layers themselves are not completely static. They too respond to the gravitational tug and then reverse their direction of flow by the Marangoni effect. The coherent surface layers act as a membrane or skin that can stretch and relax in response to the lateral forces acting on it. In doing so, the surface tension changes so as to give a force opposing the motion. This result is, in effect, an elasticity of the surface, a variation of the surface stress with surface strain. It is named "dilatational elasticity." Single pure liquids, which in the absence of a thermal gradient are not able to develop a Marangoni effect, therefore show no dilatational elasticity.

At first, the drainage of the central liquid, taking place between two membranes at the surface, is entirely induced by gravity; but once spherical bubbles are in contact, flat walls develop between them, and polyhedral cells appear in the foam. When this occurs, the rate of drainage ceases to be determined by hydrodynamic flow under gravity because capillary action becomes significant. The foam films are flat in some places and have convex curvature at the places where the liquid accumulates in the interstices between the foam cells. The line along which three films meet is called the "Plateau border" (after J.-A.-F. Plateau); and the point at which four lines meet is called the "tetrahedral angle" or the "Gibbs angle" (after J. W. Gibbs). The Plateau border is not a mathematical line but a thin vein of liquid whose cross section is approximately a curvilinear equilateral triangle; the tetrahedral angle is not a mathematical point but a small volume of liquid whose shape is approximately a tetrahedron with concave faces. The convex curvature inside the tetrahedral angles creates a capillary force that draws liquid out at the connecting Plateau borders; and the convex curvature of the Plateau borders draws liquid out of the adjoining lamellae. For convenience, let us call this the "Laplace effect"; it is also known, less happily but with logical consistency, as capillary flow. As the liquid flows, some of the surface layer is dragged along, and areas of higher tension are created. Then the Marangoni effect comes into play; the central liquid that is lost to the lamella is restored by counterflow along the surface, one effect thus creating the conditions for its reversal by the other. Ultimately, after a period that may be more or less prolonged, depending on the presence of dust, the ease of evaporation, and any departure from thermal equilibrium, the liquid film becomes so thin (20–30 nm) that the central liquid is itself affected by surface forces and so loses the property of bulk liquid. The required gradient in surface tension for the Marangoni effect to operate now cannot be so readily achieved, and the resilience of the liquid film is gradually replaced by increasing brittleness. At this stage in the life of the film, it is readily ruptured by relatively slight mechanical shocks, which it had previously been able to withstand. The foam structure, therefore, begins to collapse at the top of the foam because the older lamellae are located there. A gravitational field promotes instability of foam structure. One of the interesting features of weightlessness in space is the modification of this and other capillary phenomena with which we are familiar on Earth.

e.   Elasticity of Lamellae

The elasticity arising from the variation of the surface tension during deformation of a liquid lamella may be manifested both in equilibrium (when a surface layer under forces leading to deformation is in equilibrium with its bulk phase) and in nonequilibrium conditions. The first case refers to the Gibbs elasticity and the second to the Marangoni elasticity.[42] The Marangoni elasticity is a dynamic, nonequilibrium property, larger in value than the Gibbs elasticity that could be obtained in the same system. The elasticity is defined as the ratio of the increase in the tension resulting from an infinitesimal increase in the logarithm of the area. For a lamella with adsorbed solute on both sides, the elasticity E is given by[43]

$$E = \frac{2\,d\sigma}{d\ln A} = -\frac{2\,d\pi}{d\ln A} \qquad [\text{B.11}]$$

where σ is the surface tension and A is the area of the liquid surface. The factor 2 is required because the stretching of the lamella increases the area on both of its sides. The factor would not be required for the definition of elasticity of the surface of a solution or a liquid substrate holding an insoluble monolayer. The Gibbs elasticity of a monolayer is measured from the equilibrium π versus A curve as previously described (Section II.E.6). The Marangoni elasticity of a monolayer is determined by the dynamic (i.e., nonequilibrium) surface tension as the surface is abruptly extended, or pulsated.

The effects described by the Rayleigh–Gibbs theory depend therefore on a combination of two physical properties of the solution: The solute should be capable of lowering the surface tension of the medium; But this alone is not enough: A rate process is also required, by which a freshly created liquid surface retains its initial, high, nonequilibrium surface tension long enough for surface flow to occur. Many instances are known in which the mere reduction of surface tension by the solute does not lead to the stabilization of foam, presumably because it is not accompanied by the relatively slow attainment of equilibrium after a fresh surface is made, which is the second requirement for the ability to stabilize bubbles.

In general, the surface tension of a freshly formed solution of a surface-active solute changes with time until it reaches a final equilibrium value. Equilibrium may be reached in a fraction of a second or it may require several days. Adsorption may be considered as a two-step process involving: (a) the diffusion of the solute molecules from the bulk phase to the subsurface (i.e., the layer immediately below the surface) and (b) the adsorption of the solute molecules from the subsurface to the surface.

Recently Borwankar and Wasan[44] developed a mathematical model of adsorption of surface-active solutes at a gas–water interface that takes into account both the diffusion in the bulk phase and the energy barrier to adsorption. Some systems are diffusion controlled, that is, the activation energy barrier to adsorption is negligible in such cases. A diffusion-controlled rate of adsorption makes itself evident, therefore, by a rapid approach to surface tension equilibrium (on the order of seconds or less). Surfaces that age more slowly imply an activation energy of adsorption (see Section II.E.5). The different types of instruments developed to measure dynamic surface tension vary with respect to the age of the surface that they are designed to handle: The vibrating jet deals with surface ages of milliseconds, and conventional techniques to measure surface tension, which react to changes more slowly, can detect surface aging of several minutes, hours, or days.

These considerations apply equally to aqueous and nonaqueous solutions. A major difference between the two lies, however, in the magnitude of the effects produced by surface-active solutes. The surface tension of water is reduced from 73 to 25 mN/m quite readily by amphipathic organic solutes; but the surface tension of most organic solvents is already in the low range of 25 to 30 mN/m, so that only a small reduction remains to be achieved by an organic solute. Thus, although Marangoni effects may arise in nonaqueous solutions, they are usually much less pronounced than in aqueous solutions of soaps or detergents. Oil lamellae, therefore, have a relatively low resistance to mechanical shock; consequently oil foams are transient or evanescent, resembling the foam produced from very dilute aqueous solutions of detergents or more concentrated solutions of weakly surface-active solutes. Special solutes have been developed to stabilize nonaqueous foams for application in the field of cellular plastics. These solutes incorporate polyalkylsiloxane or perfluoroalkyl moieties in solute molecules, which are able to reduce the surface tension of organic liquid monomers by 12 to 15 mN/m and boost the Marangoni effects; and the result is obvious in an increased stability of the foam.

f. Enhanced Viscosity or Rigidity

The persistence of the liquid lamella depends not only on surface tension gradients but also on the presence of a coherent surface layer on each side of the lamella, between which layers the central portion of the sheet of liquid flows downward. Given the presence of a surface-active solute, surface layers there certainly would be. But these layers are not necessarily coherent, and if they should lack this feature, the whole liquid film, surface and central part alike, drops simultaneously. This requirement for the stability of a liquid film is often overlooked, but it explains, for instance, why the mere ability of a solute to lower surface tension is not enough by itself to ensure that it will also stabilize a liquid film. The surface layer may lack the necessary cohesion. Cohesion and elasticity of a surface film are directly related. A dilute solution of a surface-active solute has a low surface elasticity, which is evident from a π–A curve at low values of π (Fig. IIE.12). Such solutions have poor cohesion and do not stabilize bubbles well.

A single surface-active species in solution does not usually confer any increase of the viscosity, much less rigidity, in the surface layer of the solution. Although foam is capable of being produced by such a solute, the foam is of brief duration. That kind of foam is called "evanescent foam," but it can nevertheless be a cause of concern; if produced rapidly, it can reach a large expansion ratio and so flood any container. While the first stabilizing factor in foam is the elasticity of the film provided by the Marangoni effect, in special cases additional surface layer phenomena are significant, namely, gelatinous surface layers, low gas permeability, and stability of black films. These phenomena are known to occur in water with certain mixtures of solutes or with certain polymers, both natural and synthetic. Well-known examples in aqueous systems are solutions of water-soluble proteins, such as casein or albumen, as in the stable foams produced with whipping cream, egg white, beer, or rubber latex. The highly viscous surface layer is sometimes made by having present one or more additional components in the solution. An example is the surface plasticity of a mixture of tannin and heptanoic acid in aqueous solution, compared to the lack of any such effect of the two constituents separately. Surface plasticity adds enormously to the stability of foam and may be exemplified in meringue, whipped cream, fire-fighting foams, and shaving foams.

In nonaqueous liquids, particularly in bunker oils and crude oils, surface layers of high viscosity have been observed; porphyrins of high molecular weight have been indicated as a possible cause. In a hydrocarbon lubricant, the additive calcium sulfonate, for example, creates a plastic skin (or two-dimensional Bingham body) at the air surface; it also acts as a foam stabilizer.[45] Thus viscous or rigid layers in nonaqueous liquids enhance the stability of foam, just as they do in aqueous solutions.[46-48] Cellular plastics are made from a monomer foam, which on polymerization displays high viscosity, finally becoming solid; but the growth of the viscosity is not confined to the surface, nor does it even show preferential development at the surface.

Different kinds of surface viscosity can be distinguished:

(a) Innate surface viscosity: This viscosity is the resistance to flow that is innately associated with the presence of a liquid surface, whether or not there are additional sources of resistance such as those described below.[49]

(b) Surface shear viscosity: This viscosity is associated with the presence of a pellicle or skin, such as an insoluble monolayer, but not restricted to that example, at the liquid surface. A layer of denatured protein that stabilizes the foam of meringue or of whipped cream or of beer is a common example.

(c) Dilatational (or compressional) viscosity: The surface elasticity that arises from local differences of surface tension is simultaneously associated with a resistance to surface flow. The local difference of surface tension is caused by dilatation (or compression) of the surface of the solution, so the resistance to flow that results from Marangoni counterflow is known as dilatational (or compressional) surface viscosity. This type of surface viscosity is a dynamic or nonequilibrium property.

Solutions of saponins or of proteins provide surface layers that are plastic, that is, that remain motionless under a shearing stress until the stress exceeds a certain yield value, which may be greater than the small gravitational or capillary stresses usually acting on the surface layer. As already mentioned, the rate of drainage of liquid through such films is at first induced by gravity. If V is the volume of liquid drained out of the foam after time t, the rate of gravitational drainage[50] is given by

$$V = V_0 \exp(-kt) \qquad\qquad \text{[B.12]}$$

where the amount of liquid drained from the foam approaches V_0 (the total amount of liquid in the foam at $t = 0$) as t becomes large. This equation has only a single arbitrary constant and describes drainage from different types of foam reasonably well. The agreement is better, however, with a slower draining foam, for example, protein hydrolyzate, than with faster draining foams. This result is to be expected because of the way in which the equation was derived, namely, that the liquid flows down between two solid walls (i.e., infinite surface viscosity) which approach each other as the liquid between them flows out.

Where the equation is found to hold best in practice, the drainage of the central fluid ultimately stops because of the formation of gelatinous surface layers. X-ray diffraction studies of drained foam films show that a hydrous gel structure extends from the surface to a depth of about 90 nm. Water makes up the principal portion of this surface film (at least 97% by weight). We do not know sufficiently well, however, the principles of the formation of such gels as to predict the combination of solutes required to produce them, although we do have some empirical information. It is known, for example, that the addition of relatively small amounts of nonionic surface-active solutes to solutions of certain anionic detergents enhances foam stability by the formation of gelatinous surface layers. The effectiveness of these additives depends on the detergent, the more effective combinations being those in which the additive preponderates in the adsorbed surface layer. The most stable foams are found[51] with anionic–nonionic pairs having 60–90% of the nonionic in the adsorbed layer, even though the actual relative amounts of the two in the whole composition is anionic:nonionic = 100:1.

Djabbarah and Wasan[52] showed that the presence of a small amount of lauryl alcohol (equal to 1/500% of the sodium lauryl sulfate) causes significant decrease in the surface excess concentration of sodium lauryl sulfate. For example, at an SLS bulk concentration of 2.0×10^{-6} M and without LOH, $\Gamma_{SLS} = 2.47 \times 10^{-10}$ mol/cm^2. At the same SLS bulk concentration and in the presence of 6.2×10^{-9} mol/mL of LOH, $\Gamma_{LOH} = 2.90 \times 10^{-10}$ mol/cm^2, while Γ_{SLS} drops to 2.0×10^{-10} mol/cm^2. This result shows that LOH is preferentially adsorbed at the surface, replacing SLS molecules that would otherwise be there. The replacement of SLS by LOH is not a simple 1:1 replacement, as is indicated by the drastic change in molecular packing at the surface from about 0.67 nm^2/molecule without LOH to about 0.34 nm^2/molecule with LOH. Molecules containing long unbranched hydrocarbon chains and a small terminal polar group are those most likely to form a closely packed state. Molecular packing by itself does not account for high surface viscosity.[52] There is, however, a strong dependence of surface viscosity on the relative amount of the more surface-active substance at the surface, which is compatible with the suggestion that the altered composition of the surface may segregate a gelatinous three-component liquid-crystal phase while the bulk solution is still isotropic and of low viscosity. Friberg and his co-workers have pointed out that foam stability ensues when the total composition of a multicomponent system is such that two phases, an isotropic solution and a liquid crystal, are in equilibrium.[53,54] It seems likely that the liquid-crystal phase should occur at the surface where so much of the surface-active components are concentrated (Section IV.B.4h).

An important feature of the gelatinous surface film is its transition to a freely flowing film over a narrow range of temperature.[55,56] The transition is sharp and reversible and is compatible with the variation with temperature of a multicomponent isothermal phase diagram, where the exercise of another degree of freedom without change of overall composition can move the equilibrium with a sharp transition from a condition of two phases to that of a single phase. The sharp transition of the film from a gelatinous or plastic structure to a structureless Newtonian fluid as the temperature is raised is accompanied by an equally sharp transition in the foaming properties. A practical example occurs annually with certain fuel oils that foam excessively in wintertime when pumped into delivery trucks or household oil tanks but that show no sign of this unwanted property during the summer months. It was found that the surface of the oil is gelatinous below 25°F and is fluid immediately above that temperature; consequently, trouble because of foaming is encountered only in cold weather. Another example is provided by Bolles,[57] who describes a foaming problem in a hydrocarbon separation unit. The system that proved to be so troublesome showed a sharp loss of foam stability with increasing temperature. This behavior is typical of foams stabilized with gelatinous films that have a melting point characteristic of a separate phase at the surface.

g.   Black Films

When a glass or wire frame is dipped into a solution of a surface-active solute and then drawn upward, a thin liquid lamella is formed, which shows interference colors in reflected light. Upon draining, the lamella becomes thinner until it looks gray or black, indicating that its thickness is much less than the wavelength of light. Upon further draining, the lamella either breaks or reaches an equilibrium thickness consisting of two adsorbed layers at the air–liquid surface with a water core between them. The behavior of the lamella depends on the amount and the surface properties of the adsorbate. At more than a minimum concentration of solute (varying for different solutes), the black film is obtained and is stable with respect to rupture[58] (Mysels et al., 1959). These films range in thickness from about 100 nm to the thickness of a bimolecular leaflet, which in the case of sodium oleate is 4.2–4.5 nm. The bimolecular leaflet can be achieved, however, only if enough electrolyte is present in the solution to contract the electrical double layers and thus reduce to very small distances the electrostatic overlap between the two sides of the film. In certain cases, notably that of sodium oleate, the bimolecular leaflet is particularly strong and persistent. Dewar succeeded in preserving such a film for over a year in a closed container.[59] The bimolecular leaflet can be described as a two-dimensional crystal; it is no longer subject to liquid flow nor can it exhibit any phenomena that depend on surface tension effects. Foams in which the stability of black films is prolonged are characterized by an extremely fragile structure at the top of the foam, the structure persisting long after all the unbound water has drained out of the films. Gelatinous surface layers, on the other hand, usually remain thick and immobilize large amounts of water.

The stability of long-lived foams has been attributed to the stability of black films, and this view is put forth as an alternative theory of stability (Sheludko, 1966). A black film is composed of two non-Newtonian surface layers in close approach. Even a black film of limiting thinness, that is, a bimolecular leaflet, is the result of contact of two concentrated and coherent monolayers when sufficient electrolyte is present to prevent the immobilizing of water in the overlap of extended electrical double layers. We need not, however, consider black films as the only kind of film produced by the close approach of two surface layers.

h.   Influence of Liquid-Crystal Phases

A lipophilic solute capable of forming micelles, that is, one with an amphipathic structure, when dissolved in a hydrocarbon forms inverse micelles that are capable of solubilizing water. At higher concentrations, the solute and water, plus a small amount of hydrocarbon form a lamellar liquid-crystalline phase (Fig. B.2). The difference between the micellar solution and the liquid-crystalline phase is that the former is an isotropic liquid containing micelles of colloidal size; the latter is a lamellar anisotropic association of the three components in a single phase. Both structures may solubilize considerable amounts of water.

Nonaqueous foams, as well as aqueous foams, can be stabilized by the presence of a lamellar liquid crystal in conjunction with an isotropic hydrocarbon solution. If the system contains amphipathic molecules, the liquid-crystal phase may first form at the surface out of an isotropic solution, by segregation of components. The lamellar liquid crystal is surface active with respect to the hydrocarbon solution and provides a layered or structured film at the surface. This structured film has high viscosity and reduces drainage from the lamellae,

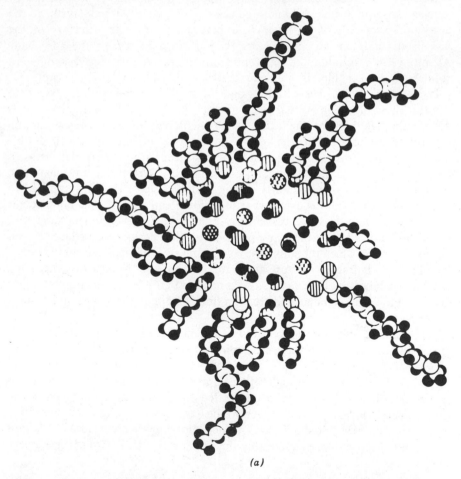

(a)

Figure B.2 The inverse spherical micelle and (b) the lamellar liquid crystal in non-aqueous media.[60]

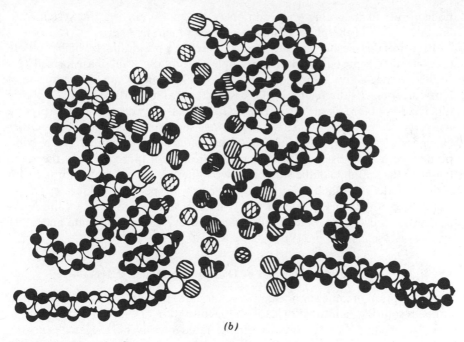

(b)

Figure B.2 *(Continued)*

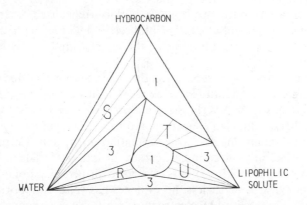

Figure B.3 Typical phase diagram for a lipophilic solute, hydrocarbon, and water in which stable foams are found in the two-phase region marked *T*.

thus conferring stability on the foam. The liquid crystal as a separate phase does not foam: Foam stability appears only in the two-phase region.

   A typical phase diagram is shown in Fig. B.3. Stable foams in such a system are found in the area *T*, a two-phase area containing isotropic solution and liquid crystal. No other portion of the phase diagram denotes systems with any foam stability. Foam films are stabilized by the presence of the liquid-crystal phase at the surface, which confers surface plasticity. An interesting feature of this nonaqueous foam is that the addition of water, by moving the system into another region of the phase diagram, acts as a foam breaker.

In view of this mechanism, the presence of surface-plastic films, which promote foam stability, is probably another example of a liquid-crystal phase formed by a soluble surface-active solute, an insoluble co-solute, and water. The addition of lauryl alcohol to sodium lauryl sulfate is well known to produce a surface-plastic film and to stabilize foam (Section IV.B.4f). The plasticity disappears sharply at a temperature of about 45°C. This sharp change suggests a phase boundary. Other examples of a similar effect of long-chain, often water-insoluble, polar compounds with straight-chain hydrocarbon groups of approximately the same length as the lipophile of the surface-active solute are lauryl alcohol for use with sodium dodecyl sulfate, lauric acid for use with potassium laurate, N, N-bis(hydroxyethyl) laurylamide for use with dodecylbenzene sulfonate, and N, N-dimethyldodecylamine oxide for use with dodecylbenzene sulfonate and other anionics (Rosen, 1978, p. 219).

i.   Mutual Repulsion of Overlapping Double Layers

The adsorption of ionic surface-active solutes into the surface layer is evident in aqueous solutions and readily leads to the formation of charged surfaces of the lamellae in foams.[61] The counterions in the liquid interlayer of the lamella are the compensating charges. When the thickness of the lamella is on the order of magnitude of 20 times the Debye thickness of the electrical double layer, the counterions adjacent to the two opposite surfaces repel each other more or less according to an exponential decline of electric potential with distance. This repulsion prevents further thinning of the lamella, and so preserves it from imminent rupture.

The mechanism of charge separation that operates in water does not function in nonionizing solvents. Until relatively recently, it was believed, therefore, that electrostatic repulsion of overlapping electrical double layers could not be a factor in stabilizing liquid lamellae in oil or other nonaqueous foams. But we now recognize that mechanisms of charge separation other than electrolytic dissociation are possible, and indeed must operate; for zeta potentials of 25 to 125 mV have been observed for various kinds of particles dispersed in nonaqueous media of low conductivity. Nevertheless, no evidence has yet been reported to suggest that foam may be stabilized by electrostatic repulsion in nonaqueous solutions.

j.   Effect of Dispersed Particles on Foam Stability

Ottewill et al.[62] found experimentally that the presence of colloidally stable, suspended, solid particles increases the tendency to form stable foams over and above that of the matrix in the absence of such particles. The increase in foam stability is linked to the increased bulk viscosity of the dispersion with solids content, which is described by a relation of the form

$$\eta_d = \eta_0(1 + k_1\Phi + k_2\Phi^2 \cdots) \qquad \text{[B.13]}$$

where η_0 is the viscosity of the liquid matrix, η_d is the viscosity of the dispersion, and Φ is volume fraction of dispersed solid. The coefficient k_2 was larger than predicted by purely hydrodynamic factors, being enhanced by the electrostatic repulsions between the solid particles, which effectively enlarges each particle and so creates a larger volume fraction of solids than is calculated from the density of the solid. In addition, the presence of a minimum in the pair interaction energy curve introduces some association between the particles with increase in volume fraction, which leads to a viscosity enhancement at the low rates of shear experienced in a slowly draining lamella. The effect of bulk dispersion viscosity on the ripples formed in the lamella surface by thermal fluctuations is not known with certainty, but it seems likely that this would have a damping effect on the magnitude of the ripples and so lead to further enhancement of foam stability.

Even greater stabilizing action conferred by solid particles is achieved if the particles are not wholly hydrophilic. They are then kept at the air–water surface by surface forces, in the same way as lipophilic molecules are adsorbed there (see Section IV.A.5). If some surface-active solute is also present, the foam lamellae carry these particles upward with the foam, a phenomenon on which is based the process of ore flotation. The particles at the surface of the lamellae add considerably to foam stability by absorbing mechanical shocks that would otherwise destroy the lamellae.

But a higher degree of water repellency of the surface of solid particles begins to introduce a new effect, namely, their dewetting effect on the liquid medium, which, if the number of particles is small* converts foam-stabilizing action into destabilizing action (Section IV.B.5e). The juxtaposition of profoaming and defoaming actions resulting from a relatively small change of the hydrophile–lipophile balance of the solid surface strongly resembles a parallel action on the molecular level of surface-active solutes (Section II.E.9) and suggest a similarity of mechanism.

k. Foaminess and Phase Diagrams

Hitherto the accepted theory of foaming in fractionation or distillation towers traces the cause to Marangoni flow, induced by a difference in composition, and hence of surface tension, between thin films of liquid at the bubble caps and in the bulk liquid phase from which they originate.[63] Liquid films on walls of the container, or foam lamellae, because of their extended surfaces, evaporate faster than does a bulk liquid phase; if the loss of the more volatile constituent causes the surface tension of the residual solution to increase, then the liquid of greater surface tension draws liquid away from that of smaller surface tension, and so the stability of the liquid lamella is maintained. The mental imagery is identical with the model used to describe wine tears (Section II.A.16).

This rule is far too inclusive: In normal liquids, volatility and low surface

---

*The concentration, size, and degree of water repellency are collective factors in determining how solid particles affect foam properties: All three factors contribute to observed behavior.

tension stem from a common cause, namely, relatively small forces of inter-molecular attraction; and therefore, occur together: thus, admittedly with certain exceptions (Table IIA.1), the usual behavior of solutions leads to the surface tension of the residual solution being greater after the loss of its more volatile components. According to this rule, therefore, foaming within a fractionation tower would be the common condition. In practice, the problem of foaming is less prevalent than this rule predicts.

A more selective prediction can be obtained by studying the phase diagram of the system. Surface activity, and hence a propensity toward foaminess, is specifically inherent in solutions only at certain temperatures and compositions that are related to solubility curves and other features of the phase diagram, as well as to the relative surface tensions of the components. Although Marangoni effects may arise as a result of the volatility of a component of small surface tension, these are less significant in stabilizing foam than are Marangoni effects derived from adsorption of solute at the liquid–gas surface, that is, surface activity.

Experimental observations of the foaminess of binary and ternary solutions in systems in which a miscibility gap exists show that the foaming of a solution reaches its maximum under conditions of temperature and concentration where a transition into two separate liquid phases is imminent. Figure B.4 shows one such diagram, for the system 2, 6-dimethyl-4-heptanol and ethylene glycol.[64] Superimposed on the diagram by means of solid lines are interpolated contours of equal foam stability (isaphroic lines, from *aphros*, Greek for foam). The isaphroic lines center about a point close to the critical point as a maximum and decrease in value the farther they are from it.

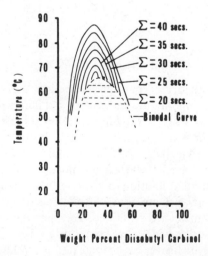

Figure B.4 Phase diagram and interpolated isaphroic lines of the two-component system, 2,6-dimethyl-4-heptanol and ethylene glycol, showing maximum foaminess at an epicenter.[64]

Two-component systems show maximum foam stability at a temperature and composition near that of the critical point; and three-component systems show maximum foam stability at compositions near that of the plait or consolute point, but only as long as the systems are maintained as homogeneous one-phase solutions. The slightest degree of separation of liquid phases produces a conjugate solution that can defoam its foamy conjugate. Both these effects, the foaming enhancement in the one-phase solution and the foam inhibition in the two-phase solutions, can be ascribed to the surface activity of the component of lower surface tension in the system, which reveals itself in the one-phase solution by adsorption at the surface; and, in the two-phase solutions, by the positive spreading of one conjugate on the other. The occurrence of surface activity in the vicinity of critical or consolute points is well established experimentally by these and many other similar types of observation.

The resemblance of the foaminess (isaphroic) contours shown on Fig. B.4 and the cosorption shown on Fig. IIB.1 is striking, even though the diagrams pertain to different systems. Later work[65] compared the cosorption contours[66] of Fig. IIB.1 with the isaphroic contours for the same system, and the resemblance confirmed the relation of surface activity and its manifestation in foaminess, to the character of the phase diagram.

Gas–liquid contacting is the basis of many production processes, in which an adequate interfacial area is required for good mass transport, but excessive foaming leads to instability, especially in equipment such as sieve trays. Industrial foaming problems were the stimulus to investigate connections between capillarity and phase diagrams. Foaming is indeed so prevalent in this type of industrial equipment that design engineers routinely provide extra capacity in gas–liquid contact towers. Problems of this kind may be anticipated and avoided by studying the phase diagram of the system.

Foaming in distillation and fractionation towers, for example, by which the liquid is carried into spaces intended for the vapor, is called "foam flooding," and is often encountered. Degasification after gas absorption or during "stripping" of a monomer in emulsion polymerization are prone to the same effect. The cause of the foam stability may occasionally be traced to the presence of minute concentrations of unintended contamination by substances of strong surface activity; but usually the pro-foaming solute is a legitimate component of the system, present in substantial concentration. Such a component is not a conventional surface-active solute but becomes surface active under certain conditions of temperature and concentration in a medium in which these conditions are conducive to decreasing its solubility. These conditions may occur in the process of fractionation, stripping, and so on; even so the surface activity thus elicited is minimal, capable of stabilizing foam only for short times. Although such evanescent foams last no more than several seconds, they are the source of severe foam-flooding problems in distillation or fractionation towers, where rapid evolution of vapor may build up a large volume of dynamic foam. In the Girdler sulfide process for producing heavy water, liquid water and gaseous hydrogen sulfide are brought into contact at high temperatures and at pressures of several atmospheres over sieve trays.[67-69] In certain parts of the exchange tower the conditions are such that the overall composition approaches the phase boundary at which liquid hydrogen sulfide separates from its conjugate aqueous solution. In accordance with observations of the more accessible systems described above, the hydrogen sulfide in aqueous solution would be expected to be surface active, and indeed foam does appear. The problem thus created was sufficiently severe to require drastic reduction of throughput in the exchange tower, and was sufficiently abstruse to delay remedial action for over a year. Again, in accord with generalizations drawn from cognate systems, the conjugate liquid hydrogen sulfide phase would be expected to inhibit foam, as in fact it was found to do.

l. Summary of Foam Stabilization

Summarizing the discussion thus far, we have pointed out three distinct mechanisms of foam stabilization: the Marangoni effect, the formation of gelatinous surface layers, and the immobilizing of interstitial liquid by electrostatic effects. These mechanisms accord with the observed fact that two distinct ranges of foam stability differing considerably in magnitude exist for two different types of foam. The relatively unstable foams (e.g., champagne foam) result from the Marangoni effect alone; they drain rapidly to a critical thickness on the order of 20 to 30 nm, at which point the Marangoni effect can no longer operate and the lamellae are brittle and soon rupture. The very stable foams (e.g., beer foam) begin their lives in the same way, being preserved from instantaneous destruction only by the existence of the Marangoni effect. This effect is soon replaced, however, either by the slow growth of gelatinous surface layers or by the overlap of diffuse double layers, which takes over the task of preserving the lamellae by immobilizing the fluid inside so that gravitational and capillary stresses are both insufficient to thin them.

5. MECHANISMS OF ANTIFOAMING ACTION

a. Impairment of Foam Stability

We distinguish between impairment of foam stability and the inhibition of foam formation. Foam stability may be impaired by the presence of certain substances of low molecular weight that by their presence in the surface layer reduce its coherence. Such a substance does not need to be surface active to affect the surface layer if enough of it is mixed with the solution to act as a cosolvent. This effect has an obvious explanation because the surface-active solute that is responsible for foaminess may be rendered more soluble, and hence less surface active, in a mixture of solvents than in one solvent alone.

The term "cosolvent" implies a concentration of 10% or more of a second solvent. Occasionally, the addition of such a large proportion of, for example, methanol, ethanol, or acetone to an aqueous solution may be practicable and would thereby solve a troublesome foaming problem. Customarily, a much smaller quantity of agent is desirable. Excluding this case then, foam stability may still be impaired by solubilizing small amounts of nonionic organic liquids in the aqueous solution, using the foam-producing solute as a solubilizing agent. This operation has previously been mentioned as a means of increasing the stability of foams by the formation of plasticized (gelatinous) surface layers, but the same operation may produce quite different results, depending on the nature of the solubilizate. If the molecular structure of the nonionic agent is such that it does not form a coherent surface layer, its presence will impair rather than enhance the stability of the foam. Thus, for example, while a small quantity of solubilized dodecanol will increase the foam stability of an anionic agent in water, a like quantity of solubilized octanol will have a contrary effect. The octanol which is less surface active than dodecanol, is not present in the surface in amount sufficient to create a liquid-crystal phase; and also, by virtue of its more rapid diffusion to the surface, reduces the dynamic surface tension more rapidly and thus allows less time for the Marangoni effect to function.[46] Nevertheless, some coherence of the surface layer and some Marangoni effect still remain, so although the foam is rendered less stable, its formation is not entirely inhibited. This result is "foam impairment."

b. Surface and Interfacial Tension Relations

Certain chemical agents are well known to do more than merely impair the stability of foam—they totally inhibit the forming of foam. These foam inhibitors are always insoluble materials, usually liquids, although hydrophobic solid particles, such as Teflon, waxes, and silane-treated inorganic oxides also are used.

Not every insoluble substance can act as a foam inhibitor: There are additional requirements. The nature of these additional requirements are readily comprehended if the mechanism of the foam-inhibiting action is first understood. Such an understanding of the underlying mechanism was inferred gradually from observations and experiments with foam-producing systems to which various agents were added so that their effects could be studied.[47] The solution of practical problems came before theoretical explanations, the latter being elicited only rather slowly and not without several hypotheses later found to be erroneous. At first, so spectacular was the phenomenon and so unfamiliar were the principles of surface activity, that one often heard it described as magic. As in so many other cases, the magic was banished by science.

Consider the successive operations that are required. First, the droplet or particle of insoluble agent has to be admitted to the surface; that is, when the agent approaches the vicinity of the surface as a result of its random movement in the foamy liquid, the medium should withdraw so that the agent has an exposed surface toward the vapor phase. A factor that would be helpful to promote this action would be a slight difference in density between the agent and the medium so that a buoyant force directed toward the surface would be added to the random movements of the droplet that are the result of thermal currents, stirring, or Brownian motion. This by itself would not be enough; droplets of the agent would indeed float to the surface, but they might still remain covered with a thin liquid film of the medium. The medium would not withdraw and expose the surface of the droplet unless, by so doing, the net surface free energy were reduced, that is, only if the surface tension of the agent σ_a were less than the sum of the surface tension of the medium σ_m and the interfacial tension σ_{int}. This condition requires that the agent have a relatively low surface tension–too low to be wetted by the foamy medium. Robinson and Woods[48] were the first to point out this condition, which they defined by means of a function named the *entering coefficient E*:

$$E = \sigma_m + \sigma_{int} - \sigma_a \qquad [\text{B}.14]$$

The agent enters the surface only if the value of E is positive. But the term "entering coefficient" seems to be an unnecessary neologism when we already possess a well-known coefficient to serve as a criterion for wetting, namely, the spreading coefficient S.[70] The condition is stated more expressively if we say that the droplet of the agent enters the surface only if it is dewetted by the medium, that is, if the medium is incapable of spreading on the agent. Let S_1 be the spreading coefficient of the medium on the agent; then,

$$S_1 = \sigma_a - \sigma_m - \sigma_{int} \qquad [\text{B}.15]$$

The agent would be dewetted by the medium if the value of S_1 were negative. The two statements of the necessary condition are, of course, identical, since $S_1 = -E$.

The droplet of the agent enters the surface by virtue of withdrawal of the medium, a process aptly described as "dewetting." A foam lamella has two surfaces, consequently dewetting can occur also on the second surface once the lamella has thinned sufficiently, either by capillary suction or hydrodynamic drainage. The second withdrawal, at the lower surface, causes the rupture of the lamella, whether due to the mechanical agitation caused by the spontaneous motion or to the poor adhesion of the medium to the droplet, which makes it ill-adapted to hold the lamella together by bridging the gap.

The surface having been entered, the next operation that may be required of a foam-inhibiting agent is for the droplet of agent to spread spontaneously across the surface of the medium. Let S_2 be the spreading coefficient of the agent on the medium; then,

$$S_2 = \sigma_m - \sigma_a - \sigma_{int} \qquad [\text{B}.16]$$

The agent spreads spontaneously if the value of S_2 is positive. The second condition is perfectly compatible with the first condition; S_1 being negative does not prevent S_2 from being positive. The only prohibition is for both S_1 and S_2 to be positive; therefore, *if S_2 is positive, S_1 must be negative.* Thus, if the interfacial tension between the medium and the foam-inhibiting agent is sufficiently low, the droplet will spread after it enters the surface of the lamella. Fowkes (personal communication of unpublished work) found that the spreading action is less effective than dewetting in causing the rupture of lamellae and may even fail to inhibit foam. But this distinction of function is not universal: the action of a spray, or even the presence of the vapor, of a volatile liquid such as diethyl ether or acetone, when introduced as a foam breaker above the surface of a foam already formed, does depend on the spreading of the droplets over the surface of the bubbles, with their consequent rupture.

The dewetting mechanism is used in the application of dispersed droplets of polydimethylsiloxane as a foam inhibitor in lubricating oils. The surface tensions of liquid hydrocarbons are between 25 and 30 mN/m; a liquid able to enter on such a low-energy substrate is required to have a surface tension lower than these by several millinewtons per meter. Such liquids are usually volatile, which makes them unsuitable in many applications. Only special polymers such as polydimethylsiloxane or perfluorinated hydrocarbons combine the usually disparate properties of low surface tension and low volatility.

c.   Foam Inhibition–Aqueous Systems

The above principles are applied in compounding a foam-inhibiting agent for a particular use. An aqueous solution or dispersion that is the source of troublesome foam usually has dissolved organic materials that act as the profoamer. Their presence is evident from the surface tension, which in such a case is less than that of pure water at the same temperature. Occasionally, when imperfectly wetted solid particles are present, the stabilizing of the foam may be caused by the solid particles holding entrapped air between their hydrophobic surfaces, but even then some dissolved profoamer is usually also present to initiate film formation. If surface tension is less than that of water by no more than 20 mN/m, the selection of a foam-inhibiting agent of abnormally low surface tension, for example, a polydimethylsiloxane, is hardly justified. A less unusual and, incidentally, less expensive agent serves just as well. One could begin by trying glyceryl esters in the form of lard, pork fat, or butter fat. These common substances have many of the desired qualities: insolubility in water, lower density than water, relatively low surface tension, and relatively low interfacial tension against aqueous solutions. For many practical purposes, they are effective foam inhibitors[71] where the systems being treated have surface tensions greater than that of the fat. If the surface tension of the aqueous system is too low, however, the fat is emulsified and is then not effective as a foam inhibitor, at least by the mechanism described above. The system may stop foaming if enough fat is emulsified to deplete the solution of its profoamer; but this result has the disadvantage, for some purposes, of incorporating a quantity of fat in the system.

If readily available natural fats prove to be ineffective agents, a number of synthetic esters of polyhydric alcohols, which are marketed under various trade names, may be tried. These substances extend the range of choice of materials and offer variations in the balance of the lipophilic and hydrophilic portions of the molecule. The lipophilic portion confers low surface tension; the hydrophilic portion confers low interfacial tension; both within the limits of a suitable balance. Completely lipophilic substances, for example, paraffin oil or kerosene, besides being very insoluble in water, are not even able to spread as a monolayer on a water surface. Their interfacial tension against water is large, which means that they cannot be wetted by water. They will, therefore, enter a water surface (S_1 negative) but will not spread over it (S_2 negative). If the substance has now a small degree of hydrophilia added to it, for example, fatty alcohols or fatty acids, the interfacial tension is the first property to respond, and it does so by becoming smaller. We now find that although S_1 is still negative, S_2 may become positive; in terms of behavior, the substance will enter the aqueous surface and also spread on it. These materials are of the class in which excellent foam inhibitors are to be found. Let the substance be a little more hydrophilic, and the interfacial tension is still further lowered to a very small value so that S_1 may become positive. These materials are not dewetted by water and consequently are unable even to enter the surface. Still more hydrophilic substances, for example, sucrose or glycerol, are soluble without limit in water (Table IIC.2).

Evidently, an optimum balance is desirable between the lipophilic and hydrophilic portions; but the optimum varies with the nature of the medium and with the nature of the profoamer dissolved therein. The more lipophilic the profoamer, the more lipophilic is the optimum foam-inhibiting agent. It seems, therefore, that a certain constant difference in the HLB scale must be maintained between a profoamer and its optimum foam inhibitor. The point has not been investigated experimentally, although it would be well worth the effort. If it were verified, we would have a basis for selecting the optimum foam inhibitor once the HLB of the profoamer were ascertained. At present, no single test has been widely accepted whereby the HLB of a surface-active agent may be measured, but such a test is the objective of a number of workers (Shinoda and Friberg, 1986). One of the fruits of success in this endeavor would be to diminish the empiricism that at present characterizes the search for an optimum foam inhibitor.

d.   Compounded Foam Inhibitors

A practical discovery of emulsion technology is that a mixture of emulsifying agents is frequently more effective to stabilize an emulsion than a single ingredient. As a result of this knowledge, the practice was derived of compounding agents to a desired HLB by mixing ingredients lying on each side of the average. Foam-inhibiting agents can be compounded in the same way, as long as all ingredients are miscible. A paraffin oil, for example, would not inhibit the foaming of an aqueous system; the surface tension, which is about 30mN/m, may not be low enough for the medium to dewet the droplet so that it could enter the surface. To make it more effective, the interfacial tension must be reduced. One way to do this is to add a surface-active agent to the aqueous medium, but this offers too narrow a margin to work with because a very small amount may be enough to reduce the surface tension of the medium as well as the interfacial tension, thereby emulsifying the paraffin oil and keeping it from entering the surface. A better way is to compound an oil-soluble surface-active agent with the paraffin oil. Such agents usually do not lower the surface tension of organic liquids but exhibit their surface activity by reducing the interfacial tension at the oil–water interface.

Substances suitable for this purpose include alkyl benzenes, natural fats, fatty acids, and metallic soaps. Even the moderate hydrophilia conferred by aromatic substituents in an alkane hydrocarbon is sufficient to reduce the oil–water interfacial tension; more hydrophilic substituents, such as a carboxylate group, make highly effective agents for this purpose. An oil-soluble polymeric silicone can also be used as the second ingredient, thereby reducing the cost of using an uncompounded silicone as the sole agent.

Another recent example of the same principle is the use of homopolymers of propylene oxide or butylene oxide and copolymers of ethylene, propylene, and butylene oxides, when compounded with metal salts of higher fatty acids, for the inhibition of foam in proteinaceous glues.[72]

The same principles can be applied *mutatis mutandis* to the problem of foam in nonaqueous media. For example, the foaming of lubricating oil in aircraft at high altitudes was an urgent problem during World War II. A foam-inhibiting agent for lubricating oil was made of glycerol, selected because of its low volatility and insolubility in the oil, with 2% of a surface-active agent (e.g., Aerosol OT) dissolved in it.[73] The function of the Aerosol OT is to reduce the surface tension of the glycerol below that of the lubricating oil and also to reduce the interfacial tension between the glycerol and the oil; when these tensions are sufficiently low, the composition functions as a foam-inhibiting agent. Other glycerol-soluble surface-active agents can be substituted for the Aerosol OT, such as a block copolymer of polyoxyethylene and polydimethylsiloxane.

e. Silicone Foam Inhibitors

Polydimethylsiloxanes are close to being universal agents for inhibiting foam because they combine two properties seldom found together, namely, involatility and low surface tension; in addition, they are chemically inert and are insoluble in water and in lubricating oil. They are effective in concentrations in the range of 10 ppm or less, whereas other foam-inhibiting agents are often used in the range of 100 to 1000 ppm of bulk fluid. On the other hand, the silicone polymers cost 10 to 20 times that of the common organic substances.

Reduction of the particle size of the composition to approximately the thickness of a foam lamella is a requirement for its effectiveness as a foam inhibitor. Unfortunately, the potentially more effective agents are precisely those most likely to coalesce on standing. Silicone antifoams are sometimes provided by manufacturers in the form of emulsions in an aqueous phase, which will then mix readily with the aqueous foamy liquid. This form meets the need for small droplets by predispersing the fluid before it is added to the foam-producing medium. Ultimately the low concentrations used are effective in preventing coalescence because of the small probability of two particles colliding.

Silicones are used as foam inhibitors[74] in a wide range of industrial foaming problems, including those in distillation towers. The standard polydimethylsiloxane molecule may be modified to decrease its solubility for certain applications and thereby improve its effectiveness as a foam inhibitor; for example, poly-trifluoropropylsiloxane defoamers are used with some organic systems in which the polydimethylsiloxanes are too soluble to be effective.[75]

f. Hydrophobic Particles

Silicone oil, polydimethylsiloxane, is an effective foam inhibitor of oil foams, and is used to prevent the foaming of hydrocarbon lubricants in aircraft engines.[76] But it is far from being equally effective to control foaming in aqueous solutions. It was found, however, that after adding finely divided silica in proportions of 3 to 6% the mixture acts as a highly effective antifoam, even when only a few parts per million is put into the foamy liquid. Scores of patents have been issued describing progressively better methods to incorporate silica into the formulation, with more effective foam-inhibiting action marking each advance in the art of preparation. The accumulated evidence reported in those published patents leaves no doubt of the reality of the "activation" by the presence of the silica: It contributes nothing to our understanding of why it should occur. The improvement of the antifoam by the addition of silica is, however, now so well established that few, if any, commercial formulations designed to suppress or destroy foam of aqueous systems by means of silicone oil are marketed without the added silica.

Silica normally has a hydrophilic surface, which is to say it is perfectly wetted by water. But the surface of silica can be drastically altered to become hydrophobic by chemical treatment. Since finely divided particles necessarily have a large surface area, and as we are discussing silicas with $200-400$ m^2 of surface area per gram, this alteration of the surface character implies marked alterations of the gross behavior of the material. A particle of hydrophobic silica is not wetted by water; it floats on the surface of the water, just as would an oiled needle, in spite of its density being greater than that of water. Water may be said to roll off its back, just as it rolls off a duck's back, and fundamentally for the same reason. When mixed with a hydrophobic liquid, such as silicone oil, however, hydrophobic silica is readily wetted by the oil and disperses easily. In such media it is now the turn of hydrophilic silica to show reluctance to be wetted; it does not allow the oil to penetrate freely along the surface and especially into the narrow spaces between particles. If many such hydrophilic particles are stirred into silicone oil, they create an interconnected or network structure throughout the oil and so give it the viscoelastic properties of a jelly.

Ross and Nishioka[77] prepared suspensions of silica in polydimethylsiloxane. These suspensions were of two types, designated alpha and beta, depending on how much heat or agitation is put into the system during its preparation. The α suspension is very viscous and elastic and resembles a jelly; when ball-milled or heated to 150°C for a few hours, or to a higher temperature for a shorter time, the jelly like structure breaks down irreversibly and the suspension becomes more fluid. The suspension is now sensibly different both to the eye and to the touch. This is the β suspension.

The α and β suspensions differ in the following respects:

(a) The obvious and immediately apparent difference is the jelly like consistency of the α suspension compared with the fluid, though still viscous, β suspension. If, for example, the concentration of silica is made more than about 5%, the α suspension is too stiff to be stirred conveniently by hand or by ordinary laboratory mixing equipment, but on conversion to the β suspension it flows like a heavy lubricating oil.

(b) On diluting the α suspension with n-hexane, the suspended silica becomes unstable and soon separates and sinks to the bottom. This behavior reveals that the suspension lacks true stability and is maintained only by virtue of its jelly like consistency. The β suspension on the other hand is evidently well stabilized because on dilution with hexane to a low viscosity the aggregates of silica remain suspended indefinitely. This difference in stability argues in favor of steric stabilization of the aggregates in the β suspension and the lack of such stabilization of those in the α suspension; that is, the presence of a thick adsorbed layer of polymer around aggregates in the β suspension.

(c) Tested as foam-inhibiting agents, the α suspension hardly differs from polydimethylsiloxane taken by itself without added silica, whereas the β suspension destabilizes foam, an effect well recognized in the commercial compositions of polydimethylsiloxane as being conferred by silica.

(d) No surface-active solute is present in this system to stabilize the suspension, which must therefore be stabilized by adsorption of the medium itself. The usual techniques to measure adsorption in a three-component system, consisting of a solid, a solvent and a solute, cannot be applied. The properties of polydimethylsiloxane, which is involatile and spreads as a monolayer on water, suggest the use of the Langmuir film balance as an instrument of investigation. The quantitative determination of bound polymer on the silica substrate can be calculated directly from the differences between pressure–area isotherms of monomolecular layers of polydimethylsiloxane with and without dispersed silica. Bound polymer on the α suspension is about 0.10 g/gram of silica; on the β suspension it is about 1.0g/gram of silica.[78]

(e) When spread as a compressed monolayer on a water surface, the α suspension stabilizes single bubbles blown under it exactly as does the monomolecular layer spread from polydimethylsiloxane taken by itself without added silica. A monomolecular layer spread from the β suspension does not stabilize a bubble at any degree of compression.

These differences demonstrate that the silica in the α suspension is only partly converted to hydrophobic silica, and in that partially converted form it adds nothing at all to the foam-inhibiting properties of the polydimethylsiloxane. The following commercial preparations of the polydimethylsiloxane containing silica, all marketed as foam inhibitors, were tested by the preceding five criteria. Each one was found to correspond to the β suspension by the previous tests: (a) Dow

Corning Antifoam M, (b) Dow Corning Antifoam MSA, (c) Rhodorsil Anti-mousse 454.

The results of the measurements of bubble stability under a monolayer are particularly revealing with respect to the mechanism of rupture of foam lamellae. Trapeznikoff and Chasovnikova[79] showed that polydimethylsiloxane when spread as a monolayer on water actually stabilizes a water lamella. A spread layer of polydimethylsiloxane on water stabilizes bubbles blown under it in a manner similar to the bilayers of Trapeznikoff and Chasovnikova. Significantly, a film of the α suspension on water also stabilizes bubbles, whereas a film of the β suspension does not. The similarity in behavior of bubbles created under a spread film of polydimethylsiloxane taken by itself without added silica and a spread film of α suspension, and the dramatic difference displayed when the β suspension is substituted for either of them, demonstrates the strong destabilizing effect produced by the hydrophobic silica particles of the β suspension in the polydimethylsiloxane film.

The mechanism by which dispersed solid hydrophobic particles are so effective in destroying bubbles depends on the degree of hydrophobia of the particle. An oily surface makes a film of water recede from it, and in the same way the surface of a hydrophobic solid particle makes water recede, or dewet, where the two make contact. The withdrawal of the water of the bubble lamella from the surface of the particle may by itself be enough of a mechanical shock to rupture the lamella and release the enclosed gas, or the rupture may be due to the poor adhesion between the water and the hydrophobic particle, causing the water to fall away from the particle as it withdraws.[78] Roberts et al.[80] published high-speed photographs that seem to demonstrate the latter mechanism for rupture of soap lamellae by insoluble droplets. Garrett,[81] Dippenaar,[82] Kurzendoerfer,[83] and Aronson[84] have also demonstrated that various chemically inert dispersed solids act as antifoams and have proposed that the particles rupture the foam film by a bridging-dewetting mechanism. "Bridging" means that the particle touches both surfaces of the lamella; but whether just one or both sides of the lamella have to be dewetted in order to cause rupture remains uncertain.

6.   METHODS OF MEASURING FOAM PROPERTIES

a.   Stability

The measurement of foam stability is made either on static or dynamic foams. A static foam is one in which the rate of foam formation is zero: The foam once formed is allowed to collapse without regeneration by further agitation or input of gas. A dynamic foam is one that has reached a state of dynamic equilibrium between the rates of formation and decay. The typical measurement of a dynamic foam is the volume of foam at the steady state; the typical measurement of a static foam is its rate of foam collapse. Dynamic-foam measurement is applicable to evanescent or transient foams; static-foam measurement is applicable to foams of high stability, such as are generated from solutions of detergents or proteins.

Stability of dynamic foams is measured by passing gas through a porous ball or plate into a solution, which may be contained in either a cylindrical or a conical vessel. Gas is passed through a suitable volume of the solution at measured rates, V/t (cm^3/s), and the steady-state volume of foam, v, is measured. The depth of the liquid layer about the porous plate should be adequate to ensure a result independent of that depth. The ratio of the flow rate of the gas, whether incoming or outgoing, to the steady-state volume of the foam has the units of time and represents the average lifetime of gas in the foam.[85] It is designated Σ and is used as a unit of foam stability. Frequently, however, the ratio is not constant through an extensive range of flow rates, or the foam is so stable that it floods the container even at low flow rates. The substitution of a conical container instead of a cylindrical one improves those situations and extends the use of the method.[86]

The stability of static foams is fundamentally the rate at which the total area of the liquid lamellae disappears. These measurements can be obtained photo-graphically[87] or by digital image analysis. An indirect technique is available that uses the equation of state of a foam as its theoretical basis:

$$3V \, \Delta P + 2\sigma \, \Delta A = 0 \qquad [\text{B.6}]$$

where ΔP is the change of external pressure and ΔA is the change of the area of the liquid lamellae during the decay of a static foam. Values of absolute area of the liquid lamellae in the foam are obtainable as follows. The area of liquid lamellae in foam is

$$A(t) = A_0 + \Delta A \qquad [\text{B.17}]$$

where $A(t)$ is the area at time t and A_0 is the initial area. When the foam has completely collapsed

$$A_\infty = A_0 - \frac{3V\Delta P_\infty}{2\sigma} = 0 \qquad [\text{B.18}]$$

where P_∞ is the final pressure in the head space when the foam is gone; therefore

$$A_0 = \frac{3V \, \Delta P_\infty}{2\sigma} \qquad [\text{B.19}]$$

and

$$A(t) = \frac{3V}{2\sigma}[\Delta P_\infty - \Delta P(t)] \qquad [\text{B.20}]$$

Interfacial area of a foam can be measured, therefore, simply by monitoring the change in pressure external to the foam in a container of constant volume and constant temperature, if the total volume of the system and the surface tension of the foamed liquid are known. The value ΔP_∞ can be obtained by letting the foam

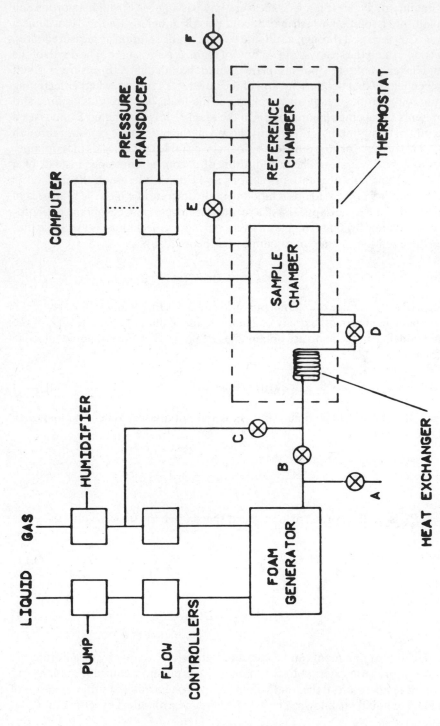

Figure B.5 Schematic of a device to determine the stability of foam by measuring the decay of area with time.[88]

decay for a sufficiently long time or by injecting a small quantity of antifoam through a septum in the container.

Values for the area of a decaying foam by use of Eq. [B.20] were found to agree with photographic estimates of the area.[34]

Nishioka[88] developed an apparatus to measure foam stability, using a mechanical foam generator, an Oakes Foamer with 2-in. mixing head, and a pressure transducer (Dynisco Model PT14-03) to measure the growth of the pressure, which was about 1.5 torr over a period of 10 h. Figure B.5 is a schematic of the equipment, which consists of controllers to regulate the flow of the liquid and gas into the generator, the foam generator, the foam-measuring system, and the data-collecting system (available from H. and N. Instruments). The results obtained by Nishioka by this equipment are reproducible with an error or 3–4%. The foams produced by the generator initially have about 2×10^5 cm^2 of area, which decays to about 0.2×10^5 cm^2 within an hour.

b. Surface Rheology

The presence of a highly viscous layer at the surface of an aqueous soap solution can be observed by sprinkling flowers of sulfur on the surface and attempting to move them by blowing gently. On a perfectly fluid surface, such as that of pure water, the particles may be moved readily; but on certain soap solutions they move only a short distance and spring back when the blowing stops. The phenomenon does not occur with solutions of a single surface-active solute but arises when sparingly soluble cosolute is also present. Current belief is that the surface is a separate liquid-crystal phase produced within the adsorbed layer of solute. While the *surface* may show extreme non-Newtonian flow behavior, the *bulk solution* retains its low Newtonian-type viscosity.

Various instruments to measure the rheology of the surface have been described of which the simplest is the torsional pendulum[6] (Fig. B.6). A circular knife edge, supported by a torsion wire, is placed so as just to touch the surface of the solution. The torsion wire is supported at the end of a shaft that can be given a rotational twist so as to cause the knife edge to oscillate in the plane of the surface. The damping of the oscillation is followed by observing a light beam reflected from a mirror attached to the wire. The amplitude of the oscillation decreases semilogarithmically when the knife edge is immersed in a perfectly fluid surface. If the surface is plastic, the damping decreases more rapidly. At certain portions of its oscillation the shearing stress produced by the pendulum is less than the yield point; the result is much as if the free oscillation were arrested from time to time by an external force. While this instrument can detect and even give relative values of the surface plasticity, the measurements do not lend themselves to deduce rheological coefficients.

The instrument designed by Burton and Mannheimer[89–91] allows absolute values of surface shear viscosity and yield strength to be obtained. The solution is contained in a stainless-steel dish into which is placed an annular canal formed by two concentric cylinders. A small gap (approximately 0.005 in.) is left between the

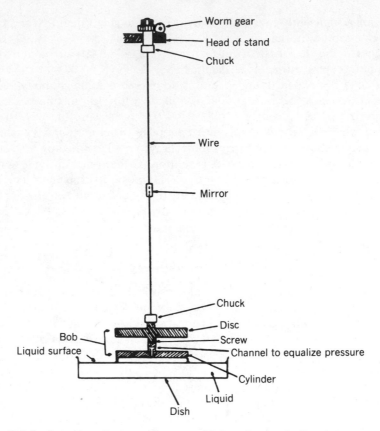

Figure B.6 Surface viscosimeter with an oscillating circular knife edge suspended by means of a torsional pendulum.

bottom of the canal walls and the bottom of the dish. The walls of the canal are held stationary as the dish is rotated on a turntable at a fixed angular velocity. A few nonwetted particles are dropped onto the surface within the canal. The velocity communicated to the surface is measured by the motion of the particles. The yield strength of a film is determined by measuring the maximum torque at which the surface flow is zero.

CHAPTER IVC

Suspensions

1. ADSORPTION FROM SOLUTION BY SOLIDS

Solid adsorbents may be classified with respect to the medium as lyophilic or lyophobic, that is, those that have a specific interaction with the medium and those that lack such an interaction. These terms usually serve to distinguish between particles of colloidal dimensions in solution (lyophilic), such as soluble polymers or micelle-forming solutes, and insoluble particles that do not attain colloidal size spontaneously (lyophobic), such as metallic oxides. The same distinction is sometimes indicated by the terms "reversible" and "irreversible" dispersions. Since the distinction between soluble and insoluble materials is clear, as is that between reversible and irreversible dispersions, neologisms that do no more than make the same distinction are redundant. They need not be wasted, however. We suggest reserving the terms "lyophilic" and "lyophobic" to characterize the affinity of the medium for the *interface*, as measured by its spreading coefficient. We shall, therefore, use *lyophilic* for interfaces, whether associated with soluble particles or not, on which the medium has a positive spreading coefficient; and *lyophobic* for those on which the medium does not spread. These terms now refer to the interface rather than to the bulk material, and consequently make it possible to say that the function of a dispersing agent is to convert a lyophobic to a lyophilic interface.

Water is readily adsorbed by and therefore spreads on lyophilic interfaces, which for that reason are called hydrophilic, whereas water does not spread on lyophobic interfaces, which are called hydrophobic. On lyophilic interfaces, specific interactions include electrostatic and electron donor–acceptor interactions, either between the interface and the medium or between the interface and a solute component; where water is the medium, the electron donor–acceptor interaction

327

on a hydrophilic interface is called hydrogen bonding. Examples of adsorbents with hydrophilic interfaces are alumina, barium sulfate, calcium carbonate, glass, ion exchange resins, quartz, silica gel, titanium dioxide and most metallic oxides, and zeolites; these can be further subdivided as electron donors (bases) or electron acceptors (acids). Examples of adsorbents with hydrophobic interfaces are bone char, carbon blacks, charcoals, graphite, organic resins and plastics, paraffin, stibnite and most metallic sulfides, selenium, sulfur, and talc. Electron donor–acceptor interactions, or the special case of such interactions known as hydrogen bonding, are stronger than interactions due to dispersion forces of attraction; but the latter are present even in the absence of the former. Dipole–dipole or dipole–induced dipole interactions are known to be less significant than dispersion force interactions, and so come fourth, and a poor fourth, in this list (Table IIIA.2).

a.   Acidic or Basic Character of Solid Substrates

The major interactions of a solid surface with an adsorbate are of two kinds: the universal interactions of the London dispersion forces and the specific interactions. London dispersion forces are long range and, even though small, their sum is a significant portion of the total interaction per unit area across an interface. The London dispersion energies of solid substrates, as determined by their interactions with various liquids, are described in Section II.A.8 and as determined from their electronic properties are described in Section III.A.2. Most of the specific interactions are Lewis base–Lewis acid reactions (electron donor–acceptor interactions) this category includes chelation, coordination, hydrogen bonding, and the complexes between nucleophiles and electrophilic sites that are so important in catalysis.[92] The energy per mole of these interactions is large but, because of their short-range character, only those pairs in proximity contribute to the total interaction. Consequently the dispersion force interactions and the donor–acceptor interactions are comparable. These two types of interactions have proved to be sufficient to account for the solution and interfacial properties of dispersions.[93]

Various empirical approaches to the problem of predicting donor–acceptor interactions are proposed (Jensen, 1980). The Drago correlation predicts the enthalpy of acid–base reactions in the gas phase or in poorly solvating solvents, by a four-parameter equation:

$$-\Delta H_{ab} = E_a E_b + C_a C_b \qquad [IIA.17]$$

The acid a and base b are each characterized by two independent parameters: an E value that measures its ability to participate in electrostatic bonding, and a C value that measures its ability to participate in covalent bonding. Both E and C are derived empirically to give the best agreement between calculation and experiment for the largest possible number of adducts. A self-consistent set of E and C values is now available for several dozen bases, allowing the prediction of ΔH for over 1500 interactions[94] (Tables IIA.5. and IIA.6).

Fowkes[92] extended Drago's approach to include the interactions taking place at an interface. For example, he has found that the silanol groups on a silica surface are strongly acidic, whereas a glass surface because of its silicate content is basic. The acidic silica surface adsorbs more of a basic adsorptive such as pyridine or polyvinyl pyridine than does a basic surface such as glass. The Drago E and C constants for a solid substrate can be evaluated by measuring the enthalpy of interaction of a test acid or base whose constants are already known. Heats of adsorption may be obtained calorimetrically. A useful technique is flow microcalorimetry in which a stream of solution containing the adsorbate at a known concentration is passed over a known weight of adsorbent previously wetted with the flowing stream of the solvent. The concentration of the adsorbate in the effluent stream is monitored as well as the heat evolved until the concentration in the effluent is the same as that in the incoming stream. Simple calculations give the total mass adsorbed and the total heat evolved for adsorption at that concentration. Repeated measurements at varying adsorbate concentrations give the adsorption isotherm and the differential heats of adsorption.

The heat of adsorption may also be obtained from adsorption isotherms measured at two different temperatures. Adsorption from solution is often described by the Langmuir equation [ID.17], which assumes that all the adsorption sites are equivalent. The assumption is valid except for substrates with a wide distribution of site energies; and a wide distribution of specific acidic or basic sites on a substrate is unlikely. The heat of adsorption is then obtained from the Clausius–Clapeyron equation applied to the two isotherms:

$$\Delta H = \frac{RT_1 T_2 \ln(K_1/K_2)}{(T_2 - T_1)} \qquad \text{[C.1]}$$

where K_1 and K_2 are the Langmuir constants in Eq. [ID.17] at absolute temperatures T_1 and T_2, respectively.

By the application of these techniques, Fowkes and his co-workers determined the E and C constants for the SiOH sites of silica, the TiOH sites of rutile, and the FeOH sites of ferric oxide. These three kinds of surface sites are all acidic. Table C.1 contains the Drago parameters determined for these oxides in terms of moles

Table C.1 E and C Parameters for Active Groups on Some Solid Surfaces

Solid	E_a(kcal/mol)$^{1/2}$	C_a (kcal/mol)$^{1/2}$
Silica	4.2 ± 0.1	1.16 ± 0.02
Rutile	5.7 ± 0.2	1.02 ± 0.03
α-Ferric oxide	4.5 ± 1.1	$0.8 \; \pm 0.2$

of active group on the surface assuming 1:1 adduct formation.[95] From these parameters the heats of adsorption of various nitrogen bases, oxygen bases, sulfur bases, aromatic bases, and basic polymers may be predicted.

b.   Nonaqueous Media

Adsorption of solute by an inorganic interface in a nonaqueous medium differs from adsorption out of an aqueous medium in that the former process lacks the entropy difference arising from the change of structure of the water (Section II.E.1). Here the significant difference of entropy resides in the withdrawal of solute from the bulk organic phase to the interface, where it has fewer degrees of freedom, which means that the change of entropy of adsorption in organic liquids is negative. Since the change of entropy is negative, the Helmholtz free energy, that is, the condition for spontaneous adsorption to occur, would be negative only if the enthalpy change had a large negative value. To obtain a large negative value for the enthalpy change, a specific chemical interaction between the solute

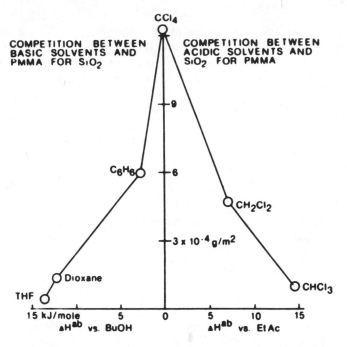

Figure C.1 Adsorption of PMMA (basic) on silica (acidic) as a function of the acidity or basicity of the solvent. The basicity is measured as the heat of acid–base interaction with *tert*-butyl alcohol and the acidity is measured as the heat of acid–base interaction with ethyl acetate. Poor adsorption occurs at the right-hand side because the acidic solvent dissolves the basic polymer too well for it to be taken out of solution by adsorption; poor adsorption occurs on the left-hand side because the basic solvent preempts the acidic surface of the silica so successfully that the basic polymer is excluded.[97]

and the substrate is required. The most significant such interaction is that between a Lewis acid and a Lewis base. Lacking such an interaction the substrate cannot compete with the solvent for the solute.

This adsorption mechanism is the one commonly found in nonaqueous systems. Dispersions of carbon black in aliphatic hydrocarbons, for example, are stabilized by aromatic hydrocarbons with long aliphatic side chains.[96] The acidic surface of the carbon black interacts with the basic aromatic group of the stabilizer. Again, amines are effective reagents for the deflocculation of carbon black in cellulose esters and oleoresinous vehicles (Fischer, 1950, p. 245), which is another example of an acid–base interaction.

A solute that is an electron acceptor (or Lewis acid) will interact with an electron donor (or Lewis base,) whether it finds that base on the substrate or in the solvent. Fowkes and Mostafa[97] measured the extent of adsorption of a basic polymer (polymethyl methacrylate or PMMA) on an acidic substrate (silica, acidity due to surface-OH groups), as a function of the Lewis acidity and basicity of six solvents used to dissolve the polymer. Carbon tetrachloride

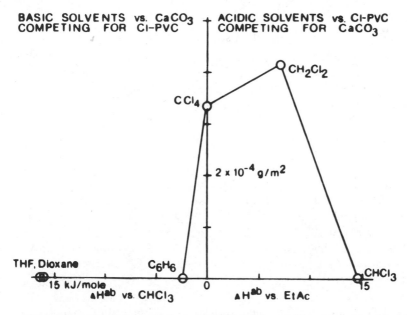

Figure C.2 Adsorption of post-chlorinated PVC (acidic) on calcium carbonate (basic) as a function of the basicity or acidity of the solvent. The basicity is measured as the heat of acid–base interaction with chloroform and the acidity is measured as the heat of acid–base interaction with ethyl acetate. Poor adsorption occurs at the left-hand side because the basic solvent dissolves the acid polymer too well for it to be taken out of solution by adsorption; poor adsorption occurs on the right-hand side because the acid solvent preempts the basic surface of the calcium carbonate so successfully that the acid polymer is excluded.[97]

was the most neutral solvent, benzene a very weak base, and dioxane and tetrahydrofuran stronger bases. The acid solvents were dichloromethane and chloroform. The basicity of the basic solvents was measured by their heat of mixing with *tert*-butyl alcohol from Drago's table (Table IIA.5) using *tert*-butyl alcohol as a model for the acid-OH groups on silica; the acidity of the acid solvents was measured by their heat of mixing with ethyl acetate, an ester comparable to the methacrylate groups of the polymer. In the PMMA-SiO$_2$ system, polymer adsorption decreases as solvent basicity increases because the solvent now successfully competes with the polymer for the acidic surface sites. Likewise, polymer adsorption decreases as solvent acidity increases because the solvent successfully competes with the acidic surface sites for the basic PMMA. These authors correlated the observed degree of polymer adsorption with the acidity and basicity of the solvent (Fig. C.1), thus showing the overriding importance of acid–base interactions in determining the outcome of the competition. Similar results were obtained for the PVC-CaCO$_3$ system (Fig. C.2). The same two polymers were used to determine the ratio of acidic to basic surface sites in a series of iron-containing pigments (iron oxides tend to have both kinds of site).[92]

c. Aqueous Media

Molecules that combine lyophilic and lyophobic moieties in their structure are spontaneously adsorbed by lyophobic substrates out of aqueous solution, as they can orient themselves at the interface with their lyophilic moiety toward the medium and their lyophobic moiety toward the substrate. Among molecules of this type we find all surface-active solutes; the free energy of the adsorption measures the spontaneity of their adsorption and their consequent action as suspending agents.

Traube's rule for the lowering of the surface tension of an aqueous solution as a function of the alkyl-chain length of the hydrophobe, is reproduced at the solid–liquid interface by implicating the degree of adsorption of solute, which is readily measured, in place of the interfacial tension, which is not readily measured. For many systems thus studied a similar rule is found: The quantity of solute that is adsorbed at a hydrophobic interface from an aqueous solution increases geometrically as the chain length of the solute increases arithmetically. Non-aqueous solutions show an inverse relation of the same sort: A lyophobic interface in a hydrocarbon solvent, such as silica in mineral oil, takes out less solute, by a constant factor, each time the chain length of the solute increases by one methylene group. The simple generalizations expressed by this form of Traube's rule are not applicable where the solute molecules are adsorbed by attraction mechanisms other than dispersion forces. For example, with cationic surface-active solutes the positive ionic charge of the surface-active ion is attracted by negatively charged substrates, such as quartz or glass, whereby the hydrophobic moiety of the solute makes the interface hydrophobic; at higher concentrations of solute a second layer of surface-active ions is adsorbed with a reverse orientation, so that the interface becomes hydrophilic and positively charged. The second layer of adsorbate would agree with the mechanism of Traube's rule, but the first layer would not.

The foregoing generalizations of adsorption behavior may be summarized in the following rules to guide prediction of relative adsorption:

(a) The extent of adsorption is usually greater from solvents in which the adsorptive is less soluble (Lundelius's rule.) A corollary to this rule is that for a given adsorbate the better solvent makes the better eluent.

(b) In the absence of specific interactions between adsorbent and adsorbate, the amount adsorbed is never large.

(c) The greater the electron donor–acceptor interaction between adsorbent and adsorbate, the greater the adsorption. This may be the reason why hydrophilic solids, such as silica gel, are more favored for adsorption separation (except special separations such as decolorization with carbon black). A corollary to this rule is that the greater the donor–acceptor interaction between solvent and adsorptive, the less the adsorption.

(d) The extent of adsorption out of aqueous solution changes in a regular manner along a homologous series. Rule (a) would warn, however, that, since solubility usually decreases (with any solvent) with sufficiently great increase in molecular weight, this corollary must be applied with discernment since the two operative factors tend to offset each other. (See the Ferguson effect, Section II.E.2).

d. Rates of Adsorption on Solid Surfaces

The rate of adsorption of a surface-active solute at the liquid–vapor surface is diffusion controlled (Section II.E.5) and reaches equilibrium in less than a minute. Rates of adsorption at solution–solid interfaces are considerably slower than the rates of arrival by diffusion. The difference between them is shown in Fig. C.3.

The time of arrival at the substrate by diffusion from a 0.5% solution is calculated for comparison with the measured amounts adsorbed from solutions of fatty acids, amines, and alcohols. The amines require an hour for complete adsorption while stearyl alcohol requires a week. Diffusion alone would have completed the monolayer in 0.02 s (Eq. [IIE.4]). The factors that can slow adsorption so markedly are the presence of cracks, pores, or capillary spaces in the solid substrate, the rate of desorption of solvent, the time required for the adsorbate to rearrange, especially at high coverage, and electrostatic repulsion. Any one of these factors can be expressed as an energy barrier for adsorption; therefore, increasing the temperature increases the rate of adsorption.

The rate of adsorption of polymers at a solid substrate is further affected by subsequent rearrangement of the polymer chain from its initial single-point attachment to loop attachment, and then to extended, multiple-site attachment.

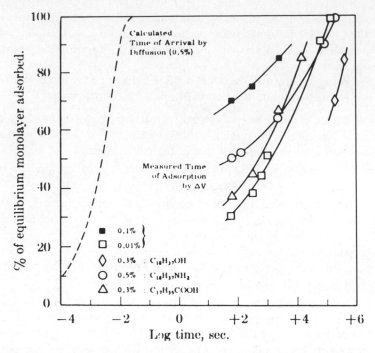

Figure C.3 Rates of adsorption on steel plates from white-oil solutions compared with rates of diffusion.[98] Black squares and white squares refer to $C_{17}H_{35}CONHCH_2$ $CHOHCH_2NH_2$.

Also, in a distribution of molecular weights, the low-molecular-weight fractions that are first adsorbed because of faster diffusion are later displaced by higher molecular weight fractions. The nature of the solvent affects both the rate of adsorption and the amount adsorbed: The rate is faster and the amount adsorbed is greater in a poorer solvent.

The rate of adsorption is the rate-limiting step in many dispersion processes. Newly created surfaces are not stabilized until sufficiently covered by adsorbate. For this reason, machinery that creates new surfaces more rapidly than they can be stabilized uses excess energy, as unstabilized particles or droplets soon flocculate. If the adsorption is slow, prolonged contact of the new surface with the solution is required. A big holding tank and a small-volume grinding unit, such as sand mills or recirculating grinding mills, meet this condition.

2.   SOLUTION ADSORPTION ISOTHERM

The Langmuir and Freundlich isotherm equations are widely applied to adsorption from dilute solution. For this application, these equations have the following forms, respectively:

$$\frac{x}{m} = \frac{aC}{K + C}$$ [C.2]

and

$$\frac{x}{m} = kC^n$$ [C.3]

where x is the amount of solute taken up by m grams of the adsorbent when present at concentration C in the solution; a, K, k and n are constants. In Eq. [C.2] a is a measure of the surface area of the solid and K is related to the adsorption potential. In Eq. [C.3] the constants k and n are relative measures of the surface area of the solid and its adsorption potential; the value of n is usually between 0.1 and 0.5. See Section I.D.4 for the use of Eq. [C.2] to determine the specific surface area of the adsorbent.

Figure C.4 shows a series of Langmuir adsorption isotherms that describe the adsorption of the selenite ion (SeO_3^{-2}) by goethite (a hydrous mineral oxide of

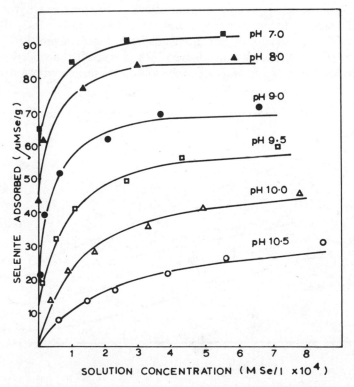

Figure C.4 Langmuir isotherms of selenite ion adsorbed by geothite at 20°C at various values of pH.[99]

iron) as a function of pH. These isotherms show that the amount of selenite ion taken up by goethite at constant pH approaches a maximum value at higher concentrations of solute. To test whether experimental data are adequately described by the Langmuir equation, the quantity $C/(x/m)$ is plotted versus C: A straight line through the points indicates a fit by the Langmuir equation; the slope of the line is $1/a$ and the intercept is K/a.

The Langmuir equation is based on the model of a uniform surface or at least assumes that the heat of adsorption does not vary with surface concentration; the Freundlich equation can be derived by assuming that the substrate is composed of a large number of homotattic patches with an exponential distribution of adsorptive potential energies.[100] As we have seen, the special models by which these equations are derived are not proved to be valid even when the resulting equation describes the experimental data. These equations are, therefore, better regarded as empirical, unless additional information besides adsorption data is available about the system.

At higher concentrations these equations are inapplicable; frequently the adsorption of solute appears to go through a maximum, then decreases and may become negative, as illustrated by the adsorption isotherms shown in Fig. C.5. For each adsorbent, ethyl alcohol is adsorbed from a benzene solution at low concentration; similarly, at low concentrations of benzene in ethyl alcohol, a preferential adsorption of benzene takes place, which also passes through a maximum. The peculiarity of a negative adsorption stems from the customary method of calculating the amount adsorbed, which is taken as proportional to the change in the concentration of solute in the solution as measured before and after

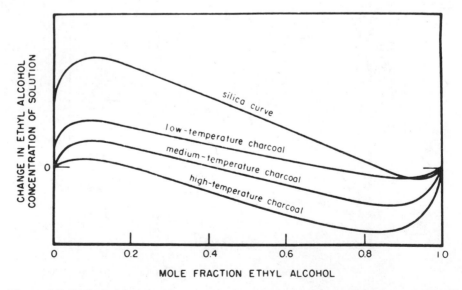

Figure C.5 Preferential adsorption of ethyl alcohol from benzene as a function of charcoal treatment.[101]

adsorption has occurred. If the solution is found to be *more* concentrated after reaching equilibrium with the adsorbent, then the solvent rather than the solute has been preferentially adsorbed. When the simultaneous adsorption of the two components of the solution is taken into account, the apparent anomaly of negative adsorption disappears.

Kipling and Tester have shown how the adsorption isotherm for the complete range of concentration, as shown in Fig. C.5, can be related to the vapor adsorption isotherms.[102] They used the Langmuir equation to describe simultaneous adsorption from solution of the two components; they also assumed that the whole surface of the adsorbent is covered by the two adsorbed components, that is, $\theta_1 + \theta_2 = 1$. Under these conditions, the fractions of the surface covered by each component are

$$\theta_1 = \frac{K_2 C_1}{K_2 C_1 + K_1 C_2} \qquad [C.4]$$

and

$$\theta_2 = \frac{K_1 C_2}{K_2 C_1 + K_1 C_2} \qquad [C.5]$$

The constants K_1 and K_2 in Eqs [C.4] and [C.5] can be derived from the vapor-adsorption isotherms of the two components, which were measured by placing the adsorbent in the vapor phase of the solution; by analysis of the mixed adsorbate both (x_1/m) and (x_2/m) were determined separately, even though mixed vapors were used. The vapor adsorption measurements provide an independent means of obtaining K_1 and K_2 for use in the mathematical description by means of Eqs [C.4] and [C.5] of adsorption from solution. The derivations of these equations require that the two components form ideal solutions. For nonideal solutions more complex relations are obtained (Ościk, 1982).

The heterogeneity of the adsorbent remains the unknown factor in all investigations of adsorption from solution. The presence of this factor nullifies all efforts to treat either the Langmuir or the Freundlich equations as anything more than empirical descriptions. Different combinations of the solute adsorption mechanism and the distribution of adsorptive potential energies can lead to either of these equations, and to several other shapes of the adsorption isotherm besides. Giles and co-workers[103] attempted to classify solution adsorption isotherms for use in diagnosing adsorption mechanisms and measuring specific surface areas of solids. The shape of the adsorption isotherm, however, is determined by an unknown combination of both lateral interactions and surface heterogeneity, and whatever the one lacks in describing the data can be supplied by the other. The system of Giles et al. is, in effect, to equate the heterogeneity to zero and to interpret all differences as due to variations in the adsorption mechanism. The system will, therefore, be successful with near-homotattic surfaces but could be completely wrong for surfaces with a wide distribution of adsorption potentials. The latter type of surface, unfortunately, is by far the more common. The system of Giles et al. can, however, be applied with more probability of success to the class of adsorption isotherms pertaining to the adsorption of dyes and other large molecules.

3. ADSORPTION OF DYES

The effect of surface heterogeneity can be suppressed by using adsorbates of large molecular size. This circumstance is, probably, the basis of the validity of dye adsorption as a technique to determine specific surface areas of solid adsorbents. The adsorption isotherm is always determined in the dilute range of concentrations where competitive adsorption of the solvent is not significant; such isotherms frequently show a saturation plateau at high equilibrium concentrations of the free dye. Sheppard and his co-workers,[104,105] who investigated the adsorption of cyanine dyes by silver halides in connection with the study of optically sensitized photographic emulsions, concluded that the saturation plateaus observed in the adsorption isotherms of a number of cyanines adsorbed by silver bromide microcrystals corresponded to the formation of a close-packed monolayer of essentially planar cations, oriented with the planes of the molecules steeply inclined to the substrate, that is, a configuration in which the edge of the molecule is presented to the substrate (edge-on adsorption.)

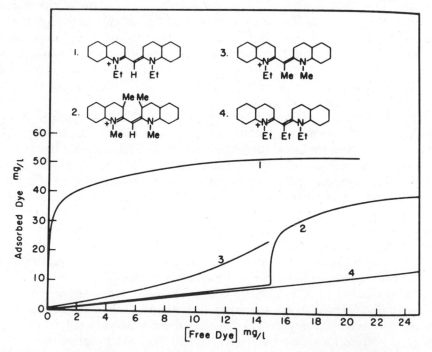

Figure C.6 Adsorption isotherms of cyanine dyes on silver bromide, showing the effect of nonplanarity of the adsorbate molecule.[106]

When the dye molecule is nonplanar or has a relatively high solubility in water, these being two distinct factors that reduce adsorption from aqueous solution, the type of isotherm shown in Fig. C.6, curve 2, frequently results.[106] This isotherm shows poor adsorption of the tetramethyl dye (curve 2 in the diagram) compared with its planar counterpart (curve 1), and curves 3 and 4 show poor adsorption of 2,2′-cyanines whose molecules have been forced from planarity by bulky substituents in the methine bridge. The discontinuity exhibited in the adsorption isotherm of the tetramethyl dye is accompanied by a change in the absorption spectrum of the adsorbed dye. In the low-concentration "foot" of the isotherm, the spectrum is that of the isolated molecule as modified by its adsorption, and probably corresponds to unassociated molecules in flat orientation with respect to the crystal surface; in the other region, the absorption maximum undergoes a bathochromic shift to a wavelength (J band) similar to that of the oriented aggregates of planar dyes in solution (micelles) For the nonplanar dye, if the adsorption plateau is identified with the completion of a monolayer, the area per molecule is found to be consistent only with edge-on adsorption, although the average intermolecular distance, 0.598 nm, is greater than for the corresponding planar molecules, as might be expected from the twisted configuration of the molecule.

For dyes such as 1,1′-diethyl-2,2′-cyanine, whose adsorption isotherm is shown in Fig. C.6, curve 1, the critical concentration at which lateral interactions become strong enough to induce cooperative edge-on orientation occurs in such a dilute solution that the isotherm appears to be continuous on the scale shown. Nevertheless, a small foot, corresponding to noncooperative adsorption, can sometimes be found in the adsorption isotherm of well-adsorbed dyes; the effect can be magnified by making the adsorption conditions less favorable, for example, by introducing a competing adsorbate or an unfavorable silver ion concentration.

With systems such as these, the large size of the adsorbate molecule so masks the smaller scale heterogeneity of the substrate that conclusions about adsorption mechanisms drawn from the shapes of the isotherms have more authority.

4. EXPERIMENTAL TECHNIQUES FOR SUSPENSIONS

Various techniques to determine particle size of particles in suspension are treated in Chapter I.D and techniques to determine the stability of suspensions are treated in Part III. The techniques described in the present section refer to physical properties of whole systems (Table C.2).

a. Flocculated and Deflocculated Suspensions

The surfaces of silica and other inorganic oxides are hydrophilic; the aqueous medium releases ions from the crystal lattice and so confers ionic charge on the surface and establishes an electric double layer extending into the solution. Thus,

**Table C.2 Generalized Summary of Physical Properties of
solid–liquid Suspensions**

Deflocculated Suspensions	Flocculated Suspension
Lyophilic interfaces	Lyophobic interfaces
$F_{sv} > \sigma_1 + F_{sl}$	$F_{sv} < \sigma_1 + F_{sl}$
Slow rate of Stokes settling	Rapid rate of Stokes settling
Small volume of sediment	Large volume of sediment
Dilatant rheology	Shear-thinning rheology
Low or zero-yield value	High-yield value
High heat of immersion	Low heat of immersion
High work of adhesion of medium on solid	Low work of adhesion of medium on solid

although the system has only two components, electrostatic repulsion between the particles can stabilize the suspensions. The "rule" that two pure components cannot form a stable dispersion does not apply to situations where the medium reacts with the dispersed phase. This consideration also explains why water dissolves sucrose: Each sucrose molecule becomes hydrated, which detaches it from the crystal lattice. A solution is the ultimate deflocculated dispersion. Nonaqueous two-component systems of an insoluble solid and a liquid are much less likely to react, and so, in spite of positive spreading coefficients, tend to flocculate. Most organic liquids have low surface tensions and will spread on inorganic solids such as ferric oxide, gold, or selenium; yet, if they lack a specific interaction such as electron donor–acceptor, deflocculated suspensions are not formed without the aid of a third component to keep particles sufficiently apart so that their kinetic energy overcomes the energy of attraction. Silica dispersed in polydimethylsiloxane can be made stable by heating the two chemicals together because of a reaction that then occurs at the silica surface (Section IV.B.5e).

b.   Sedimentation Rates and Volumes

The rate of sedimentation of a single solid spherical particle in a medium is described by Stokes' law, as long as the flow is laminar and the particle is too large to be affected by Brownian motion. The rate of settling v is given by

$$v = \frac{2a^2(\rho_2 - \rho_1)g}{9\eta} \qquad [C.6]$$

where a is the radius of the particle, η is the viscosity of the medium, ρ_1 and ρ_2 are the densities of the medium and the particle, respectively, and g is the gravitational constant. The way in which Stokes's law is used to determine particle size distributions has already been described (Section I.D.2). The law is also useful to measure the degree of coagulation of particles.[107] Kinetic units in a dispersion may be primary particles, which are ultimate working units retaining their identity throughout a dispersion process and subsequent application, or they may be aggregates or flocs. Aggregates are tightly bound clusters of primary particles, and flocs are loosely bound clusters that are broken by relatively weak mechanical forces.

A qualitative method to distinguish degrees of coagulation is to measure the specific volume of sediment (cm^3/mg). The technique is as follows: To a weighed amount of solid is added a small quantity of the liquid in which the suspension is to be made. This paste is thoroughly mixed with a spatula and more liquid is gradually added with constant stirring. The suspension is transferred to a stoppered graduate and the volume of the sediment is recorded as a function of time. Figure C.7 shows schematically the behavior of a deflocculated suspension, a flocculated suspension, and an aggregated suspension. Comparisons of the sedimentation volumes after an appropriate interval of time show the differences between the three types of suspension. Particles remain primary, or deflocculated, only if they repel each other. The repulsion, whether electrostatic or steric in origin, entails that the interface be lyophilic: that is, that

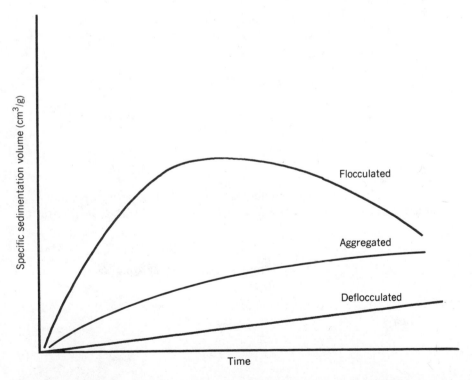

Figure C.7 Variations of the volume of sediment with time: a deflocculated suspension, a flocculated suspension, and an aggregated suspension, with time.

the spreading coefficient of the medium on the particle be positive. Such particles are free to move around each other and they settle to a densely packed sediment with minimum void volume. Flocs or agglomerates, on the other hand, which have formed irregular structures in the suspension, are less able to pack closely in the sediment; greater voids are left between them, which enlarge the volume of the sediment. These irregular structures are weakly coherent and, in time, collapse.

c.   Adsorption at the Solid–Liquid Interface

Adsorption at the solid-liquid interface is determined from the difference of concentration of surface-active solute in the medium before and after equilibration with the solid adsorbent. Known weights of solid and solution are brought together, allowed to equilibrate at constant temperature, and the resulting change of composition of the solution is measured. The solution may be separated from the solid by filtration or by centrifugation. Methods of analysis depend upon the nature of the solute; optical absorption, differential refractometry and radiotracers are favorite methods. The experiments must be so designed, by choosing the ratio of the mass of the adsorbent to the volume of the solution, that the initial and final concentrations are within the sensitive range of the analytic method.

Let the mass of solid be m and the amount of solution of mole fraction X_2^0 be n_0 (mass or volume.) At equilibrium let the composition of the solution be X_2. The surface excess amount of component 2 is

$$n_2 = n_0(X_2^0 - X_2) = \dot{n}_0 \, \Delta X_2 \qquad \text{[C.7]}$$

The adsorption by unit mass of solid, or the specific surface excess, is

$$\frac{n_2}{m} = \frac{n_0 \, \Delta X_2}{m} \qquad \text{[C.8]}$$

The specific surface excess is the preferred form to report experimental data.[108] If the specific surface area of the solid, Σ, is known, the excess surface concentration, Γ_2, is

$$\Gamma_2 = \frac{n_0 \Delta X_2}{m\Sigma} \qquad \text{moles of component 2/m}^2 \qquad \text{[C.9]}$$

It also follows for a two-component system that since $X_1 + X_2 = 1$, $\Delta X_2 = -\Delta X_1$; hence,

$$\Gamma_1 = -\Gamma_2 \qquad \text{[C.10]}$$

Equation [C.10] points out that solute and solvent compete for room on the substrate.

d. Surface Tension Titrations

Latexes made by emulsion polymerization contain 8–10% of surface-active solute, which in this context is simply referred to as "soap." The finished latex has a surface tension not much reduced from that of pure water, despite the large amount of soap in the system. Most of the soap is evidently retained, therefore, at the polymer–liquid interface, leaving very little unadsorbed soap in the solution. The latex is titrated with a standard solution, approximately 0.01 M, of the same soap that was used as stabilizer.[109] As this solution is added, the soap equilibrates between the solid–liquid interface and the serum; its presence in the serum can be detected by the change of surface tension at the liquid–air surface. As the titration proceeds, the surface tension decreases, until the concentration of free soap in the solution is equal to the CMC, which is, effectively, the largest possible concentration of soap ions in solution, and which is, therefore, in equilibrium with the largest amount of adsorbed soap that the particular system of soap + water can retain on that particular interface. (Fig. C.8). That amount of adsorbate is assumed to correspond to a saturated, close-packed monolayer. The end point of the titration occurs when minimum surface tension, corresponding to the CMC, is reached. The amount of soap added by the end point equals the quantity adsorbed at the solid–liquid interface during the titration plus the

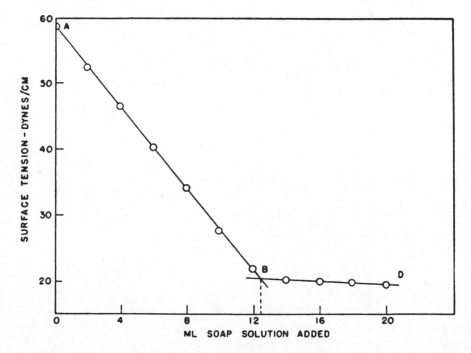

Figure C.8 Surface tension titration of a latex with soap at 50°C. The point of intersection of the lines, B, gives the quantity of added soap at which micellization occurs in the aqueous phase, and is thus the end point of the titration.[109]

quantity required to reach the CMC in the solution. The total amount of soap adsorbed on the latex particles is the sum of the amount originally present (usually negligible) and the amount added to the particles in the titration. The specific interfacial area is then calculated from the total amount of adsorbed soap per gram of polymer and the limiting area of each adsorbed soap molecule. The latter quantity can be estimated from the surface tension isotherm of the soap and the Gibbs adsorption theorem (Table IIG.1).

e.   Rheology

Whether a suspension is deflocculated or flocculated can often be determined from its rheogram. At low percent solids a suspension, whether flocculated or not, continues to behave as a Newtonian fluid with a slightly higher viscosity coefficient than that of the medium. The simplest case of the flow of a dispersion of noninteracting spheres was analyzed by Einstein to give the viscosity, η, as a linear function of volume fraction, Φ:

$$\eta = \eta_0(1 + 2.5\Phi) \qquad [C.11]$$

where η_0 is the viscosity of the medium. The Einstein equation is a limiting law,

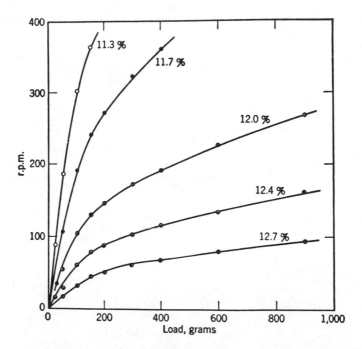

Figure C.9 Dilatant flow of a series of suspensions of red iron oxide in an aqueous solution of sodium lignin sulfonate at 10% concentration, at 30°C. The volume concentrations of solids are noted on the curves (Fischer, 1950, p. 200).

valid only as Φ tends to zero. At higher volume fractions, the flow of a deflocculated dispersion becomes more and more dilatant, as the inertia of the particles hinders their response to shearing stresses. The concentration of suspended solids at which dilatant flow is observed varies with the particle size and shape and the quantity and nature of the deflocculating agent added. Suspensions of red iron oxide in an aqueous solution of sodium lignin sulfonate at 10% concentration begin to show dilatancy at about 10% solids. Over a small concentration range, 11.3–12.7% solids, the calculated apparent viscosity increases from 2.3 to 76 P for the freshly prepared suspension measured at a low rate of shear (Fig. C.9). Most solids, however, require concentrations above 20% solids before showing dilatant flow.

Flocculated suspensions exhibit yield values at low shearing stresses, as interparticle attractive forces prevent flow until sufficient energy is imparted to the system to overcome their resistance. When the structure is broken, the rate of flow becomes linear with the applied stress. A rheogram of this type is known as a Bingham body (Chapter I.C). Oil-based paints, which are suspensions of pigments in oil, are typical Bingham bodies.

An example of the transition of a flocculated to a deflocculated system is the change of a plastic kaolin suspension to a fluid slurry, brought about by adding increasing amounts of the dispersing agent, tetrasodium pyrophosphate (TSPP.) The rheograms that accompany this change are shown in Fig. C.10.

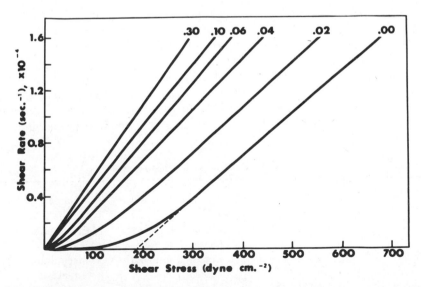

Figure C.10 Rheograms of 20 w% deionized kaolin slurries at several levels of tetrasodium pyrophosphate addition: The figures on the curves indicate percent TSPP per weight of clay. An extrapolation of the linear region of each rheogram to zero rate of shear determines an apparent yield point.[110]

f. Electrical Properties in Aqueous Media

Why measure zeta potential? Zeta potentials are used for various purposes in industrial research. As a minimum inquiry, the measurement answers the question: Is the electrical charge on the dispersed particle positive or negative? This information is often sufficient to suggest further steps in processing. The next higher level of inquiry has to do with quality control: Has the product sufficient electrostatic repulsion to maintain its stability or should more dispersing agent be provided? An aqueous clay slurry, containing tetrasodium pyrophosphate, is stabilized by electrostatic repulsion: Hence, electrophoretic measurement is useful as a quality control test in the clay-mining and papermaking industries. In R&D, electrophoretic measurements are used to screen potential suspending agents. Another question that often arises is how various processing steps affect the stability of the suspension; these changes too can be monitored by means of zeta potentials. In systems that are stabilized by a combination of electrostatic and steric repulsions, the contribution of the electrostatic component is determined by measuring the zeta potential. Finally, quantitative calculations of the stability ratio, to determine the magnitude of the repulsive barrier and hence the lifetime of the dispersion, are based on measurements of zeta potential (Chapter III.D).

Methods of measuring zeta potential use the electrokinetic phenomena that arise from the mutual effects of electric field and of tangential motion of two phases with respect to each other. An applied electric field directed along the phase boundary causes relative motion of one phase with respect to another; and conversely, motion along the charged phase boundary creates an electric potential. Small charged particles suspended in a medium move (electrophoresis) when a potential gradient is applied; or if the solid is stationary, the effect of moving the medium past the interface is to generate a potential difference (streaming potential). A value of the zeta potential is obtained from the electrophoretic mobility in the former case or from the streaming potential in the latter case. Conversely, the suspended particles can be moved, either by sedimentation or centrifugation, and the potential generated can be measured; or

Table C.3   Electrokinetic Phenomena

Solid Phase	Cause	Phenomenon	Name
Particles	Electric field	Motion of dispersed phase	Electrophoresis
Particles	Gravitational field	Potential gradient	Sedimentation potential
Tube wall or packed bed	Electric field	Motion of medium	Electroosmosis
Tube wall or packed bed	Motion of medium	Potential gradient	Streaming potential

a potential gradient can be applied to a stationary interface and the motion of the medium measured. Four distinctive phenomena are possible, as tabulated in Table C.3.

For suspensions, emulsions, or macromolecules in solution electrophoresis is the most suitable technique by which to measure zeta potential. The electric force acting on the double layer causes the particle to move at a constant velocity. The ratio of the velocity to the electric field is the electrophoretic mobility, measured as (meters/second) per (volt/meter). For aqueous systems, the electrophoretic mobility is generally independent of the electric field. The zeta potential is determined from the Smoluchowski equation (Hunter, 1981, p. 69):

$$\zeta = \frac{\eta v}{E D \varepsilon_0} \qquad [C.12]$$

where v/E is the electrophoretic mobility and ζ is the zeta potential. The same equation describes the relation between the applied average linear velocity of the medium v and the consequent potential field, E in the other electrokinetic effects.

For Eq. [C.12] to apply to the electrophoresis of nonconducting particles, the following conditions are required:

(a) The usual hydrodynamic equations for the motion of a viscous liquid are assumed to hold both in the bulk of the liquid and within the double layer.
(b) The motion is laminar and slow enough for inertia terms to be neglected.
(c) The applied field is superimposed on the field due to the electric double layer.
(d) The thickness of the double layer is small compared to the radius of curvature of the interface.

Smoluchowski's derivation essentially assumes that the applied electric field conforms to the shape of the particle, which in turn implies that the particle bends the applied field to make it become parallel to its surface.

Hückel, fresh from his triumph with the theory of interionic attraction in electrolyte solutions, also derived a relation between zeta potential and electrophoretic mobility. He assumed that the applied electric field runs straight between the electrodes, without conforming to the shape of the particle. His derivation, therefore, applies to point particles, or, in practice, to very small particles. According to Hückel (Hunter, 1981, p. 69):

$$\zeta = \frac{3 \eta v}{2 E D \varepsilon_0} \qquad [C.13]$$

Equation [C.13] gives a higher value of the zeta potential by a factor of three-halves than Eq. [C.12]. The transition between these two limiting cases for small zeta potentials was quantified by Henry.[111] The shape of the field depends on the electrical conductivities of the particle and the surrounding liquid, the size of the particle, and its electrical double layer. For an insulated sphere in a liquid of any conductivity, Henry's result may be written as

$$\zeta = \frac{3\eta v}{2ED\varepsilon_0 f(\kappa a)} \qquad\qquad \text{[C.14]}$$

where κ is the reciprocal of the double-layer thickness (Debye length):

$$\kappa^2 = \frac{e^2}{D\varepsilon_0 kT}\sum n_{i0}z_i^2 \qquad\qquad \text{[IIIB.6]}$$

$$= 10.8\sum n_{i0}z_i^2 \qquad \text{nm}^{-2} \quad \text{at } 298\,\text{K}$$

In the terms of the double layer, κ is also the reciprocal of the thickness of the double layer so that κa is the ratio between the particle radius and the thickness of the double layer. In a limiting case, κa tending to infinity, $f(\kappa a) = \frac{3}{2}$, and Smoluchowski's result is valid; that is, for large particles and high concentrations of electrolyte, Eq. [C.12] is used to calculate ζ.

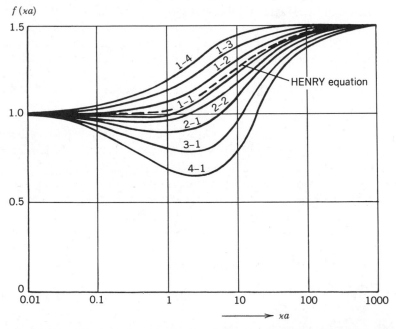

Figure C.11 The influence of the relaxation for different types of electrolyte on a negative particle, $\zeta = -50\,\text{mV}$. The dotted line is Henry's curve, calculated without taking relaxation into account (Kruyt, 1949, Vol. 1, p. 212). Diagrams for zeta potentials other than $-50\,\text{mV}$ require a complete new set of calculations, which are described by Wiersema (1964).

While the particle is moving, the spherical symmetry of the ionic double layer is distorted; thus an additional retarding force occurs as the spherical double layer tries to reform. This force, known as relaxation, is ignored in Henry's analysis. Wiersema (Wiersema, 1964) gives a detailed treatment of the relaxation effect for spherical insulating particles surrounded by a diffuse double layer. His results show that Eq. [C.14] has to be replaced by the expression

$$\zeta = \frac{3\eta v}{2ED\varepsilon_0 f(\kappa a, \zeta)} \qquad [C.15]$$

where the function $f(\kappa a, \zeta)$ approaches Henry's function, $f(\kappa a)$, under the following conditions:

(a) $\zeta \ll 25\,\text{mV}$ (any value of κa)
(b) $\kappa a \ll 1$ (any value of ζ)
(c) $\kappa a \gg 1$ (any value of ζ)

The nature of the complete expression is shown in Fig. C.11.

Table C.4. shows the limits of particle size for various concentrations of electrolyte for which either the Hückel or the Smoluchowski formulation is applicable within 10% or so, taken from Fig. C.11.

Several commercial instruments are available to measure electrophoretic mobility. Instruments marketed by Zeta-Meter, Inc. and Rank Brothers, Ltd. measure electrophoretic mobility by timing the motion of individual particles viewed through a microscope. The Laser-Zee of Pen Kem, Inc. uses a spinning prism to visually arrest the motion of a large number of independently moving particles to obtain an average velocity. Instruments that use laser Doppler velocimetry are sold commercially (see Appendix J). All these

Table C.4   Range of Applicability of the Hückel (H) and Smoluchowski (S) Equations for $\zeta = -50\,\text{mV}$

Electrolyte	Concentration	κ at 20°C	$1/\kappa$	Diameters (μm)	
	(M)	(cm^{-1})	(μm)	(S)	(H)
Pure water	—	1.0×10^4	1.0	> 600	< 1.0
Uni-univalent	10^{-5}	1.0×10^5	0.1	> 17	< 0.32
	10^{-3}	1.0×10^6	0.01	> 1.7	< 0.032
	10^{-1}	1.0×10^7	0.001	> 0.17	< 0.003
Uni-divalent	10^{-5}	1.8×10^5	0.056	> 9.4	< 0.03
	10^{-1}	1.8×10^7	0.00056	> 0.094	< 0.0003
Di-divalent	10^{-5}	2.1×10^5	0.048	> 8.1	< 0.048
	10^{-1}	2.1×10^7	0.00048	> 0.081	< 0.0005

instruments are designed for dilute suspensions. Concentrated suspensions are measured by the Mass Transport Analyzer of Micromeretics Instrument Corp., the ultrasonic analyzer of Matec Instruments, Inc. and the acoustophoretic analyzer of Pen Kem, Inc.

g. Electrical Properties in Nonaqueous Media

The inability of electrolyte to ionize in solvents of low conductivity has led to a mistaken belief that electrostatic stabilization of particles in a nonaqueous medium is nonexistent. Dispersed carbon in benzene develops appreciable zeta potentials and stability in the presence of calcium alkylsalicylate, which gives positive zeta potentials, or quaternary ammonium picrates, which give negative zeta potentials.[112,113] These two suspensions when mixed together became unstable. Such mutual antagonism is evidence of an electrostatic mechanism. The sign of the charge on the particles in these benzene suspensions is opposite to what would be expected if the charging mechanism were the adsorption of the larger ion, as it is in aqueous solution. It follows that electrostatic stabilization *is* important in stabilizing nonaqueous suspensions and that the charging mechanism is *not* the same as in aqueous suspensions.

Fowkes et al.[114] have shown by many examples that in organic media the mechanism of particle charging is the formation of ions in adsorbed films on particle surfaces, where acid–base (or donor–acceptor) interactions occur between the particle surface and the suspending agent (suspendant). Potentials develop when adsorbed suspendant ions desorb into the organic medium, where they form the diffuse electrical double layer. Zeta potentials well over $100\,mV$ result from the stronger acid–base interactions. DLVO energy barriers often exceed $25\,kT$, leading to stability ratios of 10^8 or more (Chapter III.D).

Basic suspendants, for example, are adsorbed as neutral molecules on acidic sites on the surface of the particle, where proton transfer confers on them a positive charge. During the dynamic equilibrium of adsorption and desorption, some of the charged suspendant is desorbed, leaving the particle negatively charged. The charged suspendant molecule is often thereafter incorporated into an inverse micelle, where its electrical charge finds a hydrophilic environment in the form of solubilized water. This feature of the charging process is important, for a bulky micelle keeps the charge of the counterion at a distance from the oppositely charged surface, thus reducing the force of attraction sufficiently to prevent the counterion being drawn into the Stern layer. Acidic or basic polymers, therefore, are most effective suspendants of particles in nonaqueous media. The mechanism is illustrated in Fig. C.12. Adsorption of the basic suspendant is depicted in the transition from Fig. C.12(a) to C.12(b). Desorption of the basic suspendant, now carrying a bound proton, is depicted in the transition from Fig. C.12(b) to C.12(c). The particle is now negatively charged. Carbon particles that can acquire either a negative or a positive charge from basic or acidic suspendants, respectively, have both acidic and basic sites on their surfaces (amphoteric).

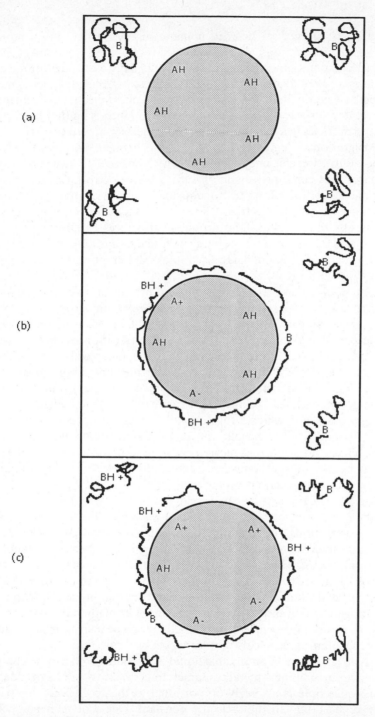

Figure C.12 Mechanism of electrostatic charging of dispersed acidic particles (with acidic sites HA) by basic dispersants (with basic sites B) in solvents of low dielectric constant.[114]

Electrically charged particles in nonaqueous media figure in a number of technological applications, such as the electrophoretic development of latent images (electrophotography), electrophoretic displays, electrodeposition of special coatings, and removal of particulate contaminants from nonaqueous media by an imposed electric field. Liquid development of an electrostatic image on a photoreceptor surface, such as a selenium drum, depends on the electrophoresis of charged particles in a nonaqueous medium. Electrophoretic displays are based on the migration of highly reflecting, charged pigment particles to a viewing electrode. Filtration efficiency is improved with positively charged microporous membranes that combine submicron sieving with electrostatic adsorption.

Harmful electrical effects in nonaqueous media are found in automotive applications in which electrified surfaces attract dispersed particles from crankcase oils, thus blocking flow through small orifices. Electric fields develop at metal surfaces by an exchange of charge with acid or basic components in the oil. These electric fields become large when tanks are emptied, as counterions are carried away with the oil.

The cause of a number of explosions and fires in oil refineries, depots, and tankers, although never established with absolute certainty, is probably traced to the discharge of electrostatic fields. Splash filling is a frequent cause of electrification inside a tank. Other sources of electrification are mixing, blending, and agitating. These operations are frequently accompanied by emulsification of the tank contents and the stirring up of water bottoms. Explosions may even occur inside grounded containers. Klinkenberg and van der Minne (1958) report that a large tank in Shell's refinery at Pernis exploded 40 min after the start of a blending operation in which a tops-naphtha mixture was being pumped into a straight-run naphtha. On the following day a second attempt was made to blend these materials and again an explosion occurred 40 min after starting the pumps. This striking and unusual coincidence could only be explained by assuming that both explosions were caused by static electricity.

A large electrical potential can develop on removing mobile countercharges from the vicinity of an interface such as a pipe wall or a sedimenting water droplet. In aqueous solutions, the high conductivity prevents the growth of any large electric potentials; but in poorly conducting media, electric potentials can become quite large. These large potentials spark on discharge and can ignite explosive vapor mixtures. A method to remove charges is to increase the conductivity of the liquid by the addition of antistatic additives. The conductivity cannot be increased by means of electrolytes in media of low dielectric constant as ionization does not take place; but conductivity can be increased by the addition of a combination of chemicals that forms charged micelles.

The half-value time of an organic liquid is the time taken for the charge in a liquid, completely filling a metal container, to decrease to half its original value. The half-value time is inversely proportional to the specific conductivity and directly proportional to the dielectric constant. Safety of refinery operations requires a conductivity of at least $5 \times 10^{-3} \Omega^{-1} cm^{-1}$. Table C.5 lists the conductivities and half-value times of several liquids. The half-value times vary by

Table C.5   Half-Value Times of Various Liquids

Liquid	Conductivity $(\Omega^{-1}\,cm^{-1})$	Half-value time (s)
Highly purified hydrocarbons	10^{-17}	12,000
Light distillates	10^{-16}–10^{-13}	1,200–1.2
Crude oil	10^{-11}–10^{-9}	0.012–0.00012
Distilled water	10^{-6}	4.8×10^{-6}

From Klinkenberg and van der Minne (1958, p. 28).

nine orders of magnitude from microseconds to hours. Electric fields in aqueous media dissipate instantaneously, even in distilled water, while taking all day to dissipate in petroleum distillates. High voltages are generated because of the slow dissipation if, in the meantime, the countercharges are carried far away from the charged interface.

Because of the ease with which countercharges are removed from the vicinity of a charged interface in a low-conductivity medium, the charge density can be determined directly by allowing the liquid with its entrained countercharges to flow out of a container into a receiving vessel connected to an electrometer (Fig. C.13).

The same principle can be applied to a suspension in a liquid of low conductivity. The separation of the mobile charge from the dispersed particles is brought about by passing the system through a microporous filter, which retains the charged particles while the countercharges are collected in an isolated metal

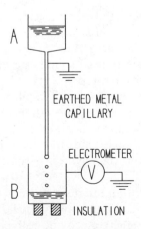

A

EARTHED METAL
CAPILLARY

ELECTROMETER

B

INSULATION

Figure C.13 Sketch of apparatus to measure the electric charging in a metal capillary (Klinkenberg and van der Minne, 1958, p. 50).

cup connected to an electrometer. The charge-to-mass ratio is calculated from the net charge collected per gram of particles. The charge per particle is calculated from the charge per gram and the particle size. Charge per particle is the end determination of most electrophoresis measurements; by this method, it is obtained directly.

Electrophoretic mobilities are difficult to measure in nonaqueous systems because low conductivities and irreversible electrodes conduce to nonuniform, time-varying electric fields. As the charges move, their distribution becomes nonuniform, generating an internal electric field. This phenomenon is not significant in aqueous media as the large number of charges per unit volume neutralizes any such field; but in media of low conductivity, these internal electric fields are superimposed on the applied electric field. Initially, the electric field is only the applied field, but during the time required to come to a steady state (Table C.5), the field decreases and is no longer known. Since the magnitude of the electric field varies in an unknown manner throughout the cell, the electrophoretic mobility cannot be calculated as the ratio of the particle velocity to the applied electric field.

The varying current flow through the system can be monitored, however, along with the velocity of the particles; and these data can be combined to yield the electrophoretic mobility under some conditions and with some assumptions.[115] The usefulness of these electrical transients depends on the fact that in low-conductivity media the particles carry a substantial part of the total current (in aqueous media the particles carry an insignificant part of the total current).

5.   POLYMER STABILIZATION OF SUSPENSIONS

Polymers can be used to stabilize suspensions either by providing electrostatic or steric repulsion. Polyelectrolytes are used in aqueous systems to confer electric charge on suspended particles. Polyacrylic acid is widely used as a stabilizer; polyacrylamide, which has many positive charges, is used as a flocculating agent, which acts by binding together negatively charge particles (Section IV.C.7).

Steric stabilization is the term used to indicate the type of stabilization of a suspension conferred by adsorbed polymers. Most examples are found in nonaqueous solvents, where electrostatic stabilization is less common. For adsorbed polymer to provide steric stabilization the molecule must be firmly anchored to the particle surface to withstand shear forces, but only at a few points so that the bulk of the molecule extends into the solvent a significant distance. This requires a careful match of polymer, particle, and solvent interactions. (Napper, 1983).

Adsorption of a polymer at an interface depends on a balance of attractions between polymer and solvent, polymer and particle, and polymer and polymer. Block copolymers can be designed to provide both polymer–particle interaction and polymer–solvent interaction. The polymer–particle interaction is usually an acid–base interaction between the surface of the particle and the groups on the polymer; that part of the copolymer is called the anchoring group. Flocculation of the suspension takes place when conditions are so altered, whether by change of temperature, concentration, or solvent composition, to reduce the solubility of the stabilizing chain in the medium. These conditions are described in the phase diagram of the stabilizing chain; hence the study of flocculation of a sterically stabilized suspension depends on solution thermodynamics.

Steric stabilization differs from electrostatic stabilization in that it is not a question of the balance of forces of attraction and repulsion between particles. The adsorbed polymer layers are so thick that particles are far enough apart to make dispersion force attraction insignificant. Repulsive forces, in the absence of electrostatics, are not long range. The long-range aspect of the repulsion is provided by the length of the polymer chain. Flocculation is not the result of the overcoming of repulsion by particle–particle attraction but is the consequence of the insolubility and precipitation of the stabilizing chain. The sensitivity to electrolyte that is characteristic of electrostatic stabilization is not displayed in steric stabilization. Further differences are illustrated in Figs C.14 and C.15, which are schematic representations of the total potential energy versus distance of separation for a pair of electrostatically stabilized particles *A* compared with a pair of sterically stabilized particles under different conditions *B*, *C*, and *D*. The

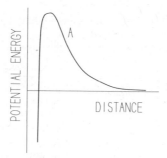

Figure C.14 Potential energy diagrams of two electrostatically stabilized particles (*A*).

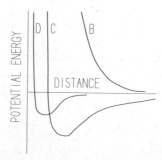

Figure C.15 Potential energy diagrams of two sterically stabilized particles: (*B*) thermodynamically stable, (*C*) in a worse than theta solvent, and (*D*) lower-molecular-weight-polymer stabilizer (or larger particle). The represented depths of the potential wells are arbitrary.

particles in *B* are thermodynamically stable; *C* represents the same particles in a worse than theta solvent, showing a potential well; and *D* represents the same particles with a polymeric stabilizer of lower molecular weight insufficient to shield the particles from their mutual dispersion force attraction. An electrostatically stabilized sol coagulates on addition of enough electrolyte and may also flocculate in a looser aggregate if a secondary minimum is present (Chapter III.C). Coagulation in the deep primary minimum of *A* is irreversible. The absence in *B*, *C*, and *D* of a deep primary minimum, is a consequence of the limit to which the adsorbed polymer can be compressed on particle–particle collision and means that the sterically stabilized sol is only capable of being flocculated in a shallow minimum, such as shown in *C*. This type of flocculation is brought about by a change of conditions, and is readily reversed by reversing the conditions.

The two types of stabilization can be combined, where it is known as electrosteric stabilization. Steric stabilization alone usually has a shallow minimum at some distance from the particle, but no primary minimum; electrostatic stabilization usually has only a primary minimum at a close distance. When the two types are combined, the particles experience repulsion at all distances and the suspension is stable.

Table C.6   Typical Anchoring Groups and Stabilizing Chains for Sterically Stabilized Dispersions

Anchoring Groups	Stabilizing Chains
Aqueous Suspensions	
Polystyrene	Polyoxyethylene
Polyvinyl acetate	Polyvinyl alcohol
Polymethyl methacrylate	Polyacrylic acid
Polyacrylonitrile	Polymethacrylic acid
Polydimethylsiloxane	Polyacrylamide
Polyvinyl chloride	Polyvinyl pyrrolidone
Polyethylene	Polyethylene imine
Polypropylene	Polyvinyl methyl ether
Polylauryl methacrylate	Poly(4-vinyl pyridine)
Nonaqueous Suspensions	
Polyacrylonitrile	Polystyrene
Polyoxyethylene	Polylauryl methacrylate
Polyethylene	Poly(1, 2-hydroxystearic acid)
Polypropylene	Polydimethylsiloxane
Polyvinyl chloride	Polyisobutylene
Polymethyl methacrylate	*cis*-1,4-Polyisoprene
Polyacrylamide	Polyvinyl acetate
	Polymethyl methacrylate
	Polyvinyl methyl ether

From Napper (1983, p. 29).

The most effective steric stabilizers are block or graft copolymers that contain both anchoring groups and stabilizing chains. The anchoring groups are usually insoluble in the medium and have strong affinity for the particle surface. These groups provide strong adsorption (chemisorption) of the polymer molecule on the particle surface, so as to prevent its being displaced during particle collisions. The stabilizing chains have strong affinity for the medium, so as to bring the polymer molecule into solution, whence adsorption proceeds. Some typical polymers for aqueous and nonaqueous suspensions are listed in Table C.6. Copolymers specifically designed for use as steric stabilizers are available from DuPont under the tradename Elvacite® and from ICI Americas under the tradename Solsperse®.

Homopolymers are also able to impart steric stabilization, for example, the naturally occurring hydrophilic polymers used as protective colloids, but they are much less effective than synthetic copolymers. The use of a homopolymer requires that the dispersion medium be a poor enough solvent to ensure strong adsorption of the stabilizer onto the particle but a good enough solvent to impart effective steric stabilization. The combination of requirements usually requires a mixture of a good solvent and a poor one, adjusted to be near the solubility limit of the polymer.

The behavior of the suspension with respect to stability is described by the phase diagram of the stabilizing polymer moiety, since its solubility determines the stability of the suspension. Typical phase diagrams of polymer solutions are shown in Fig. C.16, representing polymers of infinite molecular weight in solvents that show phase separation at theta temperatures. The theta conditions of a

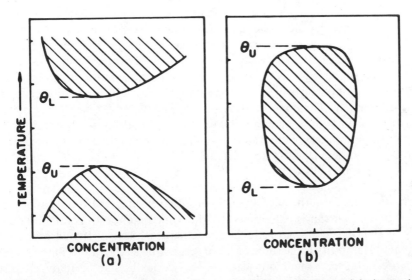

Figure C.16 Schematic diagrams of a polymer of infinite molecular weight in a solvent that shows phase separation occurring at the UCST and the LCST. (*a*) Nonaqueous polymer solution; (*b*) water-soluble polymer.[116]

polymer solution occur when the polymer–polymer interactions are the same as the polymer–solvent interactions. On one side of a theta condition, the polymer is soluble (one phase) and on the other side, it is insoluble (two phases.) Generally, polymers have both an upper critical solution temperature (UCST) and a lower critical solution temperature (LCST).

In principle, all sterically stabilized nonaqueous dispersions exhibit two flocculation temperatures, known as the upper critical flocculation temperature (UCFT) and the lower critical flocculation temperature (LCFT).[117] These are the same as the critical solution temperatures of the stabilizing polymer moiety. Although the theta conditions vary with concentration, the flocculation temperatures are almost insensitive to concentration. The region of stability of nonaqueous suspensions is approximately described on the phase diagram by that part of the one phase region between the upper and lower critical solution temperatures; and for aqueous suspensions, the region of stability is below the LCFT and above the UCFT.

The practical handling of sterically stabilized suspensions is facilitated by the reversibility of the flocculated condition, which allows the operator to go back and forth between flocculated and deflocculated states. An important variable to affect those conditions is the change of medium. Two or three components may be combined in the solvent, and the addition of one component or another can be used to flocculate or deflocculate the suspension. The suspension is flocculated at the solvent composition at which the stabilizing polymer moiety precipitates, and is deflocculated when the stabilizing polymer moiety is soluble.

The limits of stability of a sterically stabilized suspension are obtained by altering the cardinal conditions of temperature, solvent composition, or pressure that control the solubility of the stabilizing polymeric moiety. These limits may be known from the critical points of the polymer–solvent phase diagram, which are easily determined by nephelometry (cloud points). The limits may also be obtained directly by measuring the change in the number of particles as conditions are altered. Specifically, the suspension is titrated with a nonsolvent until coagulation occurs, as shown by a sudden increase in turbidity or decrease in the number of particles. Temperature or pressure changes can also be used in the same manner. These methods can be supplemented by measuring osmotic pressure as a function of temperature and particle concentration, to obtain the interaction parameters of the system. The Langmuir film balance and the method of crossed cylinders (Jacob's box, Section III.A.4) give similar information. For details see Napper (1983).

The source of steric repulsion between dispersed particles is the increase in free energy resulting from the overlap of adsorbed layers. The thermodynamic stability of the suspension is determined by the work required to concentrate molecules as two particles approach. The required thickness of the adsorbed layer for practical stability is the same order of magnitude as the radius of the particle; therefore only very small particles can be sterically stabilized by fatty-acid derivatives, which are far too short to be comparable to the size of colloids. Naturally occurring hydrophilic polymers, such as starch, gelatin, milk casein, egg albumen, lecithin, or gum arabic have been exploited for centuries as steric stabilizers. A well-known example is the silver halide "emulsion" used in photography, which is stabilized with gelatin. Another example is the use of such substances as "protective colloids." The availability of synthetic polymers of high molecular weight has extended the range and type of solid–liquid suspensions, especially nonaqueous, that can be sterically stabilized. These suspensions are increasingly important in various technical areas, such as coatings and paints, reinforcement of plastic and rubber, lubricants, and food additives. Sterically stabilized systems include paints, inks, milk and other dairy products, and biocolloids such as tissue and blood cells and protein solutions.

Linear polymers are commonly used as steric stabilizers, as polymers when cross-linked are insoluble. The primary force between the particles is their dispersion attraction; the adsorbed layers of polymer do not attract each other— the polymer, after all, is soluble in the medium. Therefore no repulsion occurs between particles until the adsorbed layers interpenetrate. Steric repulsion can be described as either an effect of the configurational entropy of the interpenetrating polymer layers or as an effect of the increase of osmotic pressure in the intervening solution between the particles on close approach. These models have been developed as quantitative theories;[118] for practical use, however, cruder approximations are adequate.

To stabilize a suspension, the thickness of the adsorbed layer should be such that the total free energy of attraction per particle, ΔG^d_{121} (Section III.A.2), is less than kT, the average kinetic energy per particle, so that no sticking occurs on collision. For two spherical particles of the same size at a short distance of separation:

$$\Delta G^d_{121} = -\frac{A_{121}af}{12H} \qquad\qquad \text{[IIIA.33]}$$

where a is the radius of the particle, f is the retardation correction, H is the shortest distance between the two particles, and A_{121} is the Hamaker constant for particles of material 1 immersed in a medium of material 2. The critical distance of separation, H_c, below which the dispersion would rapidly flocculate, is given by

$$H_c = \frac{aA_{121}f}{12kT} \qquad\qquad \text{[C.16]}$$

Values of the Hamaker constant, A_{121}, can be obtained as previously described (Section III.A.2). An example of the use of Eq. [C.16] is given by Pugh et al.[119] for carbon-black particles dispersed in dodecane. For these materials, $A_{121} = 2.8 \times 10^{-20}$ J, and the characteristic wavelength (required to calculate the retardation correction f) is 90 nm. The requirement for stability is that a particle of radius a must have a minimum film thickness of adsorbate of $H_c/2$; thus, to stabilize particles of carbon black of 200 nm radius in dodecane a film thickness of 12.5 nm is required, and particles of 500 nm radius require a film thickness of 25

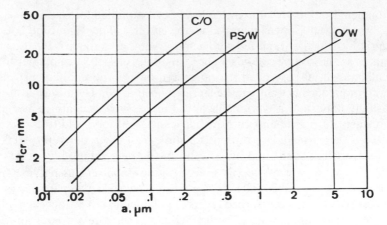

Figure C.17 Critical separation distance H_c of equal spherical particles of radius a to prevent flocculation at 25°C. C/O = carbon in oil suspensions; PS/W = polystyrene latex in water; and O/W = oil in water emulsion.[120]

nm. Clearly only adsorbed polymers can provide the required film thicknesses for steric stabilization. Results of similar calculations made for some typical suspensions are shown in Fig. C.17.

Practical applications of this concept must take into account that the polymer needs to be firmly attached to the substrate or it will be sheared off in collisions. Specific interactions between functional groups on the polymer and the substrate are required. The strongest of these is acid–base interaction, such as between basic amides and surface acid groups. A widely used oil additive that combines both high molecular weight and basic functionality is polyisobutylene succinamide, marketed as OLOA 1200 by Chevron Chemicals Co.

6. ELECTRICAL EFFECTS IN DRY SUSPENSIONS

If two solids are brought into contact and then separated, they are generally found to be electrically charged; this is the phenomenon of "contact electrification." The phenomenon has also been called "triboelectrification" since the charging of the two solids is often brought about by rubbing, but the rubbing is not necessary. Contact electrification is a key element in such processes as self-clinging wrapping materials, spray painting, adhesion, and xerography. It is also responsible for explosions in mines and flour mills, where the electric charge accumulated in moving dust clouds eventually discharges with an electric spark. The distinguishing characteristic of all these phenomena is that charges and countercharges may be separated because of low conductivity. We have discussed the same effects in suspensions of solids in media of low conductivity (Section IV.C.4g).

Contact electrification was first announced by Volta in 1797 while studying the potential difference generated by dry metals in contact. From measurements of these electric potentials, Volta developed an ordered work function series by means of his law of successive contacts. He attributed the electric potentials to an attraction of electricity by matter, varying with the difference in the nature of the substance, which produces unequal potentials and then sets up an opposition to their equalizing. In modern terms the electrons in the conduction band, also called the Fermi level, of one metal spontaneously transfer to the other metal; that is, from a low work function solid to at a high work function solid. Further separation of charges is brought to an end by the electric potential it has itself created. This potential is called the Volta potential.

For interfaces between a solid and a liquid, that is, the wet contact, charge exchange is almost always due to ionic motion, either by the adsorption of charged surface-active solutes or by the desorption of ionic components of the solid surface. The determination of the distribution of ionic charges at a solid–liquid interface is the study of the electrical double layer (Chapter III.B). The electric potential measured across a wet contact is called the Nernst potential.

A Voltaic pile is constructed by replacing every second dry contact of a pile of plates of two alternating metals, with a conducting wet contact, for example,

$$Zn|wet|Cu|dry|Zn|wet|Cu|dry|Zn|wet|Cu$$

The change of the Gibbs free energy that determines the electromotive force of the cell can be indifferently ascribed either to the difference in chemical potential of the electrons in the metals, or to the cell reaction, whatever it might be, by means of which the electrons are transferred. The cell reaction is simply the mechanism by which electrons are transported from one electrode to the other.[121]

Contact electrification is most often analyzed with the model of dry contacts. For every dry contact there is a conceptually equivalent wet contact; an analysis based on dry contacts does not exclude an alternate analysis based on wet contacts. In practice, in the presence of dirt and moisture, the actual state of the dry contact is difficult to determine.

Quantitative measures of the nature and extent of charge exchange at metal–insulator and insulator–insulator interfaces are difficult to obtain. Useful experimental techniques are:

(a) Dry-contact charging with various metals, with or without an applied electric field, in a vacuum or not.
(b) Vibrating condensor method.[122]
(c) Cascading metallic beads on an insulator surface.[123]
(d) Spectroscopic measurement of electrons emitted during separation of surfaces.[124]
(e) Field-effect capacitor.[125,126]
(f) Contact-charge spectrograph.[127]

A general review of contact electrification is given by Lowell and Rose-Innes.[128]

Several investigators[129-132] have reported that the charge transferred into a polymer by a contacting metal depends on the work function of the metal. This observation lead to the hypothesis that polymers could be arranged into a "triboelectric series," analogous to the metallic work function series. The work function of a polymer is assigned the value of the work function of the metal that does not exchange charge with it. The order of the materials in the triboelectric series is likely to be the same as the order of ionization potentials, electron affinities, acid or base strengths, reduction potentials, or Hammett sigma values.[133-135] The charging is believed to be governed by the Fermi level of the metal and the highest occupied molecular orbital (HOMO) for electron transfer to the metal or the lowest unoccupied molecular orbital (LUMO) for electron transfer to the insulator. Typical proposed triboelectric series are listed in Tables C.7 and C.8.

Table C.7   Triboelectric Series for Various Insulators

Positive	Negative
Wool	Teflon
Nylon	
Cellulose	
Cotton	
Natural silk	
Cellulose acetate	
Polymethyl methacrylate	
Polyvinyl alcohol	
Polyethylene glycol-co-terephthalic acid	
Polyvinyl chloride	
Polyacrylonitrile-co-vinyl chloride	
Polyethylene	

From Harper (1967, p. 352).

Table C.8   Triboelectric Series for Various Polymers according to Davies

Polymer	Work Function (eV)
Polyvinyl chloride	4.85 ± 0.2
Polyimide	4.36 ± 0.06
Polytetrafluoroethylene	4.26 ± 0.05
Polycarbonate	4.26 ± 0.05
Polyethylene terephthalate	4.25 ± 0.10
Polystyrene	4.22 ± 0.07
Nylon 66	4.08 ± 0.06

From ref. 130.

The absolute value of the work function for insulators and even the order of the materials in a triboelectric series is uncertain, partly because of ambient contamination, but also depending on the nature and number of contacts. Even two pieces of the same material exchange charge; for instance, one rubber rod rubbed across another.[128] When the charge exchange is not great, an insulator does not always retain the same place in a triboelectric series.

The Duke–Fabish model[136–138] of the nature of electron states in insulators attempts to account for the deficiencies of the simple idea of a triboelectric series. They propose: (a) each metal interacts with only a narrow range of electron states in the insulator; the amount of charge exchanged depending on the distribution of acceptor or donor energies; (b) injected charge is trapped to a depth of 1–$2\ \mu m$; (c) the trap states are not electronic levels of the molecule but some lower energy states stabilized by intermolecular relaxation; and (d) polymer films can have both acceptor and donor states and the number of these states accessible to the interface is limited.

7. FLOCCULATION OF SUSPENSIONS

Suspensions can be destabilized by several different mechanisms: double-layer compression, specific-ion adsorption, enmeshment or sweep flocculation, hetero-flocculation, uneven distribution of charge, polymer bridging, and changes of temperature, pressure or solvent composition.[139]

Aqueous suspensions stabilized by electrical charges are destabilized by increasing the ionic strength in the medium, which reduces the electrostatic repulsion. The effectiveness of adding an electrolyte for this purpose increases with the charge on the counterion, in accordance with the Schulze–Hardy rule. The Schulze–Hardy rule applies only to model systems with one kind of counterion of known charge and concentration. In most systems, however, a variety of solute species is present, particularly so with polyvalent ions, as these tend to form complexes with ions of opposite charge or even with nonionic solutes[140] (Section III.C.3). Freezing an electrocratic dispersion concentrates the electrolyte in the remaining liquid and flocculates the dispersion by double-layer compression. This method of flocculation is useful to obtain an equilibrium serum, as no electrolyte need be added to the system.

Changing the charge on the particle by adsorption of potential-determining ions has a much larger effect on stability than any change of ionic strength obtained by adding indifferent electrolyte. Changes of pH are particularly effective as H^+ and OH^- are often potential-determining ions (See Fig. C.4). Another example is a suspension of phosphate-stabilized kaolin, which is flocculated by acid and deflocculated by base. The magnitude of ionic charge on the phosphate ion is directly determined by the pH. Again, suspensions stabilized with polyacrylic acid are flocculated by acid and deflocculated by base. The stability of a silver iodide sol is determined by the relative concentrations of Ag^+ and I^- in solution (see Fig. IIIB.3). The surface charge need not be reduced to zero for flocculation to occur; it need only be lowered enough to allow close approach of particles. Adding excess potential-determining ions of opposite sign may repeptize the sol by reversing its charge.

In the treatment of waste water, added aluminum or iron salts hydrolyze to give a fluffy, amorphous hydroxide that traps the flocculated particles of the suspension as it settles. The aluminum salt commonly used is $Al_2(SO_4)_3 \cdot xH_2O$ where x is about 14. It is called papermakers' alum or filter alum.

Heteroflocculation refers to the flocculation of one colloid by another. A classic example is given by two electrocratic colloids of opposite charge, such as arsenic trisulfide (negative) and ferric hydroxide (positive). More recently, heteroflocculation has been studied with monodisperse latexes prepared without added stabilizers, at a pH between their isoelectric points.[141,142]

When a polymer with many charged groups, a polyelectrolyte, is added to a suspension of particles with opposite charge, it is strongly adsorbed, reversing the charge on *patches* of the surface. The polymer behaves in this way because its charge density is greater than that of the surface of the particle; thus the whole charge on the particle can be neutralized without coating the entire surface. The resulting patchwork of uneven distribution of charge allows direct electrostatic attraction between the polyelectrolyte-coated patch on one particle and the uncoated surface on another. The most common polyelectrolytes are copolymers with polyacrylamide: Acrylic acid is a possible monomer to make an anionic polyelectrolyte, and a tertiary ammonium ester of acrylic acid is a possible monomer to make a cationic polyelectrolyte.

The flocculation of negative particles by cationic polymers is primarily caused by charge neutralization as demonstrated by the fact that at the optimum concentration of flocculant, the electrophoretic mobility of the particles is close to zero.[143] Optimum flocculation occurs when sufficient charge has been neutralized, regardless of the molecular weight of the polymer.

Higher molecular weight linear polymers can be adsorbed by more than one particle at a time, forming a polymer bridge. If the polymer is a polyelectrolyte, some reduction of electrostatic repulsion also occurs and greatly enhances the rate of adsorption. The kinetics of this process are complex. The critical factor is the frequency of particle collisions. In a concentrated suspension particles may collide before adsorbed polymer has reached an equilibrium configuration. The process is well suited to flocculate dilute suspensions. The optimum polymer dosage varies widely and need not increase proportionately with percent solids.

Suspensions stabilized by adsorbed polymer are destabilized by altering the conditions of solubility of the polymer by changes of temperature, pressure, or solvent composition.

8. NUCLEATION AND CRYSTAL GROWTH

Precipitation occurs only from supersaturated solutions. Homogeneous solutions are supersaturated at solute concentrations greater than the equilibrium concentration of the two-phase system consisting of excess solute and saturated solution. Typically, supersaturated solutions are produced by sudden changes in temperature or pressure, or by mixing of two soluble materials that react to form an insoluble one. The supersaturated solution is unstable because it is at a higher free energy than that of the two-phase system. Nevertheless, supersaturated solutions may be stable for a significant time. This metastability is maintained by a barrier that is the energy required to form the first crystal nucleus. Small crystal nuclei have a high specific free energy because small size implies large curvature, and a particle of large curvature is at a high chemical potential (Kelvin equation [II.A.81]). Once a few nuclei form, however, precipitation proceeds rapidly since energy is released both by the precipitation from supersaturated solution and by the reduction of the curvature of the surface as the crystal grows. If a substrate appropriate for crystallization is introduced into a supersaturated solution, as for example, a roughened surface or a few "seed" crystals, the energy barrier to nucleus formation is removed, and precipitation follows at once. Precipitation on nascent nuclei is called homogeneous precipitation and precipitation on extraneous nuclei is called heterogeneous precipitation.

Theories of homogeneous nucleation are as yet incomplete, but the general idea is that as the degree of supersaturation increases, the time to onset of homogeneous nucleation decreases and the number of nuclei formed increases. That is, for a solution just barely beyond saturation, a long time passes before any nucleation occurs and only a few nuclei form; for a highly supersaturated solution, nucleation occurs quickly and a large number of nuclei form. After the first burst of nucleation, subsequent precipitation occurs on the already formed nuclei. Hence, from solutions just barely beyond saturation, a few large crystals form, and from highly supersaturated solutions, many small crystals form.

La Mer and co-workers[144,145] demonstrated these concepts. Figure C.18 shows the buildup of reaction product beyond saturation, followed by the formation of nuclei in a burst, and then precipitation on the already formed nuclei. The ascending curve in region I describes a reaction that continuously generates molecules in solution. The concentration of molecules increases steadily, passes the point of saturation *A*, and reaches a point *B* at which the rate

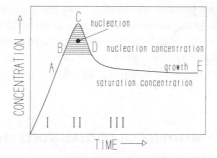

Figure C.18 The formation of a monodisperse system by controlled nucleation and growth.[144]

of self-nucleation becomes appreciable; the rate of nucleation rises rapidly to the point C, but the partial relief of supersaturation is so rapid and effective that the time of nucleation (region II) is brief and no new nuclei are formed after the initial burst BCD. The nuclei produced grow uniformly by a diffusion process (region III). A uniform suspension can be obtained by this process if the homogeneous nucleation is confined to a single burst, which can be done if the rate of generation of molecules is slow.

Kinetic theories for crystal growth are also incomplete. The rate of crystal growth depends on two separate rates: the rate of diffusion of material to the surface and the rate of incorporation of material into the crystal. The way in which the particle size distribution changes with time is a clue to the mechanism of crystal growth. If the incorporation of material is proportional to the area of the particle (as in diffusion-controlled reactions), the absolute width of the particle size distribution decreases as the average particle size grows, and the relative width of the size distribution decreases even more rapidly. If the rate of incorporation of material into the particle is the same for all sizes (as in the Smith–Ewart model for emulsion polymerization), the size distribution narrows even more markedly. If the rate of incorporation of material on to the particle is proportional to the particle volume, the relative width of the particle size distribution remains constant. Common to these three mechanisms is the decrease in polydispersity during growth, that is, the size distribution becomes narrower. Ultimately this tendency is reversed by the process of Ostwald ripening, unless the solubility is too low for that process to be significant. For this reason, uniform dispersions made by precipitation are of highly insoluble materials, such as sulfur, selenium, and ferric oxide.

The surface energies of various possible crystal faces influence the habit of the crystal, that is, its external form. Surface planes of higher energy grow most rapidly, but they are superseded in time by the slower growing planes of lower surface energy. For example, the {111} faces of sodium chloride first appear as an octahedral crystal; on each of its six vertices the slower growing {100} faces gradually develop and finally produce the stable cube crystal. The high energy of the {111} faces may be reduced by adsorption: thus, the presence of urea stabilizes the {111} faces of sodium chloride and so allows the octahedral habit to be retained. Similarly, the growth of crystals may be inhibited by surface-active solute adsorbed on all faces. An industrial application is the prevention of scaling in pipes by process water that has become supersaturated by evaporation, by adding agents to retard further growth of crystals. The growth of calcium carbonate scale, for example, is inhibited by adding polyphosphates, phosphonates, or polyacrylic acid.

9. MODEL COLLOIDAL SYSTEMS

The properties of a suspension are the sum of all the individual interactions between its particles. When these are of uniform size and chemistry, the sum is over identical interactions and the macroscopic properties are then linearly related to the properties of individual particles. Such systems come close to the simplest theoretical models and are therefore used to verify theories. Theories of light scattering, adsorption on surfaces, forces across phase boundaries, and stability of dispersions have all been tested by comparison with the behavior of model systems. Two types of systems approaching ideal have been synthesized: the first, suspensions of uniform particles of known size and surface composition; the second, uniform non porous surfaces. Suspensions of uniform particles have been used to test models of Brownian motion, theories of rapid coagulation, theories of light scattering, and theories of rheology. The uniform surfaces of soap films and adsorbed films on flat mica surfaces have been used to measure forces across phase boundaries. Carefully prepared uniform solid substrates have been used to test models of adsorption of gases.

Uniform dispersions and uniform surfaces also find practical applications: for example, as supports for agglutination tests for pregnancy and rheumatoid factors, as calibration standards for electron microscopes and particle counters, as tests for filter efficiencies, as "seeds" for laser Doppler velocimetry, as packing for chromatographic columns, as catalysts, in ceramics, as pigments with enhanced optical properties, and as media for recording devices.

a.   Particles of Uniform Size

Dispersions of uniform size may form spontaneously, as in some protein solutions, some micellar solutions, and some microemulsions. Such dispersions of proteins were used by Svedberg in his studies with the ultracentrifuge (Nobel Prize in Chemistry in 1926). Dispersions of uniform size can be obtained by repeated fractionation. Such a dispersion of gamboge particles was isolated by Perrin for use in the experiments that verified Einstein's theory of Brownian motion (Nobel Prize in Physics, 1926). Dispersions of uniform size can be prepared by chemical precipitation. A nearly uniform dispersion of gold particles was prepared in this way by Zsigmondy to test his invention of the ultramicroscope (Nobel Prize in Chemistry, 1925). Can uniform systems be prepared from any desired substance? Probably yes, if sufficiently low solubility can be obtained.

The synthetic method, the third of those mentioned above, controls particle growth by a two-step process, of which the first is a nucleation or seeding phase; the second is a growth phase, during which the average particle size increases but the size distribution narrows. Some celebrated examples follow.

(i)   Gold Sols. In 1857 Faraday reported the results of his study on "diffused" gold.[146] By "diffused" he meant a suspension of particles rather than a molecular solution. Suspended particles too small to be seen by any microscope were detected by light scattering. He adumbrated a relation between the size of the suspended gold particles and the colors of the transmitted and scattered light. His gold sols were produced by electrical evaporation of gold wire and by precipitation of gold from a solution of gold chloride. Zsigmondy developed the

"seed" method for consistently producing uniform and stable gold sols. By demonstrating, with extremely clean systems, that the number of particles produced is proportional to the amount of seed solution used, he proved that precipitation takes place on already formed nuclei. (Zsigmondy and Thiessen, 1925). By analyzing Zsigmondy's data, Mie in 1908 interpreted quantitatively the vivid colors of colloidal gold sols, thus completing the line of research started 50 years earlier by Faraday. More recently, on the basis of his own data on light scattered by monodisperse gold sols, Doremus[147] verified Mie's classical theory for gold particles as small as 8.5nm. The kinetics of flocculation of gold sols were studied in detail by measuring fractal structures as flocs developed.[148,149]

(ii) Sulfur Sols. La Mer and co-workers prepared monodisperse sulfur sols from acidified thiosulfate solutions.[144,145] They allowed a large number of nuclei to form in a first precipitation and then had those nuclei grow by the slow addition of more reactant, keeping the supersaturation below the level at which new nuclei would form. The preparation and properties of these sols were thoroughly investigated by this school. Their sulfur sols show higher order Tyndall spectra when illuminated with white light, a dramatic effect obtained only with highly uniform systems. By this and other work they demonstrated the wide potential of the method of controlled nucleation and growth to prepare various monodisperse systems. La Mer sulfur sols were prepared by Kerker et al. for use as model systems to study Mie scattering, which gives a more precise determination of average particle size than the method of higher order Tyndall spectra. The results of the two determinators were found to agree quite closely; analysis of Mie scattering also gives the width of the distribution.[150]

(iii) Selenium Sols. Watillon et al.[151-153] produced monodisperse selenium sols of 4–50 nm by precipitation of selenium from solution on to gold nuclei. The use of gold nuclei makes for better control of ultimate particle size by eliminating homogeneous precipitation. To test the Mie theory of scattering by absorbing spheres, particle size was first determined by ultramicroscope counting and electron microscopy; the Mie theory was then confirmed by direct comparisons with observations of these model systems.

(iv) Metal Oxides. Synthetic methods to obtain a variety of metal-oxide colloids, uniform in size and shape, were developed by Matijević and co-workers.[154,155] The general techniques are precipitation from homogeneous solutions, phase transformations, and reactions within uniform aerosol droplets. The most used technique is the slow hydrolysis of metal salts at elevated temperatures in aqueous solutions, usually in the presence of complex-forming sulfate or phosphate ions. Hydrolysis is controlled by the temperature or by release of anions or cations, to ensure that only one burst of nuclei is produced, followed by slow growth on the nuclei. By this method, uniform oxides of aluminum, copper, iron (spherical, cubic, and other shapes), cobalt, and nickel

have been produced, as well as zinc sulfide, lead sulfide, cadmium selenide, and cadmium carbonate.

By the method of phase transformation, the monodisperse precipitate is changed to another monodisperse form by crystallization, recrystallization, or dissolution followed by reprecipitation. By this method, uniform suspensions of cobalt ferrite, nickel ferrite, or cobalt-nickel ferrites have been produced by reaction of cobalt and nickel oxides with ferric hydroxide. These products are significant in the corrosion of steel.

Uniform particles are produced by the reaction of uniform aerosol droplets with a reactant in the vapor phase. The advantage of this method is that surface-active solutes or other reagents are not required. Titanium oxide, alumina, titanium silicate, and polymer colloids as large as $30 \mu m$ have been produced.

(v) Silver Halide Sols. Silver iodide sols figured prominently in the development of the theory of lyophobic colloids. They were extensively investigated in the schools of Kruyt and Overbeek (Kruyt, 1952, Volume 1) and of Težak.[156] Monodisperse silver iodide and silver bromide sols were obtained by carefully diluting a solution of a complex silver halide ion to form supersaturated solutions of the silver halide.[157] All the sols prepared exhibited higher order Tyndall spectra. Particle size was about 400 nm.

Silver halide sols are well suited as model systems to test double-layer-stability theory because their potential-determining ions are either silver or halide, the concentrations of which can be adjusted, the particles have surfaces of known crystal structure, and their sizes and shapes are readily measured.

(vi) Uniform Latex Particles. Aqueous suspensions of polymers are called "latexes" (or "latices") because they evolved from the work on synthetic-rubber production, and the suspensions have a milky look. Three different synthetic routes were developed to produce uniform particles. Particles with average diameters less than about $5 \mu m$ are produced by emulsion polymerization with a coefficient of variance of about 1%. Particles with diameters between about 2 and $20 \mu m$ are produced by the swollen-emulsion-polymerization technique of Ugelstad with a coefficient of variance less than 3%. Particles with larger diameters are produced by dispersion polymerization, but are not nearly so uniform having coefficients of variance in diameters of 16–30%. The differences in the uniformity of particles produced by the different techniques are direct consequences of the different mechanisms of particle formation.

Emulsion polymerization is carried out in a solution containing emulsified monomer droplets, sufficient surface-active solute to form a large number of micelles, and a soluble initiator. The initiation of polymerization is generally believed to occur in the homogeneous solution, which contains a low concentration of monomer. The growing oligomers gradually become water insoluble and are taken up by the micelles, inside which subsequent growth occurs. Because polymer growth occurs slowly after the initiation of polymerization, the narrow distribution of micellar sizes narrows even further as the particles grow. The optimum conditions for the formation of monodisperse suspensions are determined empirically.[158–161]

Larger particles are formed by the addition of more monomer to the suspension. The added monomer swells the already formed particles and is incorporated into the existing polymer chains. The ultimate size that particles reach and still remain uniform on addition of more monomer is limited. Any slight variation in particle size is increased by Ostwald ripening, the larger particles growing at the expense of the smaller ones.

The upper-size limit for particles grown by emulsion polymerization was extended by the work of Ugelstad and co-workers.[27] After a routine initial emulsion polymerization step, a swelling agent, such as dodecyl chloride, is added, which causes the particles to swell by imbibing a large amount of monomer, while still maintaining uniform size (Section IV.A.12). Large particles grown by this technique are nearly as uniform as those grown in space under conditions of microgravity at much greater expense.

The largest latex particles are grown by suspension polymerization. In this process an emulsion of the monomer is made and an oil-soluble initiator is added to polymerize the emulsion droplets. The final average particle size and size distribution of the polymer particles is nearly identical with the average particle size and size distribution of the initial emulsion droplets.

Croucher et al. synthesized a number of thermodynamically stable non-aqueous suspensions.[162–164] In every case the particles are copolymers: One moiety forms the insoluble core of the particle and the other moiety forms the soluble outer layer that stabilizes the suspension. When the particle is in a worse-than-theta solvent, which may be brought about by changing solvent composition or by changing temperature or pressure, the outer polymer collapses against the core, and loses its ability to stabilize. The flocculation, however, is into a shallow minimum. When the solvent is then restored to better-than-theta conditions, the outer polymer reextends, enabling the particles to be easily redispersed. These model systems are ideal to study mechanisms of polymer stabilization as well as to provide useful practical materials. These uniform latex suspensions are a subset of suspensions called polymer colloids, which are used as adhesives, coatings, and sealants, as well as in paints, papermaking, and medical science. The science of polymer colloids bridges from polymer chemistry to colloid science. The subject is reviewed in monographs (Buscall et al., 1985).

b. Surfaces of Uniform Composition

(i) Soap Films. Soap films have the advantage, common to any liquid surface, of being energetically uniform as long as they are not subject to changes of temperature, area, or composition. They have proved useful for direct measurements of disjunctive (disjoining) pressures since they can be thinned to tens of nanometers without rupture. The thickness is made to vary by applied pressure, which can be measured directly.[165,166] The soap film is supported, as shown in

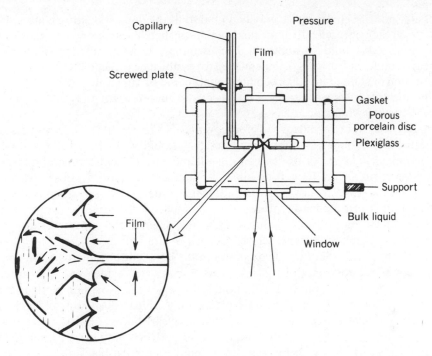

Figure C.19 Schematic cross section of the apparatus used to compress soap films. The inset shows how air is retained but liquid allowed to pass through the porous solid. A layer of liquid on the bottom of the cell prevents evaporation from the soap film and allows light to pass.[166]

Fig. C.19 (inset), by a porous porcelain ring connected to a reservoir of soap solution. When pressure is applied, the soap film thins out and the excess solution flows into the reservoir. When equilibrium is reached, the thickness of the film is measured by means of an interferometer. Thus, by controlling the air pressure within the cell, the resistance to compression between the two charged surfaces of the soap film is observed as a function of distance. The analysis of the data gives a direct measurement of the Debye length, $1/\kappa$, and a test of theories of electrostatic stabilization. For instance, a plot of the logarithm of the pressure against thickness has a slope of $-\kappa$.

(ii) Homotattic Solid Surfaces. Uniform solid substrates are the key to understanding the behavior of adsorbed vapors. To be useful, substrates must be fine powders as the amount of adsorption varies directly with the specific area; and on large flat surfaces sufficient vapor is not adsorbed to be accurately measured. Observed monolayer adsorption at constant temperature depends on both the equation of state of the adsorbed (two-dimensional) film on each patch and on the distribution of patches that compose the solid substrate. On a heterogeneous substrate, more than one combination of an equation of state and

of a degree of heterogeneity may be found to fit the data. If the true equation of state on each energetically uniform (homotattic) patch were known, then the true heterogeneity could be deduced. The usefulness of finely divided solids with nearly uniform surfaces is, therefore, to establish experimentally the actual equation of state of an adsorbed film, unobscured by substrate heterogeneity, so that thenceforth any arbitrary substrate can be analyzed with only heterogeneity as the unknown factor.

The behavior of an adsorbed monolayer bears a remarkable analogy to that of matter in the bulk. A vapor condenses to a liquid with the appearance of a discontinuity in the $p-V$ isotherm, and above a critical temperature such condensation is no longer possible. Analogous behavior is observed with adsorbed monolayers of vapors, including first-order phase transitions and two-dimensional critical temperatures. But these effects are screened from observation unless the substrate is homotattic. An excellent test for uniformity of a substrate, therefore, is whether the adsorbed film shows a first-order phase transition, as revealed by a dicontinuity in the adsorption isotherm. Even small degrees of heterogeneity of the substrate obscures this evidence. Ross and co-workers,[167-169] starting in 1947, published a series of adsorption isotherms of ethane adsorbed by cube crystals of carefully prepared sodium chloride at various temperatures above and below what is clearly a two-dimensional critical temperature. Subsequently Ross and his school (Ross and Olivier, 1964) made similar observations with krypton on graphitized carbon black.

Analysis of the adsorption isotherms of these homotattic adsorbents identified a useful equation of state as the two-dimensional analogue of the van der Waals equation of a gas. More precise data led to the use of a two-dimensional virial equation of state,[170] with refinements of the numerical analysis to determine the distribution of surface energies of adsorbents (CAEDMON).[171]

A significant outcome of this work on model solid adsorbents is that it fails to confirm that the widely used Langmuir adsorption isotherm is a valid description of vapor adsorption on homotattic substrates.

10. SOME SPECIAL DISPERSIONS

Clouds. Clouds are naturally occurring dispersions of minute droplets of water in air (aerosol). By the Stokes law a droplet of water $25\,\mu$m in diameter will fall through air at a rate of 2 cm/s. If the droplet is one-tenth that size, its rate of fall would be 100 times smaller, or about 1 cm/min. The settling of a cloud formed of such particles is unnoticeable; indeed the settling of very small particles suspended in air, such as the fine spray of waves or waterfalls and all kinds of dust and smoke, is very slow. Stokes's law, however, is not the last word on the motion of air-borne particles. Solar radiation is absorbed by the particles, which radiate heat to the surrounding air, thus reducing its density and causing it to rise, carrying the particles with it. If the particles are water, the heat of the sun causes evaporation. Water vapor is less dense than air, and moist air is lighter than dry air at the same temperature and pressure. The water droplets make the air damp and, if the density is less than that of the surrounding dry air, the cloud will rise. Moist air over a body of water rises, condenses at the lower temperature, falls, and rises again in the heat of the sun.

Smokes. The air over an industrial city contains 100–500 k of solid matter per cubic kilometer. Dust particles, even more than water droplets, absorb sunshine and generate thermal gradients in the surrounding air; consequently, in spite of their higher density, solid particles remain suspended in air and may be carried hundreds of kilometers by air currents. Finely divided particulate matter is carried into the upper atmosphere and can be transported around the world. Smoke is also responsible for "pea-soup" fogs by nucleating moisture that would otherwise remain as a vapor. These fogs carry a high concentration of dirt and sometimes sulfuric acid.

Greases. The function of a lubricating grease is to remain in contact with moving surfaces and to stay in place without dripping when the machinery is stationary. Grease is designed to flow into bearings by the application of pressure, which means that it must have a yield point. Another desideratum of a grease is to retain viscosity at temperatures up to 200°C. Grease is made from lubricating oil by dissolving lithium, calcium, or sodium soaps in hot oil; on cooling, the soaps crystallize as a mass of ramified fibers from 200 nm to 1 mm in length and from 10 to 1000 nm in width, enmeshing the oil. Nonsoap inorganic thickeners, including colloidal silica, attapulgite clay, and montmorillonite clay treated with long-chain amines, make greases that can be used at higher temperatures. Thickeners are almost always the minor component; their function is to alter the rheological behavior of the lubricant. The existence of a yield point depends on the flocculated structure of the thickener and on its particle size and aspect ratio. The lubricant is held in the solid matrix by capillarity and by adsorption, for which a large interfacial area is advantageous. The solid matrix is held together by short-range forces of attraction. Water has a large effect on the rheology of greases, suggesting that the fiber–fiber interactions are modulated by acid–base exchanges.

References for Part IV

1. Ross, S. Adhesion versus cohesion in liquid–liquid and solid–liquid dispersions, *J. Colloid Interface Sci.* **1973**, *42*, 52–61.

2. Pickering, S. U. Emulsions, *J. Chem. Soc.* **1907**, *91*, 2001–2021.

3. Griffin, W. C. Classification of surface-active agents by 'HLB', *J. Soc. Cosmet. Chem.* **1949**, *1*, 311–326.

4. Griffin, W. C. Calculation of HLB values of non-ionic surfactants, *J. Soc. Cosmet. Chem.* **1954**, *5*, 249–256.

5. Davies, J. T. A quantitative kinetic theory of emulsion type, I. Physical chemistry of the emulsifying agent, *Proc. Int. Congr. Surf. Act., 2nd, 1957* **1957**, *1*, 426–438.

6. Grist, D. M.; Neustadter, E. L.; Whittingham, K. P. The interfacial shear viscosity of crude oil/water systems, *J. Can. Pet. Tech.* **1981**, *20*, 74–78.

7. Biswas, B.; Haydon, D. A. The rheology of some interfacial adsorbed films of macromolecules. I. Elastic and creep phenomena, *Proc. Roy. Soc. London* **1963**, *271 A*, 296–316.

8. Sebba, F. Macrocluster gas–liquid and biliquid foams and their biological significance, *A.C.S. Symp. Ser.* **1975**, *9*, 18–39.

9. Ross, S. Toward emulsion control, *J. Soc. Cosmet. Chem.* **1955**, *6*, 184–192.

10. Richardson, E. G. "Emulsions," in *Flow Properties of Disperse Systems*; Hermans, J. J., Ed.; Interscience: New York, 1953, 39–60.

11. Princen, H. M. Rheology of foams and highly concentrated emulsions. I. Elastic properties and yield stress of a cylindrical model system, *J. Colloid Interface Sci.* **1983**, *91*, 160–175.

12. Clausse, M. "Dielectric properties of emulsions and related systems," in *Encyclopedia of Emulsion Technology*; Becher, P., Ed.; Dekker: New York, 1983; Vol. 1, pp. 481–715.

13. Levius, H. P.; Drommond, F. G. Elevated temperature as an artificial breakdown stress in the evaluation of emulsion stability, *J. Pharm. Pharmacol.* **1953**, *5*, 743–756.

14. Herb, C. A.; Berger, E. J.; Chang, K.; Morrison, I. D.; Grabowski, E. F. "The use of quasi-elastic light scattering in the study of particle size distributions in sub-micrometer emulsion systems," in *Magnetic Resonance and Scattering in Surfactant Systems*; Magid, L.; Ed.; Plenum: New York, 1987.

15. van der Waarden, M. The process of spontaneous emulsification, *J. Colloid Sci.* **1952**, 7, 140–150.

16. van der Waarden, M. Viscosity and electroviscous effect of emulsions, *J. Colloid Sci.* **1954**, 9, 215–222.

17. Herb, C. A. *Private communication*, 1987.

18. Briggs, T. R.; Schmidt, H. F. Experiments on emulsions II. Emulsions of water and benzene, *J. Phys. Chem.* **1915**, 19, 478–499.

19. Briggs, T. R. Experiments on emulsions III. Emulsions by shaking, *J. Phys. Chem.* **1920**, 24, 120–126.

20. Menon, V. B.; Wasan, D. T. "Demulsification," in *Encyclopedia of Emulsion Technology*; Becher, P., Ed.; Dekker: New York, 1985; Vol. 2, pp. 1–75.

21. Lissant, K. J. "Making and breaking emulsions," in *Emulsions and Emulsion Technology*, Part 1; Lissant, K. J., Ed.; Dekker: New York, 1974; pp. 71–124.

22. Bailey, P. A. The treatment of waste emulsified oils by ultrafiltration, *Filtr. Sep.* **1977**, 14, 53–55.

23. Aslanova, M. A. Demulsification and and desalting of Emba crude oils (In Russian), *Vost. Neft* (Eastern Petroleum) **1940**, No. 5–6, 59–64.

24. Canevari, G. P.; Fiocco, R. J. Treatment of Athabasca tar sands froth, U. S. Patent 3, 331, 765, 1967.

25. Higuchi, W. I.; Misra, J. Physical degradation of emulsions *via* the molecular diffusion route and possible prevention thereof, *J. Pharm. Sci.* **1962**, 51, 459–466.

26. Ugelstad, J. Swelling capacity of aqueous dispersions of oligomer and polymer substances and mixtures thereof, *Makromol. Chem.* **1978**, 179, 815–817.

27. Ugelstad, J.; El-Aasser, M. S.; Vanderhoff, J. W. Emulsion polymerization: Initiation of polymerization in monomer droplets, *J. Polym. Sci., Polym. Lett. Ed.* **1973**, 11, 503–513.

28. Ross, S.; Prest, H. F. On the morphology of bubble clusters and polyhedral foams, *Colloids Surf.* **1986**, 21, 179–192.

29. Thomson, W. (Lord Kelvin). On the division of space with minimum partitional area, *Phil. Mag.* **1887**, (5), 24, 503–514.

30. Derjaguin, B. Die elastischen eigenschaften der schäume, *Kolloid Z.* **1933**, 64, 1–6.

31. Ross, S. Bubbles and foam: A new general law, *Ind. Eng. Chem.* **1969**, 61(10), 48–58; also *Chemistry and Physics of Interfaces—II*; Ross, S., Ed.; American Chemical Society: Washington, D.C., 1971; pp. 15–25.

32. Ross, S. Cohesion of bubbles in foam, *Am. J. Phys.* **1978**, 46, 513–516.

33. Morrison, I. D.; Ross, S. The equation of state of a foam, *J. Colloid Interface Sci.* **1983**, 95, 97–101.

34. Nishioka, G.; Ross, S. A new method and apparatus for measuring foam stability, *J. Colloid Interface Sci.* **1981**, 81, 1–7.

35. Nakagaki, M. A new theory of foam formation and its experimental verification, *J. Phys. Chem.* **1957**, 61, 1266–1270.

36. Lord Rayleigh. "Foam," in *Scientific Papers by Lord Rayleigh (John William Strutt)*, Cambridge University Press: Cambridge, 1902; Vol. 3, pp. 351–362; also in 6 vols. bound as 3; Dover: New York, 1964; Vol. 2, pp. 351–362.

37. Marangoni, C. Difesa della teoria dell'elasticità superficiale dei liquidi: Plasticità superficiale, *Nuovo Cimento* **1878**, (3), *3*, 50–68.

38. Marangoni, C. Difesa della teoria dell'elasticità superficiale dei liquidi: Plasticità superficiale, *Nuovo Cimento* **1878**, (3), *3*, 97–123.

39. Marangoni, C. Difesa della teoria dell'elasticità superficiale dei liquidi: Plasticità superficiale, *Nuovo Cimento* **1878**, (3), *3*, 193–211.

40. Gibbs, J. W. "Liquid films," in *Scientific Papers of J. Willard Gibbs: Thermodynamics*; Longmans, Green: London, 1906; Vol. 1, pp. 300–314; Dover, New York, 1961.

41. Thomas, T. B.; Davies, J. T. On the sudden stretching of liquid lamellae, *J. Colloid Interface Sci.* **1974**, *48*, 427–436.

42. Rusanov, A. I.; Krotov, V. V. Gibbs elasticity of liquid films, threads, and foams, *Prog. Surf. Membr. Sci.* **1979**, *13*, 415–524.

43. van den Tempel, M.; Lucassen, J; Lucassen-Reynders, E. H. Application of surface thermodynamics to Gibbs elasticity, *J. Phys. Chem.* **1965**, *69*, 1798–1804.

44. Borwankar, R. P.; Wasan, D. T. The kinetics of adsorption of surface active agents at gas–liquid surfaces, *Chem. Eng. Sci.* **1983**, *38*, 1637–1649.

45. Mannheimer, R. J.; Schechter, R. S. Shear-dependent surface rheological measurements of foam stabilizers in nonaqueous liquids, *J. Colloid Interface Sci.* **1970**, *32*, 212–224.

46. Ross, S.; Haak, R. M. Inhibition of foaming. IX. Changes in the rate of attaining surface-tension equilibrium in solutions of surface-active agents on addition of foam inhibitors and foam stabilizers, *J. Phys. Chem.* **1958**, *62*, 1260–1264.

47. Ross, S.; McBain, J. W. Inhibition of foaming in solvents containing known foamers, *Ind. Eng. Chem.* **1944**, *36*, 570–573.

48. Robinson, J. V.; Woods, W. W. A method of selecting foam inhibitors, *J. Soc. Chem. Ind., London* **1948**, *67*, 361–365.

49. Maru, H. C.; Mohan, V.; Wasan, D. T. Dilational viscoelastic properties of fluid interfaces. I. Analysis, *Chem. Eng. Sci.* **1979**, *34*, 1283–1293.

50. Ross, S. Foam and emulsion stabilities, *J. Phys. Chem.* **1943**, *47*, 266–277.

51. Sawyer, W. M.; Fowkes, F. M. Interaction of anionic detergents and certain polar aliphatic compounds in foams and micelles, *J. Phys. Chem.* **1958**, *62*, 159–166.

52. Djabbarah, N. F.; Wasan, D. T. Relationship between surface viscosity and surface composition of adsorbed surfactant films, *Ind. Eng. Chem., Fundam.* **1982**, *21*, 27–31.

53. Friberg, S.; Ahmad, S. I. Liquid crystals and the foaming capacity of an amine dissolved in water and p-xylene, *J. Colloid Interface Sci.* **1971**, *35*, 175.

54. Friberg, S.; Saito, H. "Foam stability and association of surfactants," in *Foams, Proc. Symp., 1975* Akers, R. J., Ed.; Academic: New York, 1976, pp. 33–38.

55. Epstein, M. B.; Wilson, A.; Jakob, C. W.; Conroy, L. E.; Ross, J. Film drainage transition temperatures and phase relations in the system sodium lauryl sulfate, lauryl alcohol, and water, *J. Phys. Chem.* **1954**, *58*, 860–864.

56. Epstein, M. B.; Ross, J.; Jakob, C. W. The observation of foam drainage transitions, *J. Colloid Sci.* **1954**, *9*, 50–59.

57. Bolles, W. L. The solution of a foam problem, *Chem. Eng. Prog.* **1967**, *63*(9), 48–52.

58. Exerowa, D.; Lalchev, Z.; Marinov, B.; Ognyanov, K. Method for assessment of fetal lung maturity, *Langmuir* **1986**, *2*, 664–668.

59. Dewar, J. Soap bubbles of long duration, *Proc. R. Inst. G.B.* **1917**, *22*, 179–212.

60. Friberg, S. E.; Cox, J. M. Stable foams from nonaqueous liquids, *Chem. Ind. (London)* **1981**, *17 Jan.*, 50–52.

61. Derjaguin, B. V.; Titijevskaya, A. S. Static and kinetic stability of free films and froths, *Proc. Int. Congr. Surf. Act., 2nd, 1957* **1957**, *1*, 211–219.

62. Ottewill, R. H.; Segal, D. L.; Watkins, R. C. Studies on the properties of foams formed from nonaqueous dispersions, *Chem. Ind. (London)* **1981**, *17 Jan.*, 57–60.

63. Zuiderweg. F. J.; Harmens, A. The influence of surface phenomena on the performance of distillation columns, *Chem. Eng. Sci.* **1958**, *9*, 89–103.

64. Ross, S.; Nishioka, G. Foaminess of binary and ternary solutions, *J. Phys. Chem.* **1975**, *79*, 1561–1565.

65. Ross, S.; Townsend, D. F. Foam behavior in partially miscible binary systems, *Chem. Eng. Commun.* **1981**, *11*, 347–353.

66. Nishioka, G. M.; Lacy, L. L.; Facemire, B. R. The Gibbs surface excess in binary miscibility-gap systems, *J. Colloid Interface Sci.* **1981**, *80*, 197–207.

67. Haywood, I. R.; Lumb, P. B. The heavy water industry, *Chem. Can.* **1975**, *27*(3), 19–21.

68. Sagert, N. H.; Quinn, M. J. Influence of high-pressure gases on the stability of thin aqueous films, *J. Colloid Interface Sci.* **1977**, *61*, 279–286.

69. Bancroft, A. R. *Heavy Water GS Process: R&D Achievements*; Atomic Energy of Canada, Ltd., Report 6215, October, 1978.

70. Harkins, W. D. A general thermodynamic theory of the spreading of liquids to form duplex films and of liquids or solids to form monolayers, *J. Chem. Phys.* **1941**, *9*, 552–568.

71. Ross, S. The inhibition of foaming, *Rensselaer Polytech. Inst., Eng. Sci. Ser.* **1950**, *63*, 1–40.

72. Stephan, J. T. Combination polyglycol and fatty acid defoamer composition, U.S. Patent 2, 914, 412, 1959.

73. Robinson, J. V. The rise of air bubbles in lubricating oils, *J. Phys. Colloid Chem.* **1947**, *51*, 431–437.

74. Rauner, L. A. Antifoaming agents, *Encycl. Polym. Sci. Technol.* **1964–1972**, *2*, 164–171.

75. Whipple, C. L.; Oppliger, P. E.; Schiefer, H. M. *Abstracts of Papers*, 138th National Meeting American Chemical Society: New York, Sept. 11–16, 1960.

76. McBain, J. W.; Ross, S.; Brady, A. P.; Robinson, J. V.; Abrams, I. M.; Thorburn, R. C.; Lindquist, C. G. Foaming of aircraft-engine oils as a problem in colloid chemistry-I, *Nat. Advis. Comm. Aeronaut., Rep. ARR 4I05*, 1944.

77. Ross, S.; Nishioka, G. Experimental researches on silicone antifoams," in *Emulsions, Latices, and Dispersions*; Becher, P.; Yudenfreund, M.N., Eds.; Dekker: New York, 1978, pp. 237–256.

78. Ross, S.; Nishioka, G. Monolayer studies of silica/poly(dimethylsiloxane) dispersions, *J. Colloid Interface Sci.* **1978**, *65*, 216–224.

79. Trapeznikov, A. A.; Chasovnikova, L. V. Stabilization of bilateral films by mono-layers and thin films of poly(dimethylsiloxanes), *Colloid J. USSR* (Engl. Transl.) **1973**, *35*, 926–928.

80. Roberts, K.; Axberg, C.; Österlund, R. Emulsion foam killers in foams containing fatty and rosin acids, in *Foams, Proc. Symp., 1975* Akers, R.J., Ed.; Academic: New York 1976, pp. 39–49.

81. Garrett, P. R. The effect of poly(tetrafluoroethylene) particles on the foamability of aqueous surfactant solutions, *J. Colloid Interface Sci.* **1979**, *69*, 107–121.

82. Dippenaar, A. The destabilization of froth by solids. I. The mechanism of film rupture; II. The rate-determining step, *Int. J. Miner. Process.* **1982**, *9*(1), 1–14; 15–22.

83. Kurzendoerfer, C. P. Mechanisms of foam inhibition by trialkylmelamines, *Tr.-Mezhdunar Kongr. Poverkhn.-Akt. Veshchestvam, 7th, 1976* **1978**, *2*(I), 537–548.

84. Aronson, M. P. Influence of hydrophobic particles on the foaming of aqueous surfactant solutions, *Langmuir* **1986**, *2*, 653–659.

85. Bikerman, J. J. The unit of foaminess, *Trans. Faraday Soc.* **1938**, *34*, 634–638.

86. Watkins, R. C. An improved foam test for lubricating oils, *J. Inst. Pet., London* **1973**, *59*, 106–113.

87. Savitskaya, E. M. Analysis of the dispersity of foams, *Kolloidn. Zh.* **1951**, *13*, 309–313.

88. Nishioka, G. Stability of mechanically generated foam, *Langmuir* **1986**, *2*, 649–653.

89. Burton, R. A.; Mannheimer, R. J. Analysis and apparatus for surface rheological measurements, *Adv. Chem. Ser.* **1967**, *63*, 315–328.

90. Mannheimer, R. J. Surface rheological properties of foam stabilizers in nonaqueous liquids, *AIChE J.* **1969**, *15*, 88–93.

91. Mannheimer, R. J.; Schechter, R. S. An improved apparatus and analysis for surface rheological measurements, *J. Colloid Interface Sci.* **1970**, *32*, 195–211.

92. Fowkes, F. M. Characterization of solid surfaces by wet chemical techniques, *A.C.S. Symp. Ser.* **1982**, *199*, 69–88.

93. Fowkes, F. M. Donor-acceptor interactions at interfaces, *Polym. Sci. Technol.* **1980**, *12A*, 43–52.

94. Drago, R. S. Quantitative evaluation and prediction of donor-acceptor interactions, *Struct. Bonding (Berlin)* **1973**, *15*, 73–139.

95. Joslin, S. T.; Fowkes, F. M. Surface acidity of ferric oxides studied by flow microcalorimetry, *Ind. Eng. Chem., Prod. Res. Dev.* **1985**, *24*, 369–375.

96. van der Waarden, M. Stabilization of carbon-black dispersions in hydrocarbons, *J. Colloid Sci.* **1950**, *5*, 317–325.

97. Fowkes, F. M.; Mostafa, M. A. Acid-base interactions in polymer adsorption, *Ind. Eng. Chem., Prod. Res. Dev.* **1978**, *17*, 3–7.

98. Fowkes, F. M. Orientation potentials of monolayers adsorbed at the metal-oil interface, *J. Phys. Chem.* **1960**, *64*, 726–728.

99. Hingston, F. J.; Posner, A. M.; Quirk, J. P. Adsorption of selenite by geothite, *Adv. Chem. Ser.* **1968**, *79*, 82–90.

100. Sips, R. On the structure of a catalyst surface II, *J. Chem. Phys.* **1950**, *18*, 1024–1026.

101. Ross, S. Adsorption, *Kirl-Othmer Encycl. Chem. Technol.*, 2nd ed., Wiley-Interscience: New York, 1963–1971, Vol. 1, pp. 421–459.

102. Kipling, J. J.; Tester, D. A. Adsorption from binary mixtures; Determination of individual adsorption isotherms, *J. Chem. Soc.* **1952**, 4123–4133.

103. Giles, C. H.; MacEwan, T. H.; Nakhwa, S. N.; Smith, D. Studies in adsorption. Part XI. A system of classification of solution adsorption isotherms, and its use in diagnosis of adsorption mechanisms and in measurement of specific surface areas of solids, *J. Chem. Soc.* **1960**, 3973–3993.

104. Sheppard, S. E.; Lambert, R. H.; Walker, R. D. Optical sensitizing of silver halides by dyes I. Adsorption of sensitizing dyes, *J. Chem. Phys.* **1939**, 7, 265–273.

105. Sheppard, S. E. The effects of environment and aggregation on the absorption spectra of dyes, *Rev. Mod. Phys.* **1942**, 14, 303–340.

106. West, W.; Carroll, B. H.; Whitcomb, D. L. The adsorption of dyes to microcrystals of silver halide, *Ann. N.Y. Acad. Sci.* **1954**, 58, 893–909.

107. Kopeland, B.; Gregg, C. C. Particle agglomeration in tungsten metal powder, *J. Phys. Chem.* **1951**, 55, 557–563.

108. Everett, D. H. Manual of symbols and terminology for physicochemical quantities and units. Appendix II: Definitions and symbols in colloid and surface chemistry. Part I, *Pure Appl. Chem.* **1972**, 31, 577–638.

109. Maron, S. H.; Elder, M. E.; Ulevitch, I. N. Determination of surface area and particle size of synthetic latex by adsorption, *J. Colloid Sci.* **1954**, 9, 89–103.

110. Olivier, J. P.; Sennett, P. Electrokinetic effects in kaolin-water systems. I. The measurement of electrophoretic mobility, *Clays Clay Miner.* **1967**, 15, 345–356.

111. Henry, D. C. The cataphoresis of suspended particles. Part 1. The equation of cataphoresis. *Proc. Roy. Soc. London* **1931**, *A133*, 104–129.

112. van der Minne, J. L.; Hermanie, P. H. J. Electrophoresis measurements in benzene: Correlation with stability. I. Development of method, *J. Colloid Sci.* **1952**, 7, 600–615.

113. van der Minne, J. L.; Hermanie, P. H. J. Electrophoresis measurements in benzene: Correlation with stability. II. Results of electrophoresis, stability and adsorption, *J. Colloid Sci.* **1953**, 8, 38–52.

114. Fowkes, F. M.; Jinnai, H.; Mostafa, M. A.; Anderson, F. W.; Moore, R. J. Mechanism of electric charging of particles in nonaqueous liquids, *A.C.S. Symp. Ser.* **1982**, 200, 307–324.

115. Novotny, V. J. Physics of nonaqueous colloids, *A.C.S. Symp. Ser.* **1982**, 200, 281–306.

116. Croucher, M. D. Effect of free volume on the steric stabilization of nonaqueous latex dispersions, *J. Colloid Interface Sci.* **1981**, 81, 257–265.

117. Croucher, M. D.; Hair, M. L. Upper and lower critical flocculation temperatures in sterically stabilized nonaqueous dispersions, *Macromolecules* **1978**, 11, 874–879.

118. Clayfield E. J.; Lumb, E. C. A theoretical approach for polymer dispersant action. I. Calculation of entropic repulsion exerted by random polymer chains terminally adsorbed on plane surfaces and spherical particles, *J. Colloid Interface Sci.* **1966**, 22, 269–284.

119. Pugh, R. J.; Matsunaga, T.; Fowkes, F. M. The dispersibility and stability of carbon black in media of low dielectric constant. I. Electrostatic and steric contributions to colloidal stability, *Colloids Surf.* **1983**, 7, 183–207.

120. Fowkes, F. M.; Pugh, R. J. Steric and electrostatic contributions to the colloidal properties of nonaqueous dispersions, *A.C.S. Symp. Ser* **1984**, 240, 331–354.

121. Ross, S.; Lichtenstein, R. M. "The story of Volta potential," in *Selected Topics in the History of Electrochemistry*, Dubpernell, G.; Westbrook, J. H., Eds.; The Electrochemical Society: Princeton, NJ 1978; pp. 257-278.

122. Zisman, W. A. A new method of measuring contact potential differences in metals, *Rev. Sci. Instrum.* **1933**, *3*, 367-370.

123. Gibson, H. W.; Pochan, J. M.; Bailey, F. C. Surface analyses by a triboelectric charging technique, *Anal. Chem.* **1979**, *51*, 483-487.

124. Derjaguin, B. V.; Smilga, V. P. The present state of our knowledge about adhesion of polymers and semiconductors, *Proc. Int. Congr. Surf. Act., 3rd, 1960* **1961**, *2*, 349-367.

125. Fowkes, F. M. Interface acid-base/charge-transfer properties, *Surf. Interfacial Aspects Biomed. Polym.* **1985**, *1*, 337-372.

126. Fowkes, F. M.; Hielscher, F. H. Electron injection from water into hydrocarbons and polymers, *Org. Coat. Plast. Chem.* **1980**, *42*, 169-174.

127. Fabish, T. J.; Saltsburg, H. M.; Hair, M. L. Charge transfer in metal/atactic polystyrene contacts, *J. Appl. Phys.* **1976**, *47*, 930-939.

128. Lowell, J.; Rose-Innes, A. C. Contact electrification, *Adv. Phys.* **1980**, *29*, 947-1023.

129. Arridge, R. G. C. The static electrification of nylon 66, *Br. J. Appl. Phys.* **1967**, *18*, 1311-1316.

130. Davies, D. K. Charge generation on dielectric surfaces, *Br. J. Appl. Phys.* **1969**, (2), *2*, 1533-1537.

131. Davies, D. K. Charge generation on solids, *Int. Congr. Static Electr., 1st, 1970* **1970**, *1*, 10-21.

132. Nordhage, F.; Bäckström, G. Electrification in an electric field as a test of the theory of contact charging. *Inst. Phys. Conf. Ser.* **1975**, *27 (Static Electrif.)*, 84-94.

133. Skinner, S. M.; Savage, R. L.; Rutzler, Jr., J. E. Electrical phenomena in adhesion I. Electron atmospheres in dielectrics, *J. Appl. Phys.* **1953**, *24*, 438-450.

134. Webers, V. J. Measurement of triboelectric position, *J. Appl. Polym. Sci.* **1963**, *7*, 1317-1323.

135. Gibson, H. W. Control of electrical properties of polymers by chemical modification, *Polymer* **1984**, *25*, 3-27.

136. Duke, C. B.; Fabish, T. J. Contact electrification of polymers: A quantitative model, *J. Appl. Phys.* **1978**, *49*, 315-321.

137. Duke, C. B.; Fabish, T. J. Charge-induced relaxation in polymers, *Phys. Rev. Lett.* **1976**, *37*, 1075-1078.

138. Fabish, T. J.; Duke, C. B., Molecular charge states and contact charge exchange in polymers, *J. Appl. Phys.* **1977**, *48*, 4256-4266.

139. Halverson, F.; Panzer, H. P. Flocculating agents, *Kirk-Othmer Encycl. Chem. Technol.*, 3rd ed., Wiley-Interscience: New York, 1978-1984, Vol. 10, pp. 489-523.

140. Matijević, E. Colloid stability and complex chemistry, *J. Colloid Interface Sci.* **1973**, *43*, 217-245.

141. James, R. O.; Homola, A.; Healy, T. W. Heterocoagulation of amphoteric latex colloids, *J. Chem. Soc., Faraday Trans. 1* **1977**, *73*, 1436-1445.

142. Homola, A.; James, R. O. Preparation and characterization of amphoteric polystyrene latices, *J. Colloid Interface Sci.* **1977**, *59*, 123-134.

143. Gregory, J. The effect of cationic polymers on the colloidal stability of latex particles, *J. Colloid Interface Sci.* **1976**, *55*, 35–44.

144. La Mer, V. K.; Dinegar, R. H. Theory, production and mechanism of formation of monodisperse hydrosols, *J. Am. Chem. Soc.* **1950**, *72*, 4847–4854.

145. La Mer, V. K. Nucleation in phase transitions, *Ind. Eng. Chem.* **1952**, *44*, 1270–1277.

146. Faraday, M. Experimental relations of gold (and other metals) to light, *Phil. Trans. Roy. Soc. London* **1857**, *147*, 145–181; also *The Foundations of Colloid Chemistry;* Hatschek, E., Ed.; Benn: London, 1925, pp. 65–92.

147. Doremus, R. H. Optical properties of small gold particles, *J. Chem. Phys.* **1964**, *40*, 2389–2396.

148. Weitz, D. A.; Oliveria, M. Fractal structures formed by kinetic aggregation of aqueous gold colloids, *Phys. Rev. Lett.* **1984**, *52*, 1433–1436.

149. Weitz, D. A.; Huang, J. S.; Lin, M. Y.; Sung, J. Dynamics of diffusion-limited kinetic aggregation, *Phys. Rev. Lett.* **1984**, *53*, 1657–1660.

150. Kerker, M.; Daby, E.; Cohen, G. L.; Kratohvil, J. P.; Matijević, E. Particle size distribution in La Mer sulfur sols, *J. Phys. Chem.* **1963**, *67*, 2105–2111.

151. Watillon, A.; van Grunderbeeck, F.; Hautecler, M. Preparation et purification d'hydrosols de selenium stables et homeodisperses, *Bull. Soc. Chem. Belg.* **1958**, *67*, 5–21.

152. Dauchot, J.; Watillon, A. Optical properties of selenium sols I. Computation of extinction curves from Mie equations, *J. Colloid Interface Sci.* **1967**, *23*, 62–72.

153. Watillon, A.; Dauchot, J. Optical properties of selenium sols II. Preparation and particle size distribution, *J. Colloid Interface Sci.* **1968**, *27*, 507–515.

154. Matijević, E. Monodispersed metal (hydrous) oxides—a fascinating field of colloid science, *Acc. Chem. Res.* **1981**, *14*, 22–29.

155. Matijević, E. Production of monodisperse colloidal particles, *Annu. Rev. Mater. Sci.* **1985**, *15*, 483–516.

156. Težak, B.; Matijević, E.; Schulz, K. F.; Kratohvil, J.; Mirnik, M.; Vouk, V. B. Coagulation as a controlling process of the transition from homogeneous to heterogeneous electrolytic systems, *Faraday Discuss. Chem. Soc.* **1954**, *18*, 63–73.

157. Ottewill, R. H.; Woodbridge, R. F. The preparation of monodisperse silver bromide and silver iodide sols, *J. Colloid Sci.* **1961**, *16*, 581–594.

158. Woods, M. E.; Dodge, J. S.; Krieger, I. M.; Pierce, P. E. Monodisperse latices. I. Emulsion polymerization with mixtures of anionic and nonionic surfactants. *J. Paint Technol.* **1968**, *40*, 541–548.

159. Dodge, J. S.; Woods, M. E.; Krieger, I. M. Monodisperse latices. II. Seed polymerization techniques using mixtures of anionic and nonionic surfactants, *J. Paint Technol.* **1970**, *42*, 71–75.

160. Papir, Y. S.; Woods, M. E.; Krieger, I. M. Monodisperse latices. III. Cross-linked polystyrene latices. *J. Paint Technol.* **1970**, *42*, 571–578.

161. Tamai, H.; Hamada, A.; Suzawa, T. Deposition of cationic polystyrene latex on fibers, *J. Colloid Interface Sci.* **1982**, *88*, 378–384.

162. Croucher, M. D.; Hair, M. L. Upper and lower critical flocculation temperatures in sterically stabilized nonaqueous dispersions, *Macromolecules* **1978**, *11*, 874–879.

163. Croucher. M. D.; Lok, K. P. Stability of sterically stabilized nonaqueous dispersions at elevated temperatures and pressures, *A.C.S. Symp. Ser.* **1984**, *240*, 317–330.

164. Croucher, M. D.; Hair, M. L. Selective flocculation in heterosterically stabilized nonaqueous dispersions, *Colloids Surf.* **1980**, *1*, 349–360.

165. Derjaguin, B. V.; Martynov, G. A.; Gutop, Yu. V. Thermodynamics and stability of free films, *Colloid J. USSR* (Engl. Transl.) **1965**, *27*, 298–305.

166. Mysels, K. J. The direct measurement of the Debye length, *Phys. Chem.: Enriching Top. Colloid Surf. Sci.* **1975**, 73–86.

167. Clark, H.; Ross, S. Two-dimensional phase transition of ethane on sodium chloride, *J. Am. Chem. Soc.* **1953**, *75*, 6081.

168. Ross, S.; Clark, H. On physical adsorption. VI. Two-dimensional critical phenomena of xenon, methane and ethane adsorbed separately on sodium chloride, *J. Am. Chem. Soc.* **1954**, *76*, 4291–4297.

169. Ross, S.; Hinchen, J. J. "The evaluation and production of homotattic solid substrates of certain alkali halides," in *Clean Surf.: Their Prep. Charact. Interfacial Stud.*; Goldfinger, G., Ed; Dekker: New York, 1970, pp. 115–132.

170. Morrison, I. D.; Ross, S. The second and third virial coefficients of a two-dimensional gas, *Surf. Sci.* **1973**, *39*, 21–36.

171. Ross, S.; Morrison, I. D. Computed adsorptive-energy distribution in the monolayer (CAEDMON), *Surf. Sci.* **1975**, *52*, 103–119.

BIBLIOGRAPHY

Bibliography

Recommended general texts are marked with an asterisk.

Adam, N. K. *The Physics and Chemistry of Surfaces*, 3rd ed.; Oxford University Press: London, 1941.

*Adamson, A. W. *Physical Chemistry of Surfaces*, 4th ed.; Wiley-Interscience: New York, 1982.

Akers, R. J. Ed. *Foams*; Academic: New York, 1976.

*Alexander, A. E.; Johnson, P. *Colloid Science*; Oxford University Press: London, 1949.

Allen, T. *Particle Size Measurement*, 3rd ed.; Chapman and Hall: New York, 1981.

Attwood, D.; Florence, A. T. *Surfactant Systems, Their Chemistry, Pharmacy, and Biology*; Chapman and Hall: New York, 1983.

Aveyard, R.; Haydon, D. A. *An Introduction to the Principles of Surface Chemistry*; Cambridge University Press: London, 1973.

Bakker, G. *Kapillarität und Oberflächenspannung*; Akademische: Leipzig, 1928.

Bancroft, W. D. *Applied Colloid Chemistry*; McGraw-Hill: New York, 1932.

Barth, H. G. Ed. *Modern Methods of Particle Size Analysis*; Wiley-Interscience: New York, 1984.

Bashforth, F.; Adams, J. C. *An Attempt to Test the Theories of Capillary Action*; Cambridge University Press: Cambridge, 1883.

Becher, P. *Emulsions: Theory and Practice*; Reinhold: New York, 1957.

Becher, P., Ed. *Encyclopedia of Emulsion Technology, Vol. 1, Basic Theory*; Dekker: New York, 1983.

Becher, P., Ed. *Encyclopedia of Emulsion Technology, Vol. 2, Applications*; Dekker: New York, 1985.

Bikerman, J. J. *Surface Chemistry*, 2nd ed.; Academic: New York, 1958.

Bikerman, J. J. *Physical Surfaces*; Academic: New York, 1970.

Bikerman, J. J. *Foams*; Springer-Verlag: New York, 1973.

Bird, R. B.; Stewart, W. E.; Lightfoot, E. N. *Transport Phenomena*; Wiley: New York, 1960.

Bohren, C. F.; Huffman, D. R. *Absorption and Scattering of Light by Small Particles*; Wiley: New York, 1983.

Boys, C. V. *Soap-Bubbles, Their Colours and the Forces which Mold Them*; Dover: New York, 1959.

Buscall, R.; Corner, T.; Stageman, J. F., Eds. *Polymer Colloids*; Elsevier: New York, 1985.

Bütschli, O. *Untersuchungen uber mikrokopische Schäume und das Protoplasma* (*Foams and Protoplasm*); Engelmann: Leipzig, 1892; *Investigations on Protoplasm and Microscopic Foams*; Minchin, E. A., Transl.; Black: London, 1894.

Chattoraj, D. K.; Birdi, K. S. *Adsorption and the Gibbs Surface Excess*; Plenum: New York, 1984.

Chu, B. *Molecular Forces, Based on the Baker Lectures of Peter J. W. Debye*; Wiley–Interscience: New York, 1967.

Dahneke, B. E., Ed. *Measurement of Suspended Particles by Quasi-Elastic Light Scattering*; Wiley–Interscience: New York, 1983.

D'Arrigo, J. S. *Stable Gas-in-Liquid Emulsions, Production in Natural Waters and Artificial Media*; Elsevier: New York, 1986.

*Davies, J. T.; Rideal, E. K. *Interfacial Phenomena*, 2nd ed.; Academic: New York, 1963.

Defay, R.; Prigogine, I.; Bellemans, A. *Surface Tension and Adsorption*; Everett, D. H., Transl.; Longmans, Green: London, 1966.

Einstein, A. *Investigations on the Theory of Brownian Motion*; Dover: New York, 1956.

Everett, D. H.; Ottewill, R. H.; Eds. *Surface Area Determination*; Butterworths: London, 1970.

Fendler, J. H.; Fendler, E. J. *Catalysis in Micellar and Macromolecular Systems*; Academic: New York, 1975.

Fischer, E. K. *Colloidal Dispersions*; Wiley: New York, 1950.

Flory, P. J. *Principles of Polymer Chemistry*; Cornell University Press: Ithaca, NY, 1953.

Fowkes, F. M., Ed. *Contact Angle, Wettability, and Adhesion* (Advances in Chemistry Series No. 43), American Chemical Society: Washington, D.C.; 1964.

Fowler, R.; Guggenheim, E. A. *Statistical Thermodynamics*; Cambridge University Press: Cambridge, 1956.

Franks, F. *Polywater*; MIT Press: Cambridge, MA, 1981.

Franks, F. *Water*; Royal Society of Chemistry: London, 1983.

*Freundlich, H. *Colloid & Capillary Chemistry*, translated from the 3rd German edition by H. S. Hatfield; Methuen: London, 1926.

*Fridrikhsberg, D. A. *A Course in Colloid Chemistry*, Leib, G., Transl.; Mir: Moscow, 1986.

Gaines, G. L., Jr. *Insoluble Monolayers at Liquid–Gas Interfaces*; Wiley–Interscience: New York, 1966.

Gibbs, J. W. *The Scientific Paper of J. Willard Gibbs*, Longmans, Green: London, 1906; Dover: New York, 1961.

Goddard, E. D.; Vincent, B., Eds. *Polymer Adsorption and Dispersion Stability* (ACS Symposium Series #240), American Chemical Society: Washington, D.C., 1984.

*Goodwin, J. W., Ed. *Colloidal Dispersions*; Royal Society of Chemistry: London, 1982.

Hair, M. *Infrared Spectroscopy in Surface Science*; Dekker: New York, 1967.

Harkins, W. D. *The Physical Chemistry of Surface Films*; Reinhold: New York, 1952.

Harper, W. R. *Contact and Frictional Electrification*; Oxford University Press: London, 1967.

Hartland, S.; Hartley, R. W. *Axisymmetric Fluid–Liquid Interfaces*; Elsevier: New York, 1976.

Hauser, E. A. *Colloidal Phenomena, An Introduction to the Science of Colloids*; McGraw-Hill: New York, 1939.

Herdan, G. *Small Particle Statistics*; Elsevier: New York, 1953.

*Hiemenz, P. C. *Principles of Colloid and Surface Chemistry*, 2nd ed.; Dekker: New York, 1986.

Hirschfelder, J. O.; Curtiss, C. F.; Bird, R. B. *Molecular Theory of Gases and Liquids*; Wiley: New York, 1954.

Hunter, R. J. *Zeta Potential in Colloid Science*; Academic: New York, 1981.

*Hunter, R. J. *Foundations of Colloid Science*, Vol. 1; Clarendon Press: Oxford, 1987.

Israelachvili, J. N. *Intermolecular and Surface Forces with Applications to Colloidal and Biological Systems*; Academic: New York, 1985.

Jasper, J. J. The surface tensions of pure liquid compounds, *J. Phys. Chem. Ref. Data* **1972**, *1*, 841–1010.

*Jaycock, M. J.; Parfitt, G. D. *Chemistry of Interfaces*; Wiley: New York, 1981.

Jelínek, Z. K. *Particle Size Analysis*; Wiley: New York, 1970.

Jensen, W. B. *The Lewis Acid–Base Concepts: An Overview*; Wiley-Interscience: New York, 1980.

Kaye, B. H. *Direct Characterization of Fineparticles*; Wiley: New York, 1981.

Kerker, M. *The Scattering of Light and Other Electromagnetic Radiation*; Academic: New York, 1969.

Kipling, J. J. *Adsorption from Solutions of Non-Electrolytes*; Academic: New York, 1965.

Kitahara, A.; Watanabe, A., Eds. *Electrical Phenomena at Interfaces, Fundamentals, Measurements, and Applications*; Dekker: New York, 1984.

Klinkenberg, A.; van der Minne, J. L. *Electrostatics in the Petroleum Industry. The Prevention of Explosion Hazards*; Elsevier: New York, 1958.

Kruyt, H. R., Ed. *Colloid Science, Vol. 1: Irreversible Systems*; Elsevier: New York, 1952.

Kruyt, H. R., Ed. *Colloid Science, Vol. 2: Reversible Systems*; Elsevier: New York, 1949.

Landau, L. D.; Lifshitz, E. M. *Electrodynamics of Continuous Media*; Addison-Wesley: Reading, MA, 1960.

Lee, L. -H., Ed. *Polymer Science and Technology, Vol. 9: Adhesion Science and Technology, 1975* (in 2 vols.); *Vol. 12: Adhesion and Adsorption of Polymers, 1980* (in 2 vols.); *Vol. 29; Adhesion Chemistry—Development and Trends, 1984*; Plenum: New York.

Levich, V. G. *Physicochemical Hydrodynamics*; Prentice-Hall: Englewood Cliffs, NJ, 1962.

Lissant, K. L., Ed. *Emulsions and Emulsion Technology*, in 3 parts; Dekker: New York, 1974, 1984.

Loeb, A. L.; Overbeek, J. Th. G. Wiersema, P. H. *The Electrical Double Layer around a Spherical Colloid Particle*; MIT Press: Cambridge, MA, 1961.

Lowell, S.; Shields, J. E. *Powder Surface Area and Porosity*; 2nd ed.; Chapman and Hall: New York, 1984.

Lucassen-Reynders, E. H., Ed. *Anionic Surfactants*; Dekker: New York, 1981.

Mahanty, J.; Ninham, B. W. *Dispersion Forces*; Academic: New York, 1976.

Mandelbrot, B. B. *The Fractal Geometry of Nature*; Freeman: New York, 1983.

Manegold, E. *Emulsionen:* Strassenbau, Chemie und Technik: Heidelberg, 1952.

Matijević, E., Ed. *Surface and Colloid Science*, Vol. 1 to Vol. 9; Wiley-Interscience: New York, 1969 to 1976; Vol. 10 to date; Plenum: New York, 1978 to date.

Maxwell, J. C. *Theory of Heat*; 2nd ed.; Clarendon Press: Oxford, 1872; AMS: New York, 1972.

Maxwell, J. C. *Electricity and Magnetism*, 2 vols; Clarendon Press: Oxford, 1873.

*McBain, J. W. *Colloid Science*; Health: Boston, 1950.

McCrone, W. C.; Delly, J. G. *The Particle Atlas*; Ann Arbor Science: Ann Arbor, MI, 1973.

*McCutcheon's Emulsifiers & Detergents; McCutcheon Division, MC Publishing: Glen Rock, NJ, 1985.

*Miller, C. A.; Neogi, P. Interfacial Phenomena, Equilibrium and Dynamic Effects; Dekker: New York, 1985.

Mittal, K. L., Ed. Micellization, Solubilization, and Microemulsions, 2 vols.; Plenum: New York, 1977.

Mittal, K. L., Ed. Adhesion Aspects of Polymer Coatings, 2 vols.; Plenum: New York, 1983.

Mukerjee, P.; Mysels, K. J. Critical Micelle Concentrations of Aqueous Surfactant Systems; National Bureau of Standards, U.S. Department of Commerce: Washington, D.C., 1971.

Mysels, K. J.; Shinoda, K.; Frankel, S. Soap Films, Studies of Their Thinning; Pergamon: New York, 1959.

*Mysels, K. J. Introduction to Colloid Chemistry, 2nd ed.; Krieger: Huntington, NY, 1978.

Napper, D. H. Polymeric Stabilization of Colloidal Dispersions; Academic: New York, 1983.

Nye, M. J. Molecular Reality: A Perspective on the Scientific Work of Jean Perrin; Elsevier: New York, 1972.

Ościk, J. Adsorption; Wiley/Halsted: New York, 1982.

*Osipow, L. I. Surface Chemistry, Theory and Industrial Applications; Reinhold: New York, 1962.

Padday, J. F., Ed. Wetting, Spreading, and Adhesion; Academic: New York, 1978.

Parfitt, G. D., Ed. Dispersion of Powders in Liquids, with Special Reference to Pigments, 2nd ed.; Wiley: New York, 1973.

Parfitt, G. D.; Sing, K. S. W., Eds. Characterization of Powder Surfaces; Academic: New York, 1976.

Parfitt, G. D.; Rochester, C. H., Eds. Adsorption from Solution at the Solid/Liquid Interface; Academic: New York, 1983.

Pecora, R., Ed. Dynamic Light Scattering; Plenum: New York, 1985.

Plateau, J. Statique Expérimentale et Théórique des Liquides, 2 vols.; Gauthier-Villars: Paris, 1873.

*Popiel, W. J. Introduction to Colloid Science; Exposition: Hicksville, NY, 1978.

Rao, S. R. Surface Phenomena; Hutchinson Educational: London, 1972.

Rayleigh, Lord (John William Strutt) Scientific Papers, 6 vols. Cambridge University Press: Cambridge, 1899–1920.

Riddick, J. A., Bunger, W. B. Organic Solvents, 3rd ed.; Wiley-Interscience: New York, 1970.

Rosen, M. J. Surfactants and Interfacial Phenomena; Wiley: New York, 1978.

Ross, S.; Olivier, J. P. On Physical Adsorption; Wiley-Interscience: New York, 1964.

*Ross, S., Ed. Chemistry and Physics of Interfaces; American Chemical Society: Washington, D.C., 1965.

*Ross, S., Ed. Chemistry and Physics of Interfaces—II; American Chemical Society: Washington, D.C., 1971.

Rowlinson, J. S.; Widom, B. Molecular Theory of Capillarity; Oxford University Press: London, 1982.

Russel, W..B. The Dynamics of Colloidal Systems; University of Wisconsin Press: Madison, WI, 1987.

Sato, T.; Ruch, R. Stabilization of Colloidal Dispersions by Polymer Adsorption; Dekker: New York, 1980.

Schick, M. J., Ed. Nonionic Surfactants; Dekker: New York, 1967.

*Shaw, D. J. Introduction to Colloid and Surface Chemistry, 3rd ed.; Butterworths: London, 1980.

*Sheludko, A. *Colloid Chemistry*; Elsevier: New York, 1966.

Sherman, P. *Emulsion Science*; Academic: New York, 1968.

Shinoda, K.; Nakagawa, T., Tamamushi, B.; Isemura, T. *Colloidal Surfactants*; Academic: New York, 1963.

Shinoda, K., Ed. *Solvent Properties of Surfactant Solutions*; Dekker: New York, 1967.

Shinoda, K.; Friberg, S. *Emulsions and Solubilization*; Wiley-Interscience: New York, 1986.

Sonntag, H.; Strenge, K. *Coagulation and Stability of Disperse Systems*; Kondor, R., Transl.; Halsted: New York, 1972.

Sparnaay, M. J. *The Electrical Double Layer*; Pergamon: New York, 1972.

Stockman, J. D.; Fochtman, E. G., Eds. *Particle Size Analysis*; Ann Arbor Science: Ann Arbor, MI, 1977.

Tanford, C. *The Hydrophobic Effect: Formation of Micelles and Biological Membranes*, 2nd ed.; Wiley: New York, 1980.

Thompson, D. W. *On Growth and Form*, 2nd ed.; Macmillan: London, 1942.

Thomson, Sir W. [later Lord Kelvin] *Popular Lectures and Addresses, Vol. 1*; Macmillan: London, 1889.

van der Waals, J. D. *The Continuity of the Liquid and Gaseous States of Matter*, Physical Memoirs, Vol. 1, Part 3.; Taylor and Francis: London, 1890.

*van Olphen, H.; Mysels, K. J., Eds. *Physical Chemistry: Enriching Topics from Colloid and Surface Science*; Theorex: La Jolla, CA, 1975.

Verwey, E. J. W.; Overbeek, J. Th. G., *Theory of the Stability of Lyophobic Colloids*; Elsevier: New York, 1948.

*Vold, R. D.; Vold, M. J. *Colloid and Interface Chemistry*; Addison-Wesley: Reading, MA, 1983.

*Voyutsky, S. *Colloid Chemistry*, Bobrov, N., Transl.; Mir: Moscow, 1978.

Walters, K., Ed. *Rheometry: Industrial Applications*; Wiley-RSP: New York, 1980.

Weber, W. J., Jr.; Matijević, E., Eds. *Adsorption from Aqueous Solution* (Advances in Chemistry Series No. 79); American Chemical Society: Washington, D.C., 1968.

Weiser, H. B. *A Textbook of Colloid Chemistry*, 2nd ed.; Wiley: New York, 1949.

Wu, S. *Polymer Interface and Adhesion*; Dekker: New York, 1982.

Wiersema, P. K. *On the Theory of Electrophoresis*; Pasmans: The Hague, 1964.

Zsigmondy, R.; Theissen, P. A. *Das Kolloide Gold*; Akademische: Leipzig, 1925.

APPENDICES

APPENDIX A
Physical Constants

c	2.997925×10^8 m/sec	Speed of light in vacuum
ε_0	8.854×10^{-12} C/V m	Permittivity of free space
e	1.602×10^{-19} C	Elementary charge
	4.80298×10^{-10} esu	
g	9.807 m/s^2	Acceleration due to gravity
h	6.6261×10^{-34} J-s	Planck constant
k	1.38066×10^{-23} J/K	Boltzmann constant
	1.38066×10^{-19} erg/K	
R	8.31451 J/K mol	Gas constant
	1.98717 cal/K mol,	
	82.0575 cm^3 atm/K mol,	
	0.0820575 L atm/K mol	
	6.23637×10^4 cm^3 torr/K mol	

APPENDIX B

Units

1 calorie	$= 4.184$ joules
1 debye	$= 10^{-18}$ esu centimeter
1 dyne per centimeter	$=$ erg per square centimeter $=$ millijoule per square meter
1 dyne per square centimeter	$= 9.86923 \times 10^{-7}$ atmosphere
	$= 1 \times 10^{-8}$ bar
1 erg	$= 10^{-7}$ joule
1 electronvolt	$= 1.602 \times 10^{-19}$ joule
1 faraday	$= 9.649 \times 10^4$ coulombs per mole
1 joule	$=$ volt coulomb
1 poise	$=$ dyne second per square centimeter
0 degrees Celsius	$= 273.16$ degrees Kelvin
Dipole moment (coulomb meter)	$= 3.3349 \times 10^{-30}$ dipole moment (debyes)
Radial frequency	$= 2\pi \times$ frequency ($\omega = 2\pi\nu$)

Mathematical Formulas Used in Text

$$\exp(x) = 1 + x + x^2/2! + x^3/3! + \cdots$$

$$\ln(1 + x) = x - \frac{x^2}{2} + \frac{x^3}{3} - \frac{x^4}{4} + \cdots$$

$$\sinh(x) = 1/2\exp(x) - 1/2\exp(-x)$$
$$= x + x^3/3! + x^5/5! + \cdots$$

$$\cosh(x) = 1/2\exp(x) + 1/2\exp(-x)$$
$$= 1 + x^2/2! + x^4/4! + \cdots$$

$$\tanh(c) = \frac{\exp(x) - \exp(-x)}{\exp(x) + \exp(-x)} = x - \frac{x^3}{3} + \frac{2x^5}{15} - \frac{17x^7}{315} + \cdots$$

$$\int_0^\infty x\, dx \ln[1 - \Delta^2 \exp(-x)] = -\sum_{\nu=1}^\infty \frac{\Delta^2}{\nu^3}$$

The Laplacian operator ∇^2 is

$$\nabla^2 f = \frac{\partial^2 f}{\partial x^2} + \frac{\partial^2 f}{\partial y^2} + \frac{\partial^2 f}{\partial z^2} \qquad \text{in Cartesian coordinates}$$

$$= \frac{\partial^2 f}{\partial x^2} \qquad \text{in one-dimensional planar coordinates}$$

$$= \frac{\partial^2 f}{\partial r^2} + \frac{(2/r)\partial f}{\partial r} \qquad \text{in one-dimensional spherical coordinates}$$

The principal radii of curvature of a surface in Cartesian coordinates are

$$\frac{1}{R_1} = \frac{d^2z/dx^2}{[1 + (dz/dx)^2]^{3/2}}; \quad \frac{1}{R_2} = \frac{(1/x)dz/dx}{[1 + (dz/dx)^2]^{1/2}}$$

APPENDIX D

Electrostatic and Induction Contributions to Intermolecular Potential Energies

Charge–charge	Angle and temperature independent	$q_a q_b / 4\pi\varepsilon_0 r$
Charge–dipole	Averaged over all orientations	$-(kT/3)q_a^2\mu_b^2/(4\pi\varepsilon_0)^2 r^4$
	At the maximum	$-q_a\mu_b/4\pi\varepsilon_0 r^2$
Charge–quadrupole	Averaged over all orientations	$-(kT/20)q_a^2 Q_b^2/(4\pi\varepsilon_0)^2 r^6$
Dipole–dipole	Averaged over all orientations	$-(2kT/3)\mu_a^2\mu_b^2/(4\pi\varepsilon_0)^2 r^6$
	At the maximum	$-2\mu_a\mu_b/(4\pi\varepsilon_0)r^3$
Charge–induced dipole	Angle and temperature independent	$-q_a^2\alpha_b/8\pi\varepsilon_0 r^4$
Dipole–induced dipole	Averaged over all orientations	$-\mu_a^2\alpha_b/4\pi\varepsilon_0 r^6$

where q_a = charge on particle a (C)
  r = interparticle distance (m)
  μ_a = dipole moment of particle a (C-m)
  α_a = polarizability of particle a (m^3)
  Q_a = quadrupole moment of particle a (C-m^3)
  ε_0 = permittivity of free space, $= 8.85 \times 10^{-12}$ C/V-m

(Molecular polarizabilities are often estimated from the sum of bond polarizabilities.)

Hirschfelder, Curtiss, and Bird, 1954, pp. 25–37.

APPENDIX E

Electric Properties of Representative Molecules

	Charge, q (C)	Dipole Moment, μ (C-m)	Polarizability, α (m^3)	Ionization Potential (J)
Na$^+$	1.602×10^{-19}	—	—	—
Cl$^-$	1.602×10^{-19}	—	—	—
N$_2$	0	0	1.74×10^{-30}	2.53×10^{-18}
O$_2$	0	0	1.57×10^{-30}	2.18×10^{-18}
CO	0	3.3×10^{-31}	1.99×10^{-30}	2.29×10^{-18}
NH$_3$	0	5.0×10^{-30}	2.24×10^{-30}	1.87×10^{-18}
H$_2$O	0	6.1×10^{-30}	1.48×10^{-30}	2.01×10^{-18}
C$_6$H$_5$–OH	0	4.83×10^{-30}	—	—
C$_6$H$_6$	0	0	10.3×10^{-30}	1.54×10^{-18}
HCl	0	3.59×10^{-30}	26.3×10^{-31}	2.21×10^{-18}

Landolt-Bornstein, Volume 1, Part 3; Springer, 1951.

Lifetimes of Contributors to Colloid and Interface Science

Robert Hook	1635–1702
Sir Isaac Newton	1642–1727
Francis Hauksbee	1666–1713
Janos András Segner	1704–1777
Benjamin Franklin	1706–1790
Pierre Simon de Laplace	1749–1827
Sir John Leslie	1766–1832
Thomas Young	1773–1829
Robert Brown	1773–1858
William Henry	1774–1836
Ferdinand Friedrich Reuss	1778–1852
Claude-Louis-Marie-Henri Navier	1785–1836
Michael Faraday	1791–1867
Pierre Hippolyte Boutigny	1798–1884
Franz Ernst Neumann	1798–1895
Jean-Louis-Marie Poiseuille	1799–1869
Joseph-Antoine-Ferdinand Plateau	1801–1883
Germaine Henri Hess	1802–1850
Thomas Graham	1805–1869
Francesco Selmi	1817–1881
Jules-Celestin Jamin	1818–1886
John Couch Adams	1819–1892
Francis Bashforth	1819–1912
John Tyndall	1820–1893
Hermann von Helmholtz	1821–1894

Mathew Carey Lea	1823–1897
William Thomson, Lord Kelvin	1824–1907
Johann Wilhelm Hittorf	1824–1914
Jakob Maarten van Bemmelen	1830–1911
James Clerk Maxwell	1831–1879
Georg Hermann Quincke	1834–1924
Josiah Willard Gibbs	1839–1903
John Aitken	1839–1919
Pierre-Émile Duclaux	1840–1904
Carlo Marangoni	1840–1925
Osborne Reynolds	1842–1912
John William Strutt, Lord Rayleigh	1842–1919
Sir James Dewar	1842–1923
Joszef Fodor	1843–1901
Ludwig Boltzmann	1844–1906
Walthère Victor Spring	1848–1910
Otto Bütschli	1848–1920
Franz Hofmeister	1850–1922
Friedrich Wilhelm Ostwald	1853–1932
Louis-Georges Gouy	1854–1926
Sir Charles Vernon Boys	1855–1944
Edward Goodrich Acheson	1856–1931
Sir Joseph John Thomson	1856–1940
Percival S. U. Pickering	1858–1920
Pierre-Maurice-Marie Duhem	1861–1916
Agnes Pockels	1862–1935
Henri-Edgard Devaux	1862–1956
Frederick Stanley Kipping	1863–1949
Sir William Bate Hardy	1864–1934
Hermann Walther Nernst	1864–1941
Richard Adolf Zsigmondy	1865–1929
Heinrich Jakob Bechold	1866–1937
Wilder Dwight Bancroft	1867–1953
Georg Bredig	1868–1944
Gustav Mie	1868–1957
Raphael Edward Liesegang	1869–1947
David Leonard Chapman	1869–1958
Jean Perrin	1870–1942
Frederick M. G. Donnan	1870–1956
Marion von Smoluchowski	1872–1917
Bogdan von Szyszkowski	1873–1931
William Draper Harkins	1873–1951
Gilbert Newton Lewis	1875–1946
Leonor Michaelis	1875–1949
Carl Axel Fredrik Benedicks	1875–1958
Jerome Alexander	1876–1959
Frans Maurits Jaeger	1877–1945

Robert Whytlaw-Gray	1877–1958
Maximilian Nierenstein	1878–1946
Eli Franklin Burton	1879–1948
Albert Einstein	1879–1955
Herbert M. F. Freundlich	1880–1941
Anton Vladimirovich Dumanskii	1880–1967
Irving Langmuir	1881–1957
Hermann Staudinger	1881–1965
James William McBain	1882–1953
Hugo Rudolph Kruyt Carl Wilhelm	1882–1959
Warren Kendall Lewis	1882–1975
Wolfgang Ostwald	1883–1943
Floyd Earl Bartell	1883–1961
Peter Joseph William Debye	1884–1966
The Svedberg	1884–1971
Hendrik Sjoerd van Klooster	1884–1972
Harry Boyer Weiser	1887–1950
Sir Chandrasekhara Venkata Raman	1888–1971
Sir Hugh Stott Taylor	1890–1974
Sir Eric Keightley Rideal	1890–1974
Sergei Sergeevich Medvedev	1891–1970
Neil Kensington Adam	1891–1973
Michael Polanyi	1891–1976
George Scatchard	1892–1974
Hendrik Gerard Bungenberg de Jong	1893–1977
Sir John Edward Lennard-Jones	1894–1954
Victor Kuhn La Mer	1895–1966
Willis Conway Pierce	1895–1974
Alexander Naumovich Frumkin	1895–1976
Nikolai Albertowich Fuchs	1895–1982
Ernst Alfred Hauser	1896–1956
Erich Armand Arthur Hückel	1896–1980
Petr Aleksandrovich Rehbinder	1898–1972
Katharine Burr Blodgett	1898–1979
Ralph Alonzo Beebe	1898–1979
Foster Dee Snell	1898–1980
Georg-Maria Schwab	1899–1984
Harry Herman Sobotka	1899–1965
Fritz London	1900–1954
Curtis R. Singleterry	1900–1971
Frank Thompson Gucker, Jr.	1900–1973
Edward Armand Guggenheim	1901–1970
A.S.C. [Stuart] Lawrence	1902–1971
Arne W. T. Tiselius	1902–1971
Stephen Brunauer	1903–1986
Wilfried Heller	1903–1982

Jack Henry Schulman	1904–1967
Edward Alison Flood	1904–1979
Pierre van Rysselberghe	1905–1977
Evert Johannes Willem Verwey	1905–1981
William Albert Zisman	1905–1986
Bozo Težak	1907–1980
Lev Davidovich Landau	1908–1968
Andrei Vladimirovich Kiselev	1908–1984
Winfred Oliver Milligan	1908–1984
Robert Donald Vold	1910–1978
Paul John Flory	1910–1985
Albert Ernest Alexander	1914–1970
Myron L. Corrin	1914–1980
William James Dunning	1915–1979
Akira Watanabe	1922–1980
Frank Chauncey Goodrich	1924–1980
Geoffrey Derek Parfitt	1928–1985
John Michael Corkill	1931–1974
Raouf Shaker Mikkail	1931–1983
George Mathew Kanapilly	1934–1982

Living scientists are not included. The authors are aware that this list has many blatant omissions. Some names deserving entry are missing because vital dates were not found. The authors welcome suggestions for additional names. They request that correspondents provide vital dates.

APPENDIX G

Kendall Awardees

The ACS Award in Colloid or Surface Chemistry is sponsored by the Kendall Company (a subsidiary of the Colgate-Palmolive Co.) to recognize and encourage outstanding scientific contributions to colloid or surface chemistry in the United States and Canada.

1954	Harry N. Holmes
1955	John W. Williams
1956	Victor K. La Mer
1957	Peter J. W. Debye
1958	Paul H. Emmett
1959	Floyd E. Bartell
1960	John D. Ferry
1961	Stephen Brunauer
1962	George Scatchard
1963	William A. Zisman
1964	Karol J. Mysels
1965	George D. Halsey, Jr.
1966	Robert S. Hansen
1967	Stanley G. Mason
1968	Albert C. Zettlemoyer
1969	Terrell L. Hill
1970	Jerome Vinograd
1971	Milton Kerker
1972	Egon Matijević
1973	Robert L. Burwell, Jr.

1974	W. Keith Hall
1975	Robert Gomer
1976	Robert J. Good
1977	Michel Boudart
1978	Harold A. Scheraga
1979	Arthur Adamson
1980	Howard Reiss
1981	Gabor A. Somorjai
1982	Gert Ehrlich
1983	Janos H. Fendler
1984	Brian E. Conway
1985	Stig Friberg
1986	Eli Ruckenstein
1987	John T. Yates, Jr.
1988	Howard Brenner

APPENDIX H

Electrostatic Units*

In the older literature, the three-quantity electrostatic system is generally used. In this system the Coulomb equation is written as $F = Q_1 Q_2 / D^2$, and the Poisson equation is then $\nabla^2 \Phi = -4\pi\rho/D$, where D is the dielectric constant or the *relative static permittivity*.

In the rationalized four-quantity (part of the SI system) the Coulomb equation is

$$F = \frac{Q_1 Q_2}{4\pi\varepsilon r^2}$$

and the Poisson equation is

$$\nabla^2 \Phi = -\rho/\varepsilon$$

where ε is now the static permittivity, which equals the product of the relative static permittivity and the permittivity in vacuum: $\varepsilon = D\varepsilon_0 = \varepsilon_\pm \varepsilon_0$.

Hence, in the older literature on the diffuse electrical double layer, the factor 4π appears in the equations, and the symbol ε_r stands for the dielectric constant of the medium, which is on the order of 80 for water. When using the rationalized four-quantity system, these factors, 4π, are eliminated in the double-layer formulas, whereas the symbol ε stands for the static permittivity, which for water at room temperature is $80 \times 8.85 \times 10^{-12} \, \text{J}^{-1}\text{C}^2\,\text{m}^{-1}$ or Fm^{-1}. Factors of π will only occur in the formulas of the rationalized system when spherical problems are involved.

We follow Hunter (*Zeta Potential in Colloid Science*, p. 358, 1981) in using the symbol D to represent the dielectric constant (or relative permittivity), rather than the more modern ε_r, to emphasize that D has no units and is, therefore, a different kind of quantity than ε_0, the permittivity of a vacuum.

*From *Enriching Topics from Colloid and Surface Science*, van Olphen and Mysels, (1975 pp. x–xi).

APPENDIX J
Manufacturers of Instruments

Manufacturer	Instruments
Analytic Measuring Systems Shirehill Saffron Walden Essex CB11 3AQ England (0799)24080	Image analysis
ATM Corporation 645 South 94th Place Milwaukee, WI 53214 (414)453-1100	Particle size by sieving Pilot scale sieving
Bausch and Lomb Analytic Systems Division Rochester, NY 14625 (716)385-1000	Image analysis
Bohlin Reologi Box 742 S-72007 Lund, Sweden	Surface viscosimeters
Brinkmann Instruments Co. Cantiague Rd. Westbury, NY 11590 (516)334-7500	Langmuir–Blodgett film balance

Brookfield Engineering Laboratories, Inc. 240 Cushing St. Stoughton, MA 02072 (617)344-4310	Couette viscosimeters
Brookhaven Instruments Corp. 200 Thirteenth Ave. Ronkonkoma, NY 11779 (516)588-4100	Particle size by quasi-elastic light scattering Particle size by Fraunhofer diffraction
Cahn Instruments, Inc. 16207 S. Carmenita Rd. Cerritos, CA 90701 (213)926-3378	Surface tension and contact angle by Wilhelmy plate or fiber balance
Carl Zeiss, Inc. One Zeiss Drive Thornwood, NY 10594 (914)747-1800	Image analysis
Carlo Erba Strumentazione Strada Rivoltana 20090 Rodano (Milan) Italy (2)950591/9588161	Surface area by gas adsorption
Chem-Dyne Research Corp. P.O. Box 17968 Milwaukee, WI 53217-0968 (414)352-8179	Surface tension by maximum bubble pressure
Climet Instruments Co. P.O. Box 1760 Redlands, CA 92373 (714)793-2788	Particle counting by light scattering
Coulter Electronics, Inc. 590 West 20th St. Hialeah, FL 33012-0145 (305)885-0131	Particle size and number by resistazone detection Particle size by QELS Electrophoretic mobilities by laser Doppler velocimetry
Dantec Electronics, Inc. 6 Pearl Court Allendale, NJ 07401 (201)825-3339	Particle size and velocity by laser Doppler velocimetry
DuPont Company Instrument Systems Concord Plaza, Quillen Bldg. Wilmington, DE 19898 (302)772-5488	Particle size by sedimentation field flow fractionation

GCA Corporation
Technology Division
213 Burlington Rd.
Bedford, MA 01730
(617)275-5444

Particle size by elutriation

Gelman Instrument Co.
600 South Wagner Rd.
Ann Arbor, MI
(313)747-1800

Image analysis

Gilson Company, Inc.
P.O. Box 677
Worthington, OH 43085
(614)846-5979

General laboratory particle size equipment

Haake, Inc.
244 Saddle River Rd.
Saddle Brook, NJ 07662
(201)843-2320
Dieselstrasse 4
D-7500 Karlsruhe 41
W. Germany
(0721)406053

Capillary, cone-and-plate, Couette, and falling ball viscosimeters

Hach Company
P.O. Box 389
Loveland, CO 80539
(800)227-4224

Turbidity measurements

H & N Instruments, Inc.
412 Daniel Ave.
Newark, OH 43055
(614)927-0156

Instruments for characterizing transient and stable foams

Horiba Instruments, Inc.
1021 Duryea Ave.
Irvine, CA 92714
(714)540-7874

Particle size by centrifugation

Imass, Inc.
3 King Philip Path
P.O. Box 134
Hingham, MA 02018-0134
(617)749-5728

Contact angle goniometer

Komline-Sanderson
12 Holland Ave.
Peapack, NJ 07977
(201)234-1000

Particle charge by microelectrophoresis

Krüss GmbH Borsteler Chaussee 85-99a D-2000 Hamburg 61 W. Germany 040/511 6033	Instruments for surface and interfacial tensions Film balance Contact anglemeter
KSV Chemicals OY Valimotie 7.00380 Helsinki, Finland 358 0 556 351	Langmuir–Blodgett film balance

Leeds & Northrup Instruments
Sumneytown Pike
North Wales, PA 19454
(215)643-2000
	Particle size by Fraunhofer
	  diffraction
	Surface area by
	  gas adsorption

Macro Scientific 1055 Sunnyvale-Saratoga Road No. 8 Sunnyvale, CA 94087 (408)739-9418	Particle size by Fraunhofer diffraction

Malvern Instruments, Inc.
200 Turnpike Rd.
Southborough, MA 01772
(617)480-0200
	Electrophoretic mobilities
	  by laser Doppler velocimetry
	Particle size by quasi-elastic
	  light scattering
	Particle size by
	  Fraunhofer diffraction

Matec Instruments, Inc. 60 Montebello Rd. Warwick, RI 02886 (401)739-9030	Particle charge by ultrasonic electrophoresis

Micromeretics Instrument Corp.
5680 Goshen Springs Rd.
Norcross, GA 30093
(404)448-8282
	Mercury porosimetry
	Surface area by gas adsorption
	Particle charge by
	  electrophoresis
	Particle size by X-ray/
	  sedimentation
	Particle size by centrifugation
	Particle size by hydrodynamic
	  chromatography

Microscal, Ltd. 79 Southern Row London W10 5AL, England (01)969-3935	Heats of adsorption by flow microcalorimetry Particle size by sedimentation

Mine Safety Appliances Co.
MSA Instrument Division
400 Penn Center Blvd.
Pittsburgh, PA 15235
(412)273-5095

Particle size by sedimentation
and centrifugation

Pacific Scientific Co.
2431 Linden Ln.
Silver Spring, MD 20910
(301)495-7000

Particle size by photozone
detection
Particle size by quasi-elastic
light scattering
Fineness of grind gauges
Viscosimeters

Particle Data, Inc.
Box 265
111 Hahn St.
Elmhurst, IL 60126
(312)832-5653

Particle size by electrozone
detection

Pen Kem, Inc.
341 Adams St.
Bedford Hills, NY 10507
(914)241-4777

Rheometry of viscoelastic
dispersions
Rates of flocculation by
photozone detection
Particle charge by
acoustophoresis
Particle charge by
electrophoresis
Particle charge by laser
Doppler velocimetry

Polytec Optronics
22651 Lambert Street, Unit 108
El Toro, CA 92630
(714)770-9911

Particle size and velocity
by laser Doppler
velocimetry

Quantachrome Corp.
Six Aerial Way
Syosset, NY 11791
(516)935-2240

Surface area by gas
adsorption
Mercury contact angles
Powder packing densities
Powder sampling
Mercury porosimetry
Particle size by X-ray
sedimentation
Particle density

Ramé-Hart, Inc.
43 Bloomfield Ave.
Mountain Lakes, NJ 07046
(201)335-0560

Contact angle goniometer
Wilhelmy balance

Rank Brothers 56 High Street, Bottisham Cambridge CB5 9DA England 0223-811369	Particle charge by microelectrophoresis
Schott America Glass & Scientific Products, Inc. 3 Odell Plaza Yonkers, NY 10701 (914)968-8900	Capillary viscosimeters
Shape Technology, Ltd. 901 Park Place Iowa City, IA 52240 (319)351-3736	Image analysis
Spectrex Corporation 3594 Haven Ave. Redwood City, CA 94063 (415)365-6567	Particle size by low-angle light scattering
Vickers Instruments, Inc. P.O. Box 99 Riverview Business Park 27 300 Commercial St. Malden, MA 02148 (617)324-0350	Particle size by centrifugation Langmuir–Blodgett film balance

Wyatt Technology Corp.
820 East Halet St.
P.O. Box 3003
Santa Barbara, CA 93130
(805)963-5904
	Particle size (or molecular
	  weight) by multiple-angle
	  light scattering

Zeta-Meter, Inc. 50-17 Fifth St. Long Island City, NY 11101 (718)392-0229	Particle charge by microelectrophoresis
Zed Instruments, Ltd. 336 Molesey Rd., Hersham Surrey, KT12 3PD England 0932-228977	Surface-shear rheometry

APPENDIX K
Manufacturers of Processing Equipment

Manufacturer

Equipment

Paul O. Abbé, Inc.
152 Center Ave.
Little Falls, NJ 07424
(201)256-4242

Ball-and-pebble mills
Jar mills
Blade mixers
Blenders

Willy A. Bachofen Maschinenfabrik
Untengasse 15/17 CH-4005
Basel, Switzerland
(061)335555

Dyno-mills

Biospec Products
P.O. Box 722
Bartlesville, OK 74005
(918)333-2166

Laboratory high-speed
  stirrers

Branson Sonic Power Co.
Eagle Rd.
Danbury, CT 06810
(203)744-0760

Ultrasonic dispersion

Brinkmann Instruments Co.
Cantiague Rd.
Westbury, NY 11590
(516)334-7500

High-speed rotor/stator
  dispersers

Donaldson Co., Inc. Majac Division Box 1299 Minneapolis, MN 55440	Micronizers
Gaulin Corp. 44 Garden St. Everett, MA 02149 (617)387-9300	High-shear mixers
Glen Mills 203 Brookdale St. Maywood, NJ 07607 (201)845-4665	Hammer mills Blade-and-tooth mills Dyno-mills
J. W. Greer, Inc. South Main St. Wilmington, MA 01887 (617)658-3301 3 Tribune Dr. Trinity Trading Estate Sittingbourne, Kent, England 0795-70742	Colloid mills Homogenizers
Greerco Corp. Executive Dr. Hudson, NH 03051 (603)883-5517	High-speed mixers Colloid mills Homogenizers
Farrel Co. Emhart Machinery Group 25 Main St. Ansonia, CT 06401 (203)734-3331	Banbury mixer
Kinetic Dispersion Corp. 127 Pleasant Hill Scarborough, ME 04074 (207)883-4141	Kady mills

Microfluidics Corp.
44 Mechanic St.
Newton, MA 02164
(617)965-7255
	Size reduction by dynamic
	  interaction of two fluid
	  streams

Morehouse Industries, Inc. 1600 W. Commonwealth Ave. P.O. Box 3620 Fullerton, CA 92633 (714)738-5000	High-speed dispersers and dissolvers Horizontal mills Colloid mills Sand mills

Myers Engineering 8376 Salt Lake Ave. Bell, CA 90201 (213)560-4723/771-2184	High-speed mixers and dispersers
Oakes Machine Corp. 235 Grant Ave. Islip, NY 11751 (516)581-4130	Continuous mixers/foamers
Premier Mill Corp. 220 East 23rd St. New York, NY 10010 (212)686-8190	Planetary mixers Media mills Homogenizers Colloid mills
Sonic Corp. One Research Dr. Stratford, CT 06497 (203)375-0063	Sonic homogenizers Colloid mills
Sturtevant Mill Co. 103 Clayton St. Boston, MA 02122 (617)825-6500	Crushers, pulverizers Cyclones Fluid energy mills Hammer mills
Sweco, Inc. 6033 E. Bandini Blvd. P.O. Box 4151 Los Angeles, CA 90051 (213)726-1177	Vibratory media mills
Union Process, Inc. 1925 Akron-Peninsula Rd. Akron, OH 44313 (216)929-3333	Attritors
VirTis Company, Inc. Gardiner, NY 12525 (914)255-5000	Ultrasonic dispersion
Waring Products Division Dynamics Corporation of America New Hartford, CT 06057 (203)379-0731	Laboratory high-shear mixers

Index